QVI MAL Y PENSE
HONI SOIT
This Book
belongs to the Library of
King Edward VI'S
Grammar School.
Guildford, Surrey.

LIQUID CRYSTALS & PLASTIC CRYSTALS

Volume 2

PHYSICO-CHEMICAL PROPERTIES and METHODS OF INVESTIGATION

Ellis Horwood Series in Physical Chemistry

Series Editor: T. Morris Sugden, F.R.S.

Director 'Shell' Research Limited, Thornton Research Centre; Associate Professor of Molecular Biology, University of Warwick

Founded as a library of fundamental books on important or growing areas in physical chemistry, this series will serve chemists in industrial research, and in teaching or advanced study

Published or in active preparation

Liquid Crystals and Plastic Crystals

Vol. 1 Preparation, Constitution and Applications
Vol. 2 Physico-Chemical Properties and Methods of Investigation

Gray, G. W., *University of Hull*
Winsor, P. A., *'Shell' Research Ltd* } *Editors*

Kinetics and Mechanisms of Polymerization Reactions

Allen, P. E. M., *University of Adelaide*
Patrick, C. R., *University of Birmingham*

Polymer Physics

Hearle, J. W. S., *University of Manchester Institute of Science and Technology*

Mechanisms of Reactions in Solution

Kohnstam, G., *University of Durham*

Electrochemical Phase Formation

Thirsk, H. R.
Armstrong, R. D. } *University of Newcastle-upon-Tyne*
Harrison, J. A.

Surface Chemistry

Parfitt, G. D., *Tioxide International*
Jaycock, M. J., *Loughborough University of Technology*

LIQUID CRYSTALS
&
PLASTIC CRYSTALS

Volume 2

Physico-Chemical Properties and Methods of Investigation

Editors

G. W. GRAY
Department of Chemistry, University of Hull

P. A. WINSOR
'Shell' Research Limited

ELLIS HORWOOD LIMITED

First published in 1974 by

ELLIS HORWOOD LIMITED
Coll House, Westergate, Chichester, Sussex, England

this book bears his colophon reproduced from
James Gillison's drawing of the
ancient Market Cross, Chichester.

Distributed in:

Australia, New Zealand, South-east Asia by
JOHN WILEY & SONS AUSTRALASIA PTY LIMITED
110 Alexander Street, Crow's Nest, N.S.W. Australia

Europe, Africa by
JOHN WILEY & SONS LIMITED
Baffins Lane, Chichester, Sussex, England

N. & S. America and the rest of the world by
HALSTED PRESS a division of
JOHN WILEY & SONS INC.
605 Third Avenue, New York, N.Y. 10016, U.S.A.

Library of Congress Catalog No. 73 11505
ISBN (Ellis Horwood) 85312 004 8
ISBN (Halsted Press) 0470 32340 X

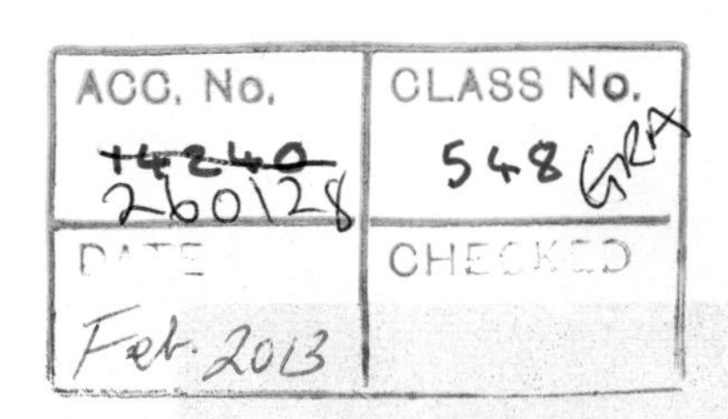

Printed in Great Britain by
Butler & Tanner Ltd., Frome and London

Contents

1

Electron Microscopy of Liquid Crystals

J. F. GOODMAN and J. S. CLUNIE

During the past decade conventional electron microscopy has been used with increasing success in the study of mesomorphic phase structure in amphiphilic systems [1]. In particular, transmission electron micrographs have been obtained which directly reveal the morphology and molecular organization in several mesomorphic phases, thus providing strong and independent confirmation of structures postulated on the basis of low-angle X-ray diffraction data. However, despite this success, the method suffers from severe limitations. The electron microscope operates under a high vacuum ($<10^{-5}$ torr) yet most mesophases in amphiphilic systems contain appreciable amounts of water. Exacting sample preparative techniques [2] are required to produce a specimen for examination which is both anhydrous and very thin (<600 Å thick), which possesses sufficient contrast and is stable in the electron beam, and which above all preserves the original structure. Furthermore, in examining mesomorphic phase structures, the microscopist is often operating at the resolution limit, not of the electron microscope (3–10 Å), but of the specimen preparative techniques currently available (ca. 20 Å). Therefore, it is important to use several different specimen preparative techniques in order to reduce the chance of being misled by artefacts peculiar to any one preparative method. Variants of two main specimen preparative methods have been generally used:

(1) surface replication of (*a*) anhydrous and (*b*) quench-frozen systems
(2) chemical fixation and staining by heavy metal reagents.

Since the electron microscopic approach, unlike the low-angle X-ray diffraction method, is not one of general applicability, the choice of system to be studied is very important, and this frequently requires some *a priori* knowledge of phase behaviour. Particularly convenient for study are those systems where the mesomorphic phases occur at room temperature since this facilitates equilibration of samples before specimen preparation is begun.

Surface Replication

The established technique of surface replication has been used to examine the surface detail of mesomorphic phase structures [3], with the method of "shadow-casting" being adopted in order to increase contrast in the replica. The procedure is carried out *in vacuo* and consists of evaporating a thin metal film (ca. 100–200 Å thick) obliquely on to the surface of the specimen which is supported on either a mica or a glass substrate. Shadows are cast by any topographical features in the specimen surface and a three-dimensional impression of the surface is thus created. A backing film of carbon is usually evaporated on to the shadowed surface in order to strengthen the replica before it is stripped from its substrate at an air/water interface and then mounted on a conventional microscope grid. For high-resolution micrographs, amorphous replicas produced by simultaneous shadowing with platinum–carbon are usually prepared, although recently it has been claimed that better resolution is obtained if tantalum–tungsten shadowing is used [4]. For accurate measurement of shadow lengths, however, shadowing with platinum alone followed by carbon backing has been preferred [3] because the resulting micro-crystalline replicas give sharper image definition.

The key step in the method is to find a mesomorphic phase system which is sufficiently stable under the high vacuum ($<10^{-5}$ torr) of the evaporator unit. Several branched-chain amphiphiles satisfy this criterion [3, 5]. For example, under high-vacuum conditions at room temperature, sodium 2-ethylhexyl sulphate and sodium hept-6-enyl sulphate exist as neat phases whereas the sodium di-2-ethylalkyl sulphosuccinates (C_4–C_{18})* exist as "inverse" middle phases. In these systems, surface replicas of the anhydrous mesomorphic phase have been prepared from dried-down aqueous and non-aqueous solution droplets. A typical micrograph of a surface replica from neat (lamellar) phase is shown in Fig. 1.1. A characteristic feature is the terraced droplet shape which gives rise to the planar optical texture observed in the polarizing microscope [6]. Also evident in the micrograph are stepped growth forms and focal conic fringes. Multiple steps are mostly observed but separate high-resolution micrographs have shown that the minimum step height determined from the measured shadow length and known shadowing angle corresponds to a bimolecular layer thickness. These bimolecular step heights are roughly equal to the X-ray long spacing for the original composition [3, 5].

One of the disadvantages of the replication method as applied to anhydrous mesomorphic specimens is that morphological changes may occur during the drying down of the dilute solution droplets. To remove these objections, and to minimize any possible specimen damage during shadowing, a quench-freeze replication method has been developed [5]. Neat phase compositions from branched-chain amphiphile + water systems containing appreciable amounts of water are first equilibrated

*$CH_2 \cdot CO \cdot OR$
|
$CH(SO_3Na) \cdot CO \cdot OR$

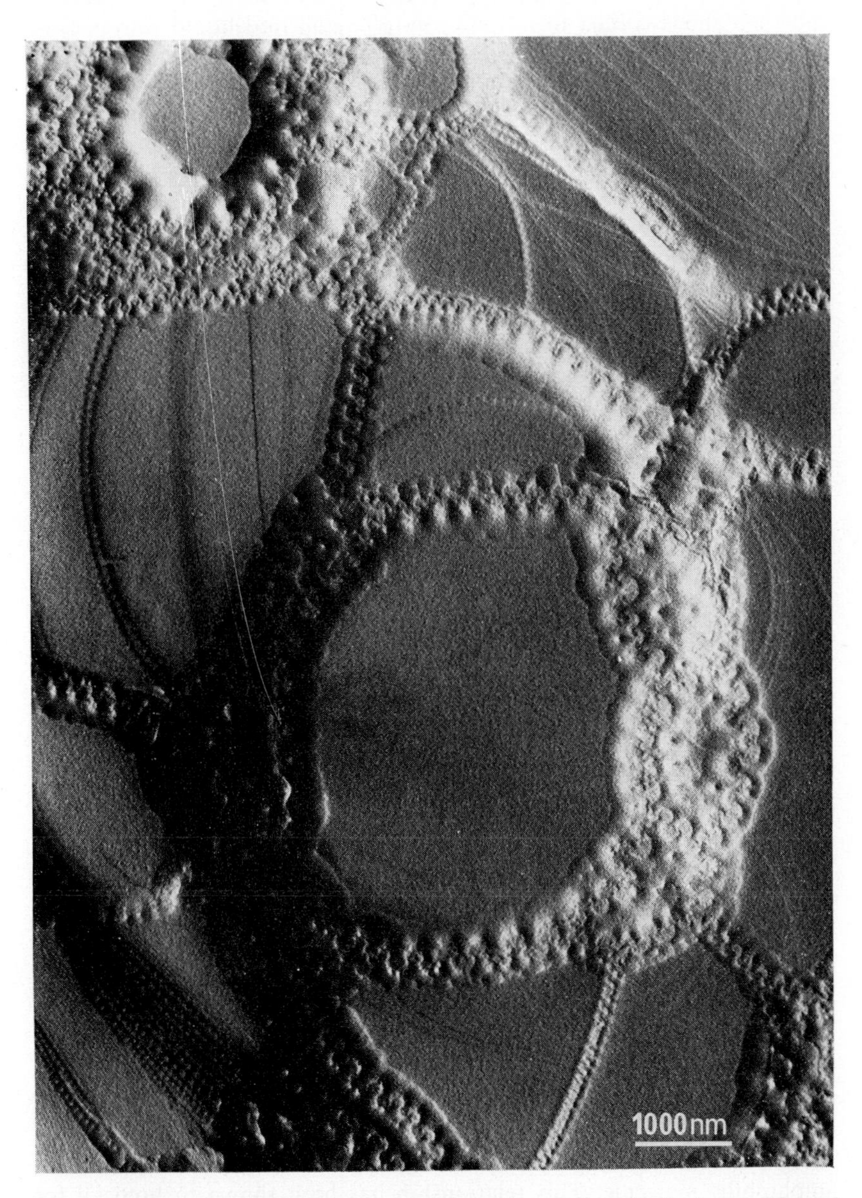

FIG. 1.1. Electron micrograph of a replica of neat (lamellar) phase obtained from a dried-down droplet of a 1% ethyl acetate solution of sodium 2-ethylhexyl sulphate. The specimen was shadowed with platinum–carbon at room temperature (1 nm = 10 Å).

isopiestically on mica at the appropriate relative water vapour pressure and then shock-cooled to $-196°C$ by plunging into liquid nitrogen, or preferably, into liquid Freon pre-cooled to liquid nitrogen temperature. If no transition to a gel or crystalline phase occurs on shock-cooling then the original mesomorphic phase structure is preserved in the form of a transparent glassy "solid". Due to the rapidity of the freezing process, the water in the sample is converted into vitreous ice which differs from the crystalline polymorphs formed at higher temperatures in that its formation from water is not accompanied by any volume expansion. Thus, no disruption of the structure occurs. However, if any crystallization does occur during shock-cooling then the resulting solid has an opaque, frosted appearance. Surface replicas of quench-frozen mesophases are prepared on a cold stage (ca. $-145°C$) after an initial freeze-etch step in which the surface is etched by a brief vacuum sublimation of ice to leave the surface features exposed [5]. A representative micrograph from a quench-frozen neat phase is shown in Fig. 1.2. The micrograph is similar to that obtained from anhydrous neat phase, showing a stratified morphology with stepped growth forms and prominent focal conic features.

There have been recent further developments in the freeze-etch technique [4, 7] in which surfaces of myelinic neat phase in phospholipid + water systems have been examined. By comparing both surface views and cross-fracture views [7], a much clearer appreciation of the three-dimensional structure is possible. In particular, replicas of freeze-etched fractured and sectioned surfaces of quench-frozen specimens have shown the lamellar structure of neat phase to be corrugated [7].

A comparison of replicas obtained from anhydrous and quench-frozen neat phase specimens suggests that in certain favourable cases complete removal of water from the system does not lead to any serious structural modification. This conclusion has been confirmed by similar studies on "inverse" middle (hexagonal II) phase in the sodium di-2-ethylalkyl sulphosuccinate + water system [1, 3]. Platinum–carbon replicas of anhydrous inverse middle phase (Fig. 1.3) clearly show the domain microstructure responsible for the characteristic angular texture of middle phase observed in the polarizing microscope [6, 8]. The tactoidal (cigar-shaped) features seen in the micrograph are a consequence of the anisotropic surface tension of the mesomorphic phase [9]. A stepped growth pattern is also clearly visible with the steps always crossing the major axes of the tactoidal features normally [3]. From high-magnification micrographs, the smallest step heights measured correspond closely to the fundamental X-ray periodicity, being approximately equal to twice the length of the amphiphilic molecule. This relationship has been shown to hold [3] for homologues from C_4 to C_{18} in the sodium di-2-ethylalkyl sulphosuccinate series. Furthermore, replicas from quench-frozen "inverse" middle phase samples in these systems [1] containing ca. 20% water (Fig. 1.4) have confirmed all the features observed in the corresponding micrographs from anhydrous specimens.

When anhydrous mesomorphic specimens are prepared by drying down

FIG. 1.2. Electron micrograph of a replica of quench-frozen neat (lamellar) phase containing 30% water from the potassium 10-*p*-styrylundecanoate + water system. The specimen was shadowed with platinum–carbon at −145°C after freeze-etching (1 nm = 10 Å).

FIG. 1.3. Electron micrograph of a replica of "inverse" middle (hexagonal II) phase obtained from a dried-down droplet of a 1% toluene solution of sodium di-2-ethyldodecyl sulphosuccinate. The specimen was shadowed with platinum–carbon at room temperature (1 nm = 10 Å).

FIG. 1.4. Electron micrograph of a replica of quench-frozen "inverse" middle (hexagonal II) phase containing 20% water from the sodium di-2-ethylhexyl sulphosuccinate + water system. The specimen was shadowed with platinum–carbon at −145°C after freeze-etching (1 nm = 10 Å).

FIG. 1.5. Electron micrograph of a replica from a pseudomorph of viscous isotropic (cubic) phase obtained from a dried-down droplet of a 1% aqueous solution of sodium 2-ethylhexyl sulphate. The specimen was shadowed with platinum–carbon at room temperature (1 nm = 10 Å).

aqueous solution droplets, the systems studied frequently pass through a series of mesomorphic phases and it is possible to finish up with a pseudomorph of one of the preceding mesophases in the drying sequence. One case has been reported [3] where the topography revealed by the replica is not that of the final anhydrous mesomorphic phase (neat phase) but is, in fact, a pseudomorph of the precursory viscous isotropic (cubic) mesophase in the sodium 2-ethylhexyl sulphate + water system. The highly developed polygonal forms (Fig. 1.5) cannot be ascribed to any equilibrium form of neat phase or middle phase, and no fine detail (e.g. steps) can be observed within the polyhedral features. Furthermore, surface replicas from a quench-frozen cubic mesophase specimen containing 24% water, in the sodium di-2-ethylhexyl sulphosuccinate + water system, have confirmed [10] that faceted droplets do indeed represent the energetically favoured morphology of cubic mesomorphic phases. Again, no indications of any growth steps were observed.

Thus, the surface replication method has revealed the morphological aspects of mesomorphic phase structure in three-dimensional perspective but, until the recent introduction of the freeze-etch and fracture technique, the nature of the underlying structure could only be inferred from the surface topography. However, detailed images of the molecular organization have been obtained by the other main method of specimen preparation, viz. chemical fixation and staining.

Chemical Fixation and Staining

Many fixation and staining procedures have been used in preparing biological specimens for examination in the electron microscope [2, 11, 12]. Several of these preparative techniques have been successfully exploited to provide direct visual confirmation of mesomorphic phase structure in amphiphile + water systems.

(*a*) Negative Staining

Because of its simplicity, negative staining [11] was one of the first methods to be used in the electron microscopic study of mesomorphic phase structure (e.g. myelinic neat phase in a naturally occurring phospholipid + water system [13]). The technique has subsequently been used by other investigators to examine neat phase dispersions (i.e. neat phase + isotropic solution mixtures) from naturally occurring [14, 15] and pure synthetic [1, 16] dialkanoyl phosphatidyl cholines (lecithins).

In the negative staining method, the mesomorphic phase as a dilute dispersion in water is treated with a heavy metal staining reagent, such as potassium phosphotungstate, and then allowed to dry out to a thin layer on a carbon support film. During drying, the negative stain sets to an amorphous electron-dense "glass" in which the amphiphilic structural units of the mesophase appear as regions of electron transparency embedded in an electron-dense background. Electron micrographs obtained from negatively stained neat phase dispersions (Fig. 1.6) frequently show a mixture of dark, stained amphiphile crystals surrounded by extensive

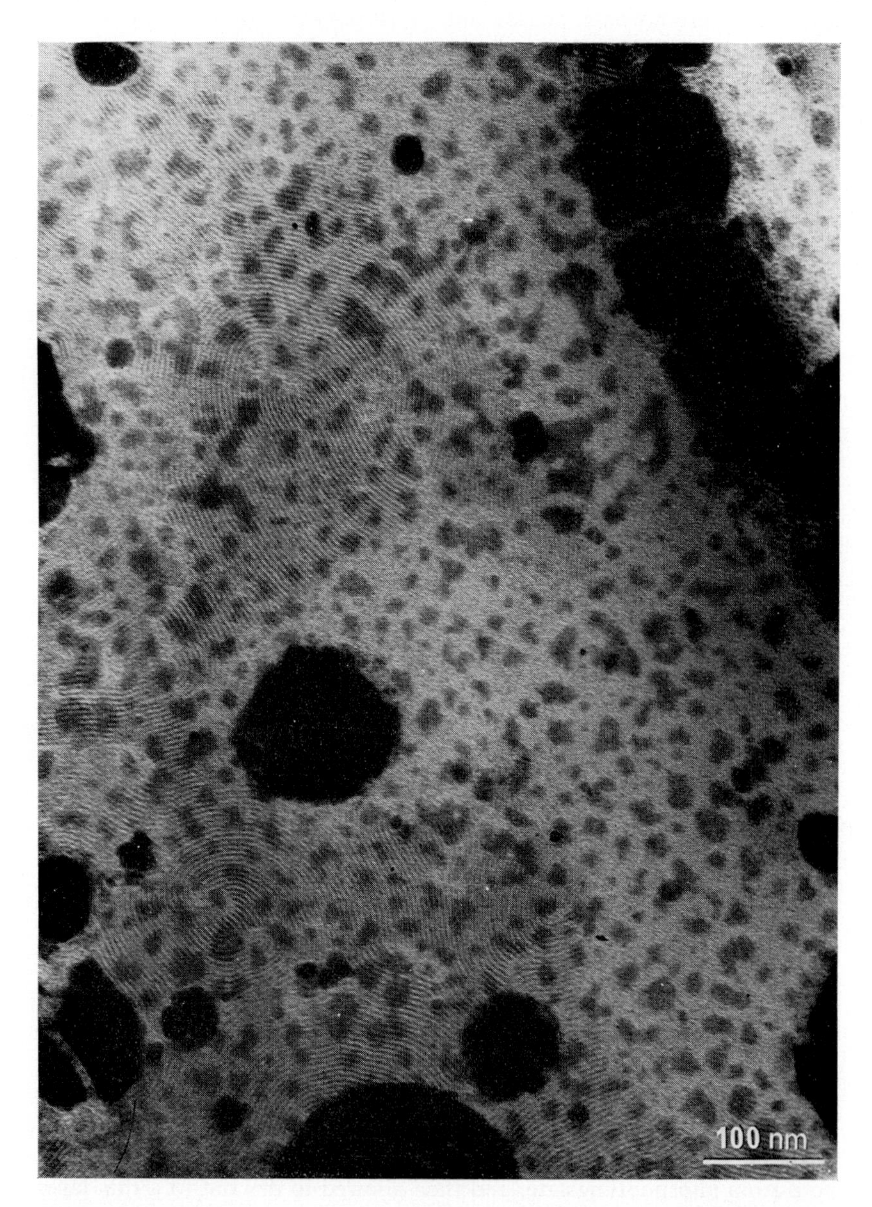

FIG. 1.6. Electron micrograph of neat (lamellar) phase [banded structure] and amphiphilic crystals [electron-dense areas] obtained from a 0·5% dispersion of didecanoyl phosphatidyl choline in water which has been negatively stained with potassium phosphotungstate at 25°C (1 nm = 10 Å).

areas of neat phase in the form of continuous, and sometimes convoluted, layers trapped during the drying process.

For an homologous series of dialkanoyl phosphatidyl cholines, the measured lamellar spacings from negatively stained neat phase preparations are in good agreement with the low-angle X-ray diffraction results from untreated specimens (Fig. 1.7). A comparison has also been made

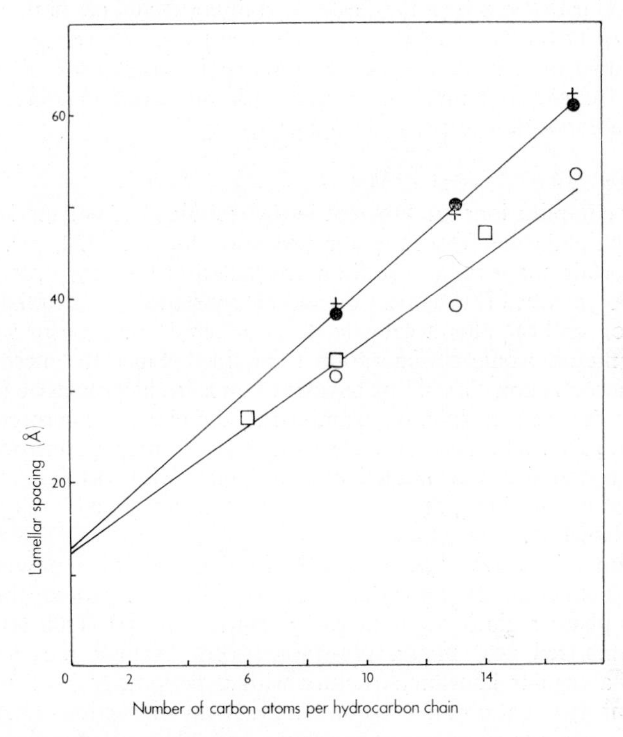

FIG. 1.7. Comparison of low-angle X-ray diffraction data and electron microscopic results for neat (lamellar) phase given by some pure dialkanoyl phosphatidyl cholines.

(+) X-ray long-spacing, d_{001}, for neat phase containing ~10% water [10]
(●) negative-staining with potassium phosphotungstate [1]
(□) tri-complex fixation with cobalt nitrate + ammonium molybdate [28]
(○) osmium tetroxide fixation followed by osmium tetroxide staining [1]

with other fixation methods in Fig. 1.7 and, although some differences in measured spacings are noted for different preparative techniques, the increments in lamellar spacing with chain length agree fairly well.

Further correlation of negative staining results with X-ray data has recently been reported for neat phase and "inverse" middle phase in a natural phosphatidylethanolamine + water system [17]. In particular, both transverse and longitudinal views of middle phase structure have been recorded. But, perhaps the most spectacular micrographs obtained by the negative staining method are those given by a complex and ill-defined system (cholesterol + lecithin + saponin + water) [14, 15, 18]. Various staining reagents have been used under different preparative conditions to reveal a variety of lamellar, spherical, tubular and helical structures, and, although comparable features were observed using fixation methods [18], no systematic X-ray diffraction data for this, or similar [19], systems have been reported.

The major drawback to the negative staining technique is that there is little control over the method. Although negative staining appears capable of providing images of hydrated mesomorphic structures [18], it is still possible for phase changes associated with variations in ionic strength, pH and composition to occur during drying.

(*b*) Polymerization and Staining

Fewer complications are inherent in the technique of polymerizing suitable amphiphilic monomers in the mesomorphic state [20, 21]. For example, certain soaps such as sodium or potassium 11-*p*-styrylundecanoate can be polymerized [21] as neat or middle phase, using γ irradiation, UV irradiation or heat plus a catalyst, to produce a polysoap in which the original mesomorphic configuration is retained. Since the mesomorphic configuration is now "fixed" by covalent bonds in the polymeric form, the structure can, in principle, be examined in the electron microscope after thin sectioning. The necessary electron contrast may be introduced by either pre-staining the unsectioned polymer with osmium tetroxide vapour or post-staining the cut sections with a heavy metal reagent such as uranyl acetate.

Electron microscopy has been used in the study of a polymerizable system (potassium 10-*p*-styrylundecanoate + water) where the mesomorphic phase regions occur at room temperature [1]. Thin sections of the polymerized neat phase, pre-stained with osmium tetroxide, have revealed a regular lamellar structure with a periodicity (30 Å) in good agreement with that obtained by low-angle X-ray diffraction (33 Å). However, thin sections from the polymerized neat phase, post-stained with a 1% solution of uranyl acetate, have yielded a different aspect of the lamellar structure (Fig. 1.8) which also shows a slightly greater periodicity (43 Å). The difference in spacing recorded for the post-stained sections may be associated with the characteristic swelling of the polysoap which occurs in the presence of water.

Unfortunately, attempted polymerization of middle phase in the potassium 10-*p*-styrylundecanoate + water system leads to structural disruption of the mesophase [1, 21]. Ultracentrifugation [10] shows the resulting polymer to have a molecular weight of ca. 170,000 and surface replicas [10] produced by drying down aqueous solution droplets have revealed a dendritic structure bearing a striking resemblance to crystals given by certain linear polymers [22].

(*c*) Osmium Tetroxide Fixation and Staining

Several fixatives, principally potassium permanganate and osmium tetroxide, have been used in studying amphiphile + water systems [23].

FIG. 1.8. Electron micrograph of a uranyl acetate-stained thin section of polymerized neat (lamellar) phase from the potassium 10-*p*-styrylundecanoate + water system. The original monomer sample contained 25% water at 25°C (1 nm = 10 Å).

To date, the most versatile and effective fixative has been osmium tetroxide vapour, and the present discussion will be concerned mainly with this preparative technique. In particular, osmium tetroxide fixation has been pre-eminently successful in preserving mesomorphic phase structure in unsaturated amphiphile systems [1], although certain saturated phospholipid + water systems have also been successfully fixed by this reagent [1, 16].

Much argument surrounds the chemistry of the fixation reaction [24, 25. 26]. With unsaturated amphiphiles, chemical fixation by osmium tetroxide vapour is considered to occur by cross-linking of the unsaturated alkyl chains via glycol osmate ester bridges while the electron-dense staining is due to accumulation of colloidal osmium dioxide in the polar head group region [27]. In essence, the insoluble black reaction product is a cross-linked polymer in which the structural organization of the mesophase has been retained in most cases. Less certain, however, is the nature of the reaction of osmium tetroxide vapour with saturated amphiphiles, usually at high temperature and low water content [1, 16]. It is possible that coulombic interactions are involved, since saturated phospholipid systems can also be satisfactorily fixed by the method of tri-complex fixation [28] with cobalt nitrate and ammonium molybdate. In this fixation method, interactions between electrolyte ions and the zwitterionic polar groups are known to occur [28].

After fixation, the material is extracted with water, dehydrated in a graded series of water/ethanol mixtures, embedded in thermosetting plastic (methacrylate or epoxy resin) and then cut into ultra-thin sections (ca. 250–400 Å thick) using an ultra-microtome. All these stages, including subsequent examination in the electron microscope where the sectioned material is bombarded by a stream of electrons, are capable of introducing artefacts of various kinds to confuse interpretation of the final image. This question of artefacts has been considered in some detail by others [25, 27].

Investigations of lyotropic mesomorphic phase structure by fixation methods have, in general, been stimulated by the pioneering studies of Stoeckenius. Following other workers in the biological field, Stoeckenius, Schulman and Prince [23] examined myelinic neat phase in the phospholipid fraction of brain tissue using osmium tetroxide vapour fixation, followed by embedding and thin sectioning. This electron microscopic study was supported by X-ray diffraction results obtained at each stage of the preparative procedure. Micrographs were obtained showing the myelinic neat phase structure in cross-section as an alternating series of light and dark bands corresponding respectively to hydrophobic and hydrophilic layers in the original structure [29]. Later investigations have centred on neat phase in pure amphiphile + water systems [1] and Balmbra *et al.* [3] have examined, at high resolution, osmium tetroxide fixed sections from a well-characterized, unsaturated amphiphile + water system in which neat phase is the only mesomorphic phase formed. A typical micrograph from an osmium tetroxide fixed and sectioned neat phase is shown in Fig. 1.9. The lamellar structure is clearly shown in

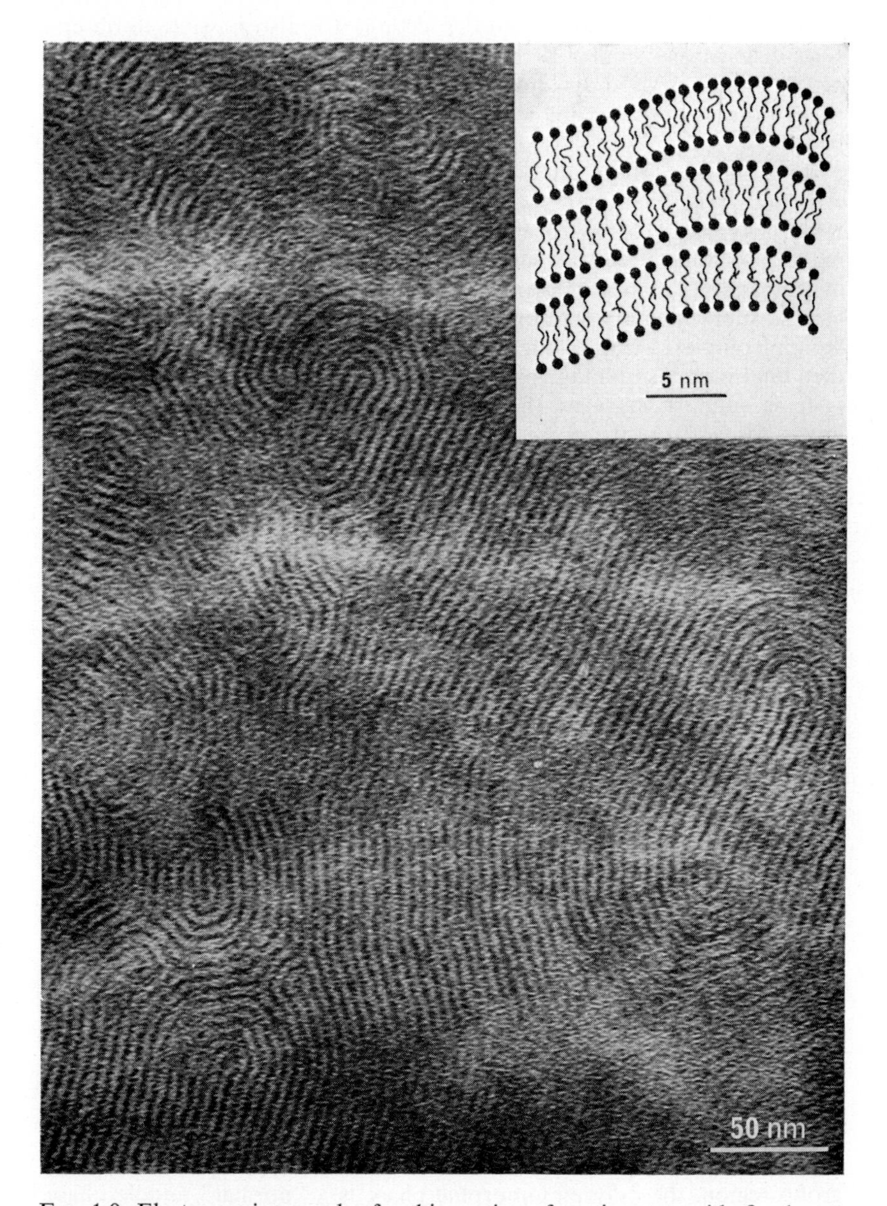

Fig. 1.9. Electron micrograph of a thin section of osmium tetroxide-fixed neat (lamellar) phase from the sodium linolenate + water system. The original sample contained 28% water at 25°C. Inset shows a schematic representation of neat phase structure in cross-section (1 nm = 10 Å).

cross-section. The layers are seen to be convoluted in a characteristic pattern (cf. Fig. 1.6) and, within the section, it is also possible to observe edge dislocations similar to those observed in conventional crystals. In general, there is good agreement between the measured lamellar periodicity in the micrograph and the X-ray long-spacing for the original neat phase [1].

In fixing neat phase an unsaturated alkyl chain is desirable but not essential. Neat phase samples from saturated synthetic phospholipid + water systems [1, 16] have been fixed with osmium tetroxide but the resulting thin sections often lack sufficient contrast when viewed in the electron microscope [1]. Nevertheless, it is possible to enhance the contrast by using a supplementary staining technique [31] whereby the osmium in the section complexes with a bidentate ligand, thiocarbohydrazide, which can then bind more osmium tetroxide vapour. The results from fully saturated systems support the view that the electron-dense contrast in the final image derives from the polar regions of the original structure.

This interpretation had been earlier suggested by Stoeckenius [30] from studies on middle phase structure. Electron micrographs obtained by osmium tetroxide fixation and sectioning of a "normal" middle phase (sodium linolenate + 54% water) were compared with those from an "inverse" middle phase (brain phospholipid + 3% water). In transverse section, the "normal" middle phase appeared as an hexagonal array of white dots on a dark background, whereas the "inverse" middle phase gave the reversed contrast, namely an hexagonal array of electron-dense dots on a light background. Typical micrographs from a "normal" middle phase (potassium oleate + 55% water at 25°C) and an "inverse" middle phase (dipalmitoyl phosphatidyl choline monohydrate at ca. 200°C.) are shown in Fig. 1.10 and Fig. 1.11 respectively. Correct interpretation of the observed image contrast enables the molecular organization in the original structure to be deduced as shown in the insets to Fig. 1.10 and Fig. 1.11. However, it has been pointed out [1] that some care must be exercised before interpreting such contrast effects in high-magnification micrographs. This is because the contrast in the middle phase image, as in all periodic images [32], can be varied with the focus, being completely reversed at the extremes of a through-focal series. It is therefore important always to select the in-focus micrograph from a through-focal series of plates.

In relating electron optical images to original mesomorphic phase structure it has proved useful to study systems which give only one mesomorphic phase. For example, in the dodecyl triallyl ammonium bromide + water system [33], where unsaturation is present in the head group region, the only mesomorphic phase is a "normal" middle phase. Hence it is possible to present a completely unambiguous interpretation of micrographs from osmium tetroxide fixed middle phase in terms of contrast and of obliquely sectioned aspects of the structure [30, 34]. It has been confirmed [33] that middle phase, in longitudinal section, yields an unconvoluted lamellar structure in which the periodicity (d) bears a simple relationship to that of the transverse section i.e. $d = \sqrt{3}a/2$, where

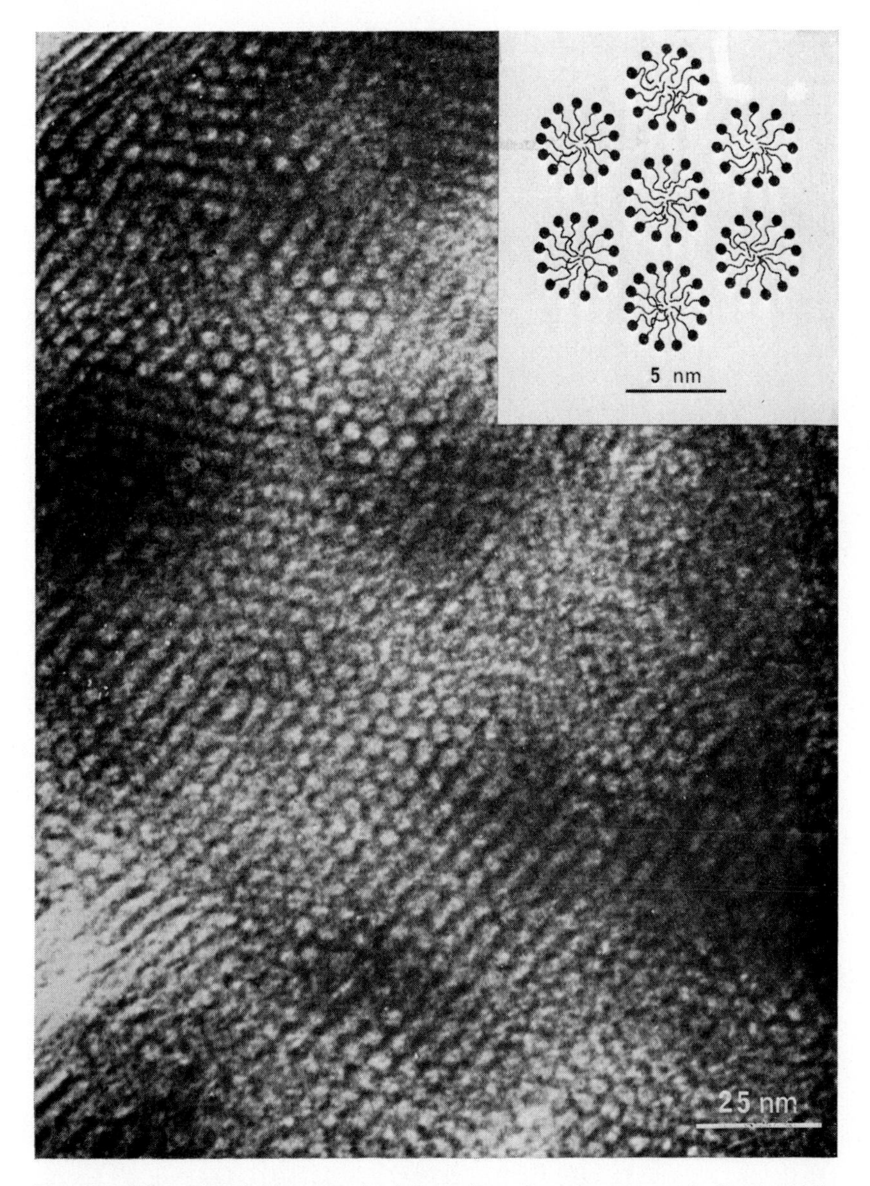

FIG. 1.10. Electron micrograph of a thin section of osmium tetroxide-fixed middle (hexagonal I) phase from the potassium oleate + water system. The original sample contained 55% water at 25°C. Inset shows a schematic representation of a "normal" middle phase structure in cross-section (1 nm = 10 Å).

FIG. 1.11. Electron micrograph of a thin section of osmium tetroxide-fixed middle (hexagonal II) phase from dipalmitoyl phosphatidyl choline monohydrate at ca. 200°C. Inset shows a schematic representation of an "inverse" middle phase structure in cross-section (1 nm = 10 Å).

a = intermicellar spacing. Frequently, the hexagonal (here white dots on a dark background) and the lamellar aspects of middle phase are observed in adjacent regions within the same micrograph, but this appears to present few difficulties when authentic neat phase + middle phase mixtures are examined [10], e.g. potassium 10-*p*-styrylundecanoate + water system in the range 49–54% water at 25°C. Micrographs from osmium tetroxide-fixed compositions in the two-phase region indicate that the segregated neat phase (d = 30 Å) and middle phase (a = 49 Å and d = 42 Å) may be readily distinguished in longitudinal section by their different periodicities.

Still more interesting is the potassium oleate + water system. Between neat phase and middle phase, there exist two other, quite distinct mesomorphic phases—complex hexagonal phase and rectangular phase [33, 35]. Direct visual confirmation of the existence of these intermediate mesomorphic phases has recently been obtained [33] by the osmium tetroxide fixation and sectioning method. Electron micrographs from through-focal series clearly show complex hexagonal phase in cross-section (Fig. 1.12) as an hexagonal array of electron-dense rings (interparticle spacing, a = 110 Å). These annular features have the dimensions expected from low-angle X-ray diffraction studies [33, 35] and represent the transverse section of an hexagonal structure consisting of indefinitely long tubes of water, filled with and surrounded by amphiphile. This structure is the inverse of that originally favoured by Luzzati *et al.* [35]. By comparison, rectangular phase has been revealed [33] as a reticular structure with two sets of electron-dense bands crossing at right angles (Fig. 1.13). This micrograph does not represent a Moiré effect due to superimposed layered lattices since the bands have a fixed periodicity and always cross at right angles. Moreover, measurements of the two-dimensional rectangular lattice yield spacings (a = 56 Å, b = 46 Å) in good agreement with the low-angle X-ray diffraction data for the unfixed mesophase [33, 35]. The electron micrographs showing rectangular phase in transverse section support the earlier tentative proposals of Luzzati *et al.* [35] that rectangular phase consists of a two-dimensional orthorhombic lattice array of indefinitely long ribbons composed of amphiphile. The nature of the cross-section of the structural units in rectangular phase remains uncertain. This is because a blurring of particle shape occurs when working near the practical resolution limit [36]. For example, at a resolution limit of 20 Å, an object which is actually square but less than 60 Å in edge-length will appear circular in the micrograph. Thus, for rectangular phase, the structural units may be either ellipsoidal (Fig. 1.13) or, less likely, rectangular [35] in cross-section.

Conclusion

Many uncertainties still surround the use of conventional electron microscopy in the study of mesomorphic phase structure in amphiphilic systems. Nevertheless, the method has provided striking visual support for structures proposed on the basis of a limited amount of low-angle X-ray diffraction data.

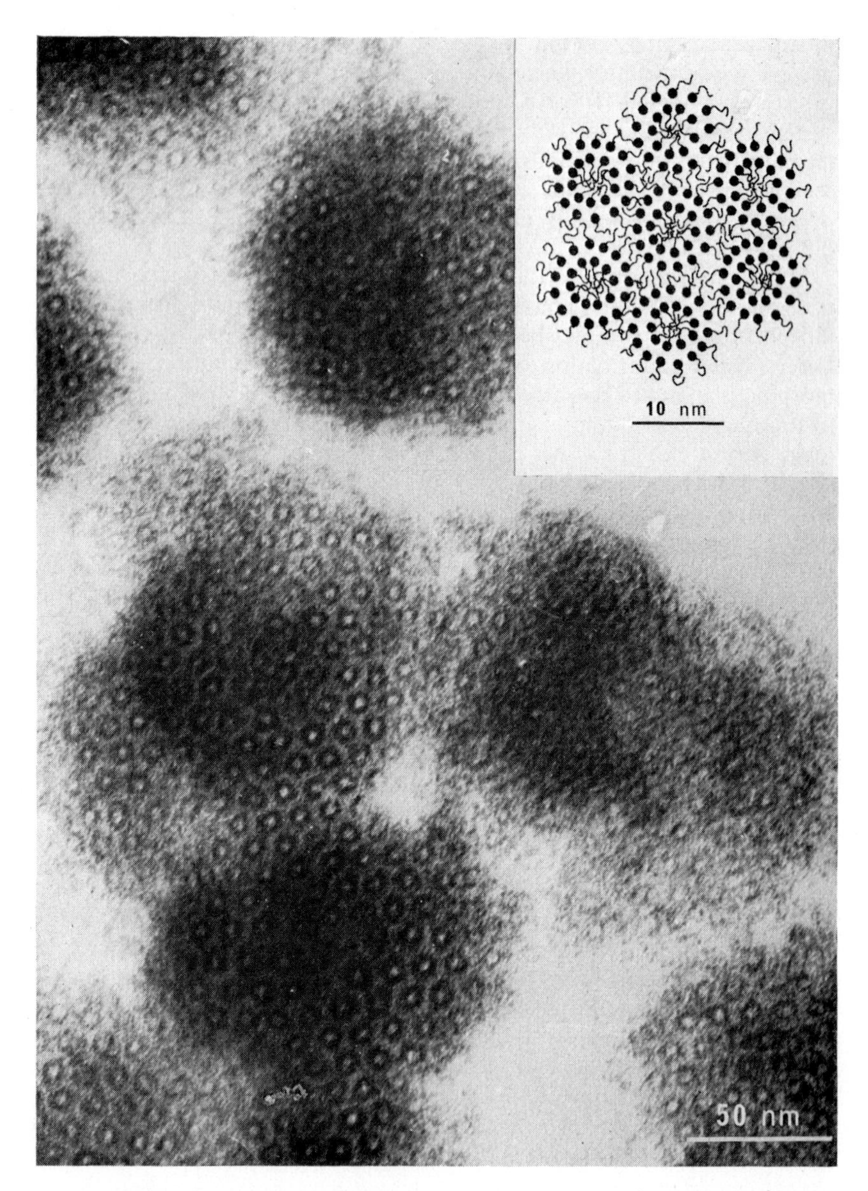

Fig. 1.12. Electron micrograph of a thin section of osmium tetroxide-fixed complex hexagonal phase from the potassium oleate + water system. The original sample contained 42% water at 25°C. Inset shows a schematic representation of complex hexagonal structure in cross-section (1 nm = 10 Å).

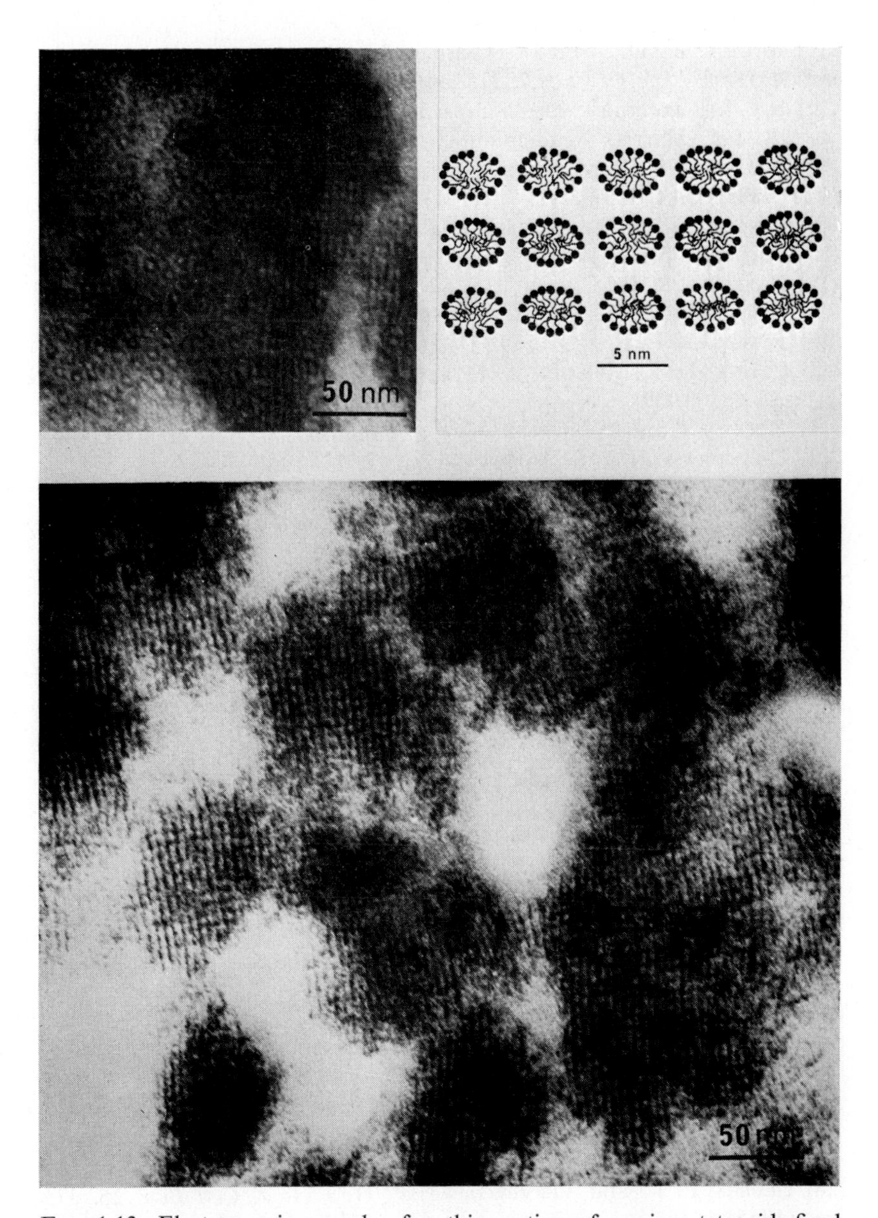

FIG. 1.13. Electron micrograph of a thin section of osmium tetroxide-fixed rectangular phase from the potassium oleate + water system. The original sample contained 45% water at 25°C. Inset (*left*) shows an area of complex hexagonal phase adjacent to an area of rectangular phase. Inset (*right*) shows a plausible representation of rectangular phase structure in cross-section (1 nm = 10 Å).

REFERENCES

[1] BUCKNALL, D. A. B., CLUNIE, J. S. and GOODMAN, J. F. *Molec. Crystals Liqu. Crystals* **7,** 215 (1969).
[2] See for example, *The Interpretation of Ultrastructure* (edited by R. J. C. Harris), Academic Press (1962).
[3] BALMBRA, R. R., CLUNIE, J. S. and GOODMAN, J. F. *Proc. R. Soc.* **A285,** 534 (1965).
[4] ZINGSHEIM, H. P., ABERMANN, R. and BACHMANN, L. (*J. Phys. E*): *Sci. Instr.* **3,** 39 (1970).
[5] BALMBRA, R. R., CLUNIE, J. S. and GOODMAN, J. F. *Molec. Crystals* **3,** 281 (1967).
[6] ROSEVEAR, F. B. *J. Am. Oil Chem. Soc.* **31,** 628 (1954).
[7] FLUCK, D. J., HENSON, A. F. and CHAPMAN, D. *J. Ultrastruct. Res.* **29,** 416 (1969).
[8] ROGERS, J. and WINSOR, P. A. *J. Colloid Interface Sci.* **30,** 500 (1969).
[9] CHANDRASEKHAR, S. *Molec. Crystals* **2,** 71 (1966).
[10] BUCKNALL, D. A. B. and CLUNIE, J. S., unpublished work.
[11] KAY, D. H. *Techniques for Electron Microscopy*, 2nd edn, Blackwell Scientific Publications, Oxford (1965).
[12] MORETZ, R. C., AKERS, C. K. and PARSONS, D. F. *Biochim. biophys. Acta* **193,** 1, 12 (1969).
[13] FERNÁNDEZ-MORÁN, H. *Circulation* **26,** 1039 (1962).
[14] BANGHAM, A. D. and HORNE, R. W. *J. molec. Biol.* **8,** 660 (1964).
[15] LUCY, J. A. and GLAUERT, A. M. *J. molec. Biol.* **8,** 727 (1964).
[16] CHAPMAN, D. and FLUCK, D. J. *J. Cell Biol.* **30,** 1 (1966).
[17] JUNGER, E. and REINAUER, H. *Biochim. biophys. Acta* **183,** 304 (1969).
[18] GLAUERT, A. M. and LUCY, J. A. *J. Microscopy* **89,** 1 (1969).
[19] BANGHAM, A. D., STANDISH, M. M. and WEISSMAN, G. *J. molec. Biol.* **13,** 253 (1965).
[20] SADRON, C. *Pure appl. Chem.* **4,** 347 (1962).
[21] HERZ, J. Thesis, Univ. of Strasbourg (1963).
[22] KARGIN, V. A. *Russ. chem. Revs* **35,** 427 (1966).
[23] STOECKENIUS, W., SCHULMAN, J. H. and PRINCE, L. M. *Kolloid-zeitschrift* **169,** 170 (1960).
[24] KORN, E. D. *Science* **153,** 1491 (1966).
[25] KORN, E. D. *J. gen. Physiol.* **52,** 257S (1968).
[26] RIEMERSMA, J. C. *Biochim. biophys. Acta* **152,** 718 (1968).
[27] ELBERS, P. F. in *Recent Progress in Surface Science* **2** (1964), p. 443 (edited by J. F. Danielli, K. G. A. Pankhurst and A. C. Riddiford), Academic Press.
[28] ELBERS, P. F. and VERVERGAERT, P. H. J. T. *J. Cell Biol.* **25,** 375 (1965).
[29] STOECKENIUS, W. *Proc. Eur. Reg. Conf. on Electron Microscopy* (Delft), Vol. II, 716 (1960).
[30] STOECKENIUS, W. *J. Cell Biol.* **12,** 221 (1962).
[31] SELIGMAN, A. M., WASSERKRUG, H. L. and HANKER, J. S. *J. Cell Biol.* **30,** 424 (1966).

[32] Grivet, P. *Electron Optics*, p. 543, Pergamon Press, Oxford (1965).
[33] Balmbra, R. R., Bucknall, D. A. B. and Clunie, J. S. *J Molec. Crystals Liqu. Crystals* **11,** 173 (1970).
[34] Eins, S. *Stud. Biophys.* (Berlin), **1,** 391 (1966).
[35] Husson, F., Mustacchi, H. and Luzzati, V. *Acta Crystallogr.* **13,** 668 (1960).
[36] Heidenreich, R. D. *Fundamentals of Transmission Electron Microscopy*, Interscience, New York (1964).

2

Optical Properties of Liquid Crystals

N. H. HARTSHORNE

The structures of the great majority of mesophases are based on parallel or approximately parallel arrangements of elongated molecules, or of elongated associations of molecules. Such arrangements are *birefringent*. In some cases the structure has in addition a twist about an axis normal to the direction of elongation of these units, and it then exhibits the property of *optical rotation*. Secondary and localized optical effects also arise when the deformability of the structure responds to the influence of centres of strong attraction for its molecules. Such centres may exist, for example, on the surfaces supporting the material.

Consider first some basic principles underlying the optical properties of mesophases, and in particular their birefringence. As the simplest case of an elongated molecule we take first an atom pair as shown in Fig. 2.1. Suppose that this is at a considerable distance from other molecules, and that it is subjected to a beam of light. The electric field associated with the

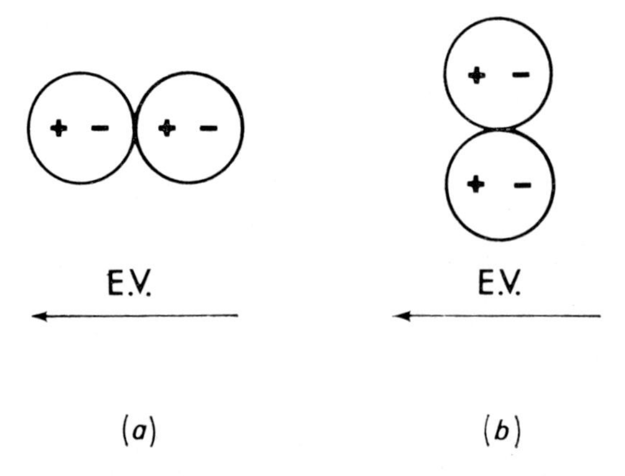

FIG. 2.1 Polarization of atom pair by light wave. E.V. = electric vector.

light waves displaces the electron systems of the atoms with respect to their nuclei so that they become electrical dipoles, the effect being termed *electron polarization*. The polarity of the dipoles undergoes continual reversals in sympathy with the oscillations of the field. At any instant the degree of polarization, i.e. the strength of the dipoles, depends not only on the phase of the oscillations, but also on the direction of the electric vector of the field (the vibration direction of the light) because of electrostatic induction effects. The two limiting cases of this directional effect are shown in Figs. 2.1(*a*) and (*b*). In (*a*) the electric vector is parallel to the line joining the atoms, so that the dipoles are in line, thus $+-+-$. Induction then results in the strength of the dipoles being greater than if the atoms were widely separated. In (*b*) the vector is at right angles to the line joining the atoms, similar charges are adjacent, thus $\begin{smallmatrix}+-\\+-\end{smallmatrix}$, and induction results in the dipole strength being less than if the atoms were far apart.

These directional effects fall off very rapidly with increase in the distance between atoms—as the third power of this distance. Therefore in a condensed structure in which diatomic molecules are arranged parallel to one another at van der Waal's distances of separation, we can as a first approximation neglect such effects between neighbouring atoms belonging to *different* molecules, because they will be much further apart than those in the same molecule which are chemically linked. The difference between the polarization of the atoms for different directions of the electric vector will thus be approximately the same as for widely separated molecules,* and since for a given density the refractive index of a medium increases with increase in the polarization of its atoms, this index will be greater when the light is vibrating parallel to the long axes of the molecules as in Fig. 2.1(*a*), than across them as in (*b*). Thus the medium will be birefringent.

The above arguments can be broadly extended to the larger and more complicated molecules that form mesophases and to their arrangements in the mesophases. In such molecules the direction of elongation is still in general the direction of maximum polarizability, or near to this direction, but the smaller transverse polarizability may vary with direction owing to the molecules having a lath-like shape (as is often the case) instead of being essentially rod-like like the simple atom pairs just considered. However, if the molecules in addition to being arranged with their long axes approximately parallel have a random lateral distribution and a random orientation around the long axes, the medium will have its maximum refractive index for light vibrating along the mean direction of these axes and its minimum one for all directions normal to this; any differences in the individual molecules between their directions of maximum polarizability and elongation, or between their polarizabilities for different transverse directions,

* The individual values of the polarization for any given direction of the vector will not, however, be independent of the density of the medium, because the polarization of an atom is increased by the field set up by the surrounding dipoles, *including those not in its immediate neighbourhood*, and the strength of this field increases with the density.

will be averaged out by the random lateral arrangement. A medium with such optical properties is said to have *positive birefringence* and to be *uniaxial*, the latter term referring to the fact that it has one *optic axis*, namely the single vibration direction corresponding to the maximum refractive index (Fig. 2.2(*a*)). Light *propagated* along this direction can

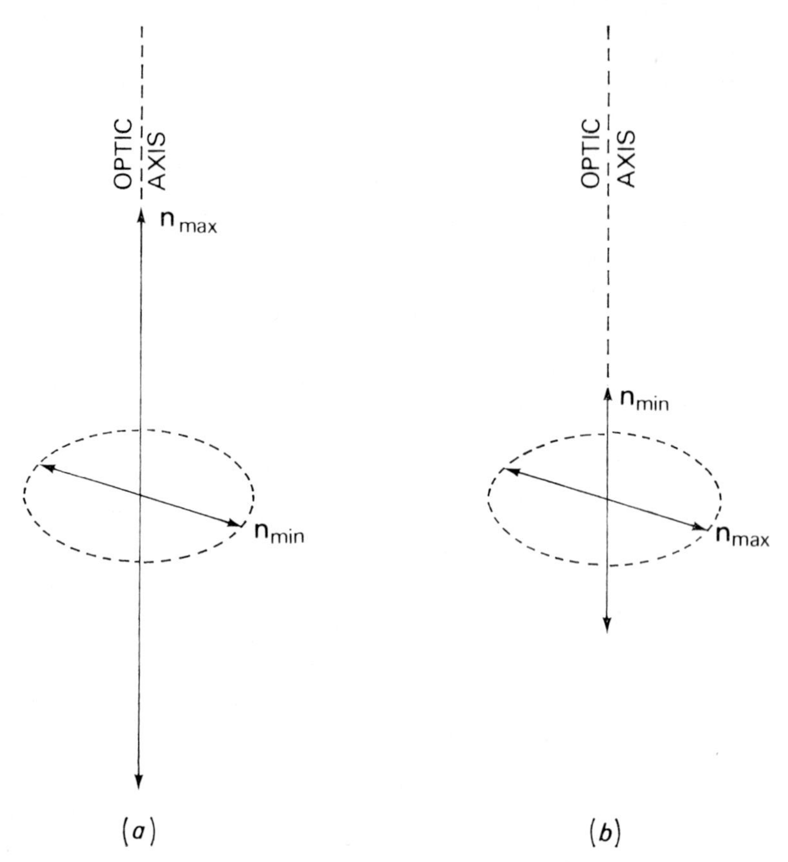

FIG. 2.2. Principal vibration directions in uniaxial media: (*a*) positive, (*b*) negative n_{max} = maximum refractive index; n_{min} = minimum refractive index.

vibrate with equal ease in any direction, and thus does not suffer double refraction.

Nematics are optically positive and are believed to have structures with a random lateral arrangement of the molecules as described above. Most smectics are also optically positive and uniaxial; the problem of their structure is complicated by *smectic polymorphism* [1, 2]; this is discussed later (p. 32).

In some, apparently few, cases the presence of highly polarizable side groups on a long molecular chain so increases the transverse polarizability that it approaches, or may even exceed, that for light vibrating along the chain. Examples are polyacrylonitrile and highly acetylated or nitrated cellulose; fibres of these substances in which the molecules are oriented approximately parallel to the fibre axis may have a greater refractive index for light vibrating across this axis than along it. As far as the writer is aware, such cases have not been found among compounds which form mesophases.* However, the regular parallel arrangements of elongated *associations* of molecules forming the "M" mesophases in systems of single-chain amphiphiles and water (p. 53) do show their maximum refractive index for light vibrating in any direction across these long units and a minimum one for that vibrating along them. This is because the average orientation of the long molecules in each unit is transverse and they radiate from the central axis. A medium with such optical properties is uniaxial and has *negative birefringence*, the single optic axis (direction of propagation for which there is no double refraction) being the same as the direction of vibration corresponding to the minimum refractive index (Fig. 2.2(*b*)). Other negatively birefringent structures will be encountered in the sequel.

Form Birefringence

It was shown [3] that a body consisting of parallel rod-shaped particles, narrower than the wavelength of light and composed of isotropic (i.e. non-birefringent) material, embedded in an isotropic medium having a different refractive index, will be positively birefringent with a greater refractive index for light vibrating along the rods than across them. Also a body in which parallel plate-like particles or layers composed of isotropic material, and thinner than the wavelength of light, dispersed in an isotropic medium of different refractive index, showed negative birefringence with a greater refractive index for light vibrating parallel to the plates or layers than for that vibrating normal to them. These effects are termed *form birefringence*. Wiener derived equations relating the birefringence to the refractive indices and volume fractions of the particles or layers and the embedding medium.

Form birefringence may play a part in determining the total birefringence in lyotropic systems. In general, however, its value as calculated from Wiener's equations is considerably less than the birefringence which can be attributed to the directional polarization effects in such molecules.

Optical Rotation

A mesophase which rotates the plane of polarization† of light has a structure in which a twist is superimposed on the usual parallel arrangement

* Many nematogens used in display devices have negative *dielectric* anisotropy (see e.g. ref. [50]). This refers to fields of low frequency; the high-frequency field of light waves produces only electron polarization, and the *optical* anisotropy of their nematic phases is positive.

† It is unfortunate that this term is used both in reference to the vectorial properties of light vibrations, as here, and also to denote the deformation of atoms by the electric field of light waves, as in the previous pages. But the context will always convey the sense in which it is used.

of long molecules or molecular associations. It may also be thought of as a pile of parallel birefringent molecular layers in each of which the molecules are arranged parallel to one another but along a direction which changes progressively from layer to layer, like the patterns on the backs of the cards in a pack which has been given a twist. The individual atoms in such a structure must lie on a series of parallel and like-handed spirals, though these spirals and their arrangement will be somewhat irregular in contrast to those in an optically active crystal, which have to conform to a definite space lattice.

These like-handed spirals in a condensed structure (crystal or mesophase) are the principal cause of its optical rotatory property, in fact the sole cause if the molecules or ions of which the structure is composed are themselves devoid of optical activity, as in crystals of sodium chlorate, sodium periodate trihydrate, etc. A simple qualitative explanation of the effect of a spiral on light transmitted along it has been given by Wooster (private communication) and uses Fresnel's theory of optical activity. A plane-polarized ray entering an optically active medium is resolved into two circularly polarized rays of opposite hands, which travel through the medium with different velocities and are therefore associated with different refractive indices. The resultant of these two components at any stage in the transmission is a linear vibration the direction of which is rotated with respect to that of the incident ray by an amount which is proportional to the distance travelled in the medium. (In the case of a birefringent crystal it is supposed that the light is travelling along an optic axis, normally the direction of the spirals, so that double refraction in the ordinary sense does not contribute to the optical behaviour.) Wooster considers plane-polarized light passing along a spiral of atoms and shows that as a result of directional inductive effects, the net deformation of the atoms will be different for the two circularly polarized components of the light. This is demonstrated in Fig. 2.3 which shows a right-handed atomic spiral and the direction of the electric vector (small arrows) in different atoms for the passage along the spiral of (*a*) right-handed and (*b*) left-handed circularly polarized light. For simplicity of presentation it has been assumed that the pitch (p) of the spiral is the same as the wavelength of the light (λ), i.e. the distance along the direction of propagation corresponding to a complete rotation of the electric vector. (In cholesterics p and λ are in fact of the same order.) As a further simplification the atoms in the figure are shown at positions corresponding to 45° intervals of rotation of the electric vector, but the existence of atoms at other positions on the spiral would not affect the following conclusions. It will be seen that in (*a*) the direction of the vector varies from tangential (*i*) to radial (*ii*) as the light travels, whereas in (*b*) alternate atoms are polarized radially and tangentially in both stages (*i*) and (*ii*) of the passage of the light. Thus it is to be expected that in (*b*) directional effects will largely cancel out, whereas in (*a*) the net polarization will be increased or reduced depending on the angle made by the row of atoms at any point in the spiral with the axis of the spiral. Thus if p is large, adjacent atoms will be more side by side than in line with respect to the direction of the electric vector for both

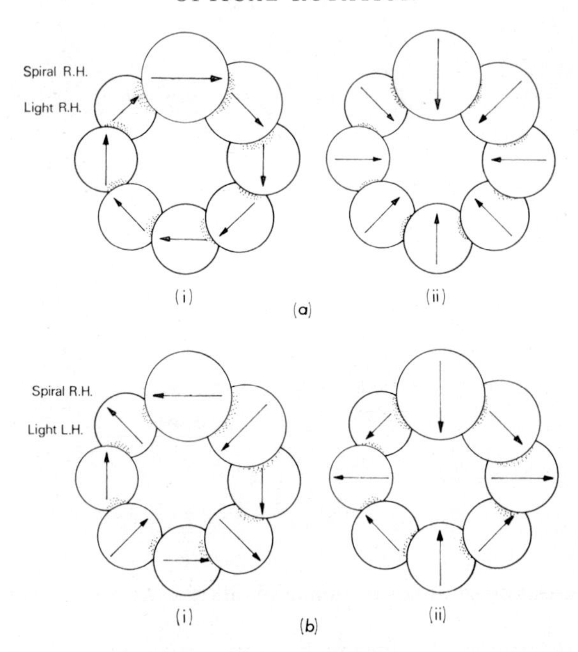

FIG. 2.3. Polarization of atoms in atomic spiral by circularly polarized components of light (W. A. Wooster).

tangential and radial directions of this vector (Fig. 2.4(I)) and so the polarization will be reduced (see Fig. 2.1(*b*)), i.e. the right-handed component of the light will be the faster one. If p is small (Fig. 2.4(II)), the opposite is the case, and for a tangential direction of the vector the polarization will be increased. It can be shown that this effect will predominate over the one arising when the vector is radial (Fig. 2.3(*a*),(*ii*)), and so the right-handed component of the light will be the slower one. These conclusions must be reversed if a left-handed spiral of atoms is considered.

In cholesterics, p is large and of similar magnitude to λ, so the spirals are very drawn out and Fig. 2.4(I) would seem to apply. In crystals, however, p is very small (a few Å) so Fig. 2.4(II) is a more appropriate picture. But the conditions assumed in Fig. 2.3 ($p = \lambda$) then no longer apply. However, it can be shown that the effect of reducing p will be only to reduce the difference in polarization between the two cases corresponding to Figs. 2.3(*a*) and (*b*) without reversing it. In agreement with these considerations, the rotations given by crystals are mostly quite small, rarely exceeding about 25° for a thickness of 1 mm, whereas those for cholesterics may be hundreds of complete turns for this thickness.

It would not be justifiable to expect more from this explanation of optical rotation than very qualitative, and somewhat tentative, conclusions like those above. Cholesterics show certain optical properties not accounted for by it, in particular a reversal in the sign of the rotation at a certain

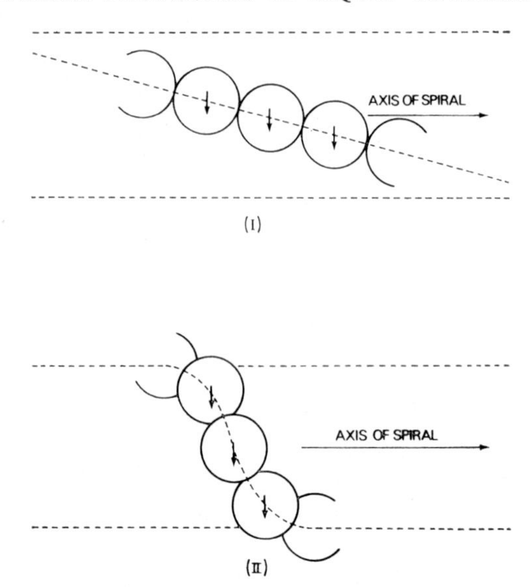

FIG. 2.4. Polarization of atoms in atomic spiral: (I) pitch large, (II) pitch small.

wavelength determined by the pitch of the structure, and the selective reflection of circularly polarized light around this wavelength. These properties call for the much more sophisticated treatment dealt with later.

Apparatus and Methods

The optical properties of mesophases have been studied mostly by the combined techniques of hot-stage and polarization microscopy. These enable substances under microscopic observation to be heated to the temperature ranges within which their mesophases are stable, and the transition points between phases to be determined. By means of the polarization equipment the directions of "fast" and "slow" vibration and the sign of the birefringence of a specimen can be determined in whatever texture it presents, and from these observations the orientation of the molecules can usually be deduced, perhaps in conjunction with evidence yielded by other techniques, e.g. X-ray analysis. The equipment also enables optical rotations to be detected and measured.

The first systematic studies of mesophases by microscopical techniques were made by Lehmann [4]. Major contributions to the understanding of mesomorphism were later made by G. Friedel, whose comprehensive paper in 1922 [5] is a classic on the subject, particularly on the microscopy of mesophases.

In what follows it is assumed that the reader is familiar with the principles of polarization microscopy and its application to the study of birefringent materials. Those wishing for further information on these subjects should consult one of the standard works, examples of which are given in references [6] and [7].

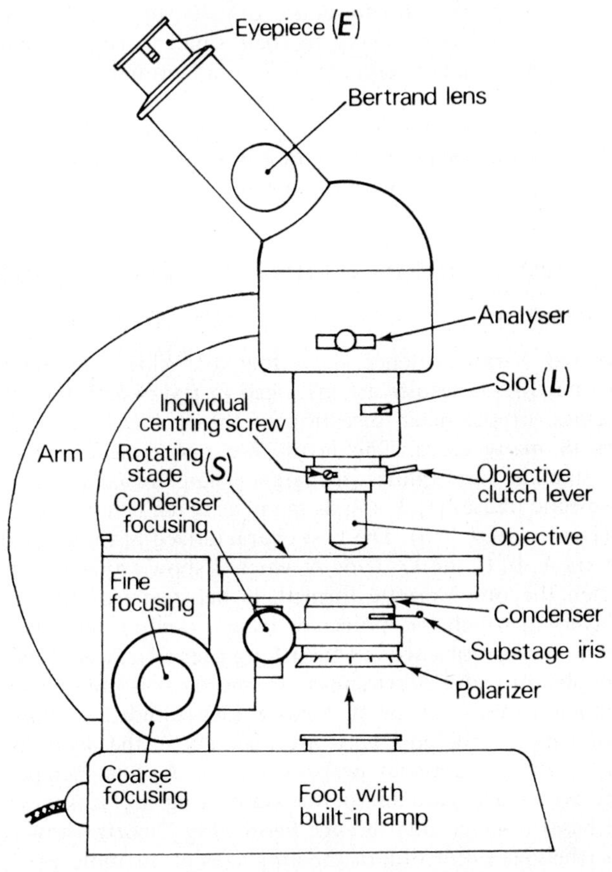

FIG. 2.5. The polarizing microscope.

Fig. 2.5 shows the general arrangement of the components of a polarizing microscope with a modern type of stand. For attachment to the microscope, hot stages with an upper limit of 300–350°C are commercially available, e.g. the Kofler stage (Reichert Optische Werke), the Leitz "cold/hot stage, −20 to 350°C" (E. Leitz), and the Mettler "FP2" and "FP52" stages (Mettler Analytical and Precision Balances). This range of temperature covers the regions of stability of practically all mesophases. However, these stages require considerable working distances between specimen and objective and condenser, and so the study of high-aperture interference figures for determining the optical sign of a phase is not possible. For obtaining such figures at moderate temperatures the Leitz "cold/hot stage, −20 to 80°C", with which ordinary high-power objectives can be used, is suitable. The writer has constructed stages with an upper limit of 100 to 150°C, consisting of a thin electrically heated metal plate, bolted direct to the microscope stage and incorporating a thermocouple to measure temperature [7]. Satisfactory interference figures have been obtained, but to attempt to reach higher temperatures with such devices would probably be to court disaster to ordinary high-power objectives and condensers owing to their small working distances from the heated object. Higher temperatures could probably be reached with safety by use of the special medium-aperture objectives and condensers (numerical apertures up to 0·65) supplied by E. Leitz and Carl Zeiss (both the Oberkochen and Jena concerns) for conoscopic work with the universal stage and which have large working distances.

THERMOTROPIC MESOMORPHISM OF SINGLE COMPOUNDS

Smectic Mesophases

Optical and X-ray evidence show that the chief characteristic of a smectic is that the molecules are arranged in layers with the long axes of the molecules, or the mean direction of these, normal to the planes of the layers in many cases. The layers are one molecule thick, or two molecules thick in amphiphiles (see later). There are a number of different types of smectic phases [1, 2, 45], as many as seven being claimed by some workers (Demus *et al.* [44]). The best characterized of these are designated by the letters A, B, C and D. Type A which is shown by many smectogens, and is often the only smectic formed, is believed from X-ray evidence to have layers in which the molecules have a random lateral distribution and random orientations around their long axes ([1], second reference). In substances showing also other smectics, smectic A is the one stable at the higher temperatures. In Type B, X-ray evidence indicates that the molecules have a more ordered lateral arrangement in the layers ([1], second reference). Optically uniform preparations of A- and B-type phases in which the layers are parallel to the supporting surfaces normally give a centred basal uniaxial interference figure (Fig. 2.6(*a*)), showing that the optic axis (the mean direction of the long axes of the molecules) is normal to these surfaces. This orientation of the layers may arise spontaneously when the substance is heated to give the smectic either between a slide and cover slip cleaned without the use of any agent which might etch the glass, or on a freshly cleaved and uncovered mica surface. In the latter case the characteristic stepped or layered drops ("gouttes à gradins") discovered by Grandjean may be formed. Friedel [5] gives photomicrographs of such a preparation. By inserting a suitable compensator into the slot of the microscope (L, Fig. 2.5), e.g. a unit retardation ("Red I") plate, if the preparation is thin and the interference figure shows no colours, the optical sign may be determined, and is positive in the great majority of cases (Fig. 2.6(*b*)). Some amphiphiles with comparatively short lipophilic chains give negative smectics (p. 59).

Lehmann and Friedel [5] used the term *homeotropic* to describe preparations in which the optic axis is normal to the plane of the preparation. The term is applicable to both smectics and nematics (p. 38). Such preparations are characterized by appearing dark between crossed Nicols *in parallel light*, and for that reason were also termed "pseudo-isotropic" by Lehmann.

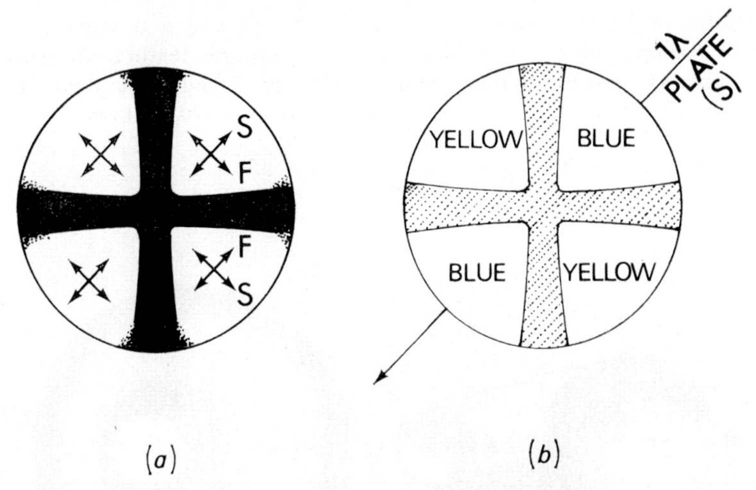

FIG. 2.6. (*a*) Basal interference figure of positive uniaxial medium, showing "slow" (S) and "fast" (F) vibration directions in different quadrants. (*b*) The same after insertion of unit retardation plate (length "slow"). (The case depicted is that for a thin section of material and using an optical system of medium numerical aperture.)

C-type phases have layers in which the molecules are tilted, and some at least present an off-centre biaxial interference figure corresponding to a small optic axial angle in preparations in which the layers are parallel to the supporting surfaces (Taylor *et al.* [46]). The biaxial character is thought to be due to an anisotropy in the tilt of the molecules for different directions.

D-type phases [1], which so far seem to occur in very few cases, have the puzzling property of presenting an isotropic texture for parallel light and no interference figure in convergent light, thus indicating a truly isotropic structure. No satisfactory explanation of this has been offered, and it seems doubtful whether this phase can in fact be smectic in the sense of having a stratified structure.

When a smectic A-type phase (and frequently other smectic polymorphs) are confined between surfaces with which they form strong local attachments, e.g. between a slide and cover slip which have been previously roughened by treatment with hydrofluoric acid, the layers become contorted while preserving their constant thickness and parallelism. From microscopic study* it may be inferred that the layers form series of parallel curved surfaces of the geometrical type known as the *cyclides of Dupin*, and of the possible arrangements satisfying the above conditions this is the one having the lowest potential energy. It is known as the *focal conic texture* for reasons given below. A Dupin cyclide (Fig. 2.7) may be described as a hollow circular ring having a circular cross-section the

* Largely carried out on such textures by G. Friedel [5], whose microscopical researches on the contorted phases revealed their essentially stratified nature, before this was confirmed by X-ray analysis.

diameter of which varies from a maximum (PQ) to a minimum (RS). Fig. 2.8 shows the geometrical basis of a focal conic texture, the parallel and equally spaced cyclides being numbered serially. As their cross-sectional diameters increase the central hole (Fig. 2.7(*b*)) shrinks to a point

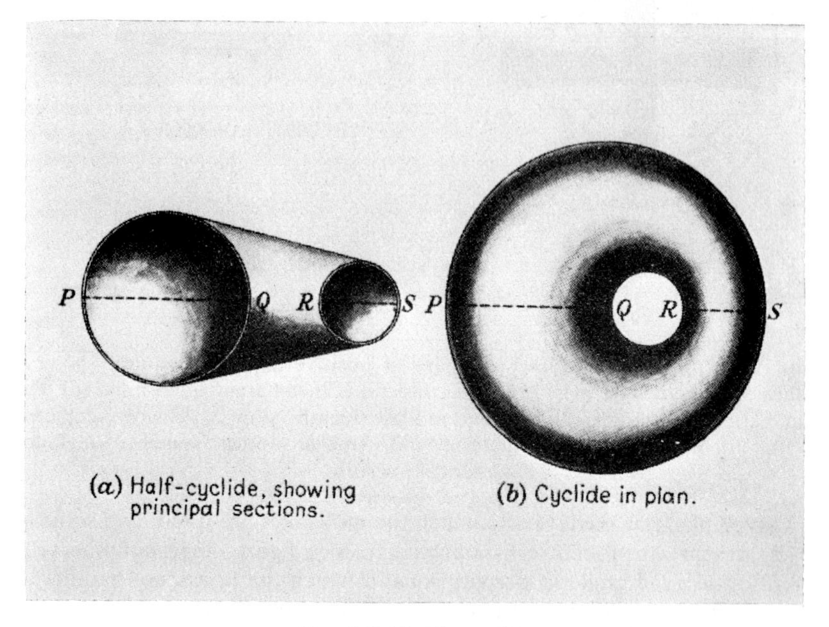

FIG. 2.7. Dupin cyclide.

(cyclide No. 1) and thereafter is replaced by "dimples" as at E and F (Fig. 2.8(*a*)). In the direction of decreasing cross-sectional diameters (cyclides 1 to 8) the minor one shrinks to a point (B, cyclide No. 5) and thereafter the cyclides become crescents, e.g. VXV′ (cyclide No. 8, Fig. 2.8(*b*)). The loci of points at which the cyclides show an abrupt change in direction are the hyperbola DFOEC (Fig. 2.8(*a*)) and the ellipse AVUTB (Fig 2.8(*b*)). These curves are at right angles to one another. The hyperbola passes through one of the foci of the ellipse (O), and the focus of the hyperbola is one of the apices of the ellipse (B). An ellipse and a hyperbola related in this way are termed a pair of focal conics and this gives the name to the structure. A further geometrical feature is that the hyperbola is the locus of the apices of a series of cones of revolution of which the ellipse is a common section. Thus C and E are the apices of cones of revolution CAB and EAB respectively.

In a smectic showing a well-developed focal conic texture the ellipses and hyperbolae can be seen as dark lines since they mark structural and therefore optical discontinuities in the medium. One of the best examples is the so-called *polygonal texture* (Fig. 2.9), obtained in general when the preparation is fairly thick and not too viscous [5]. With the microscope focused on either the upper or lower surface of the preparation the field

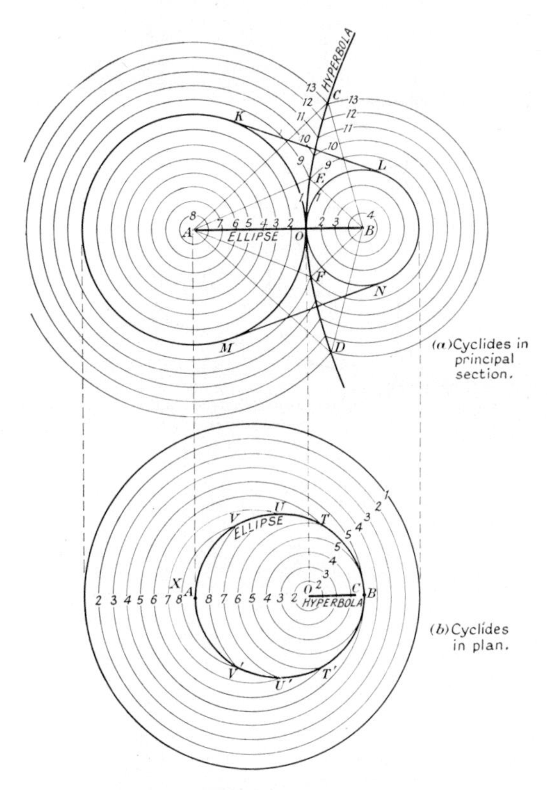

FIG. 2.8. Focal conic structure.

is seen to be divided into polygonal areas which may have any number of sides and in each of which there appears a family of ellipses. The spaces between the larger ellipses are occupied by smaller ellipses, and the spaces between these by still smaller ones and so on. Fig. 2.9(*a*) shows the larger ellipses in an imaginary case of a four-sided polygon ABCD and portions of the ellipses in neighbouring polygons. The major principal axes of all the ellipses in a given polygon pass, when produced, through one point (P in the figure). If the upper surface of the preparation be focused and the analyser (without the polarizer) be inserted, or alternatively, if the lower surface be focused and only the polarizer be used, each ellipse is seen to be crossed by a straight "brush" or isogyre the direction of which is parallel to the vibration direction of the polar, and the narrowest point of which coincides with one of the foci of the ellipse (L, M, N, O in the figure, in which the microscope is supposed to be focused on the upper surface).

If the focus of the microscope be gradually altered from the upper to the lower surface it is possible to make out that each ellipse is partnered by one branch of a hyperbola which starts at the focus crossed by the brush (L, M, etc.) and that the hyperbolae belonging to ellipses in the same polygon all meet on the lower surface at one point (K in Fig. 2.9(*b*))

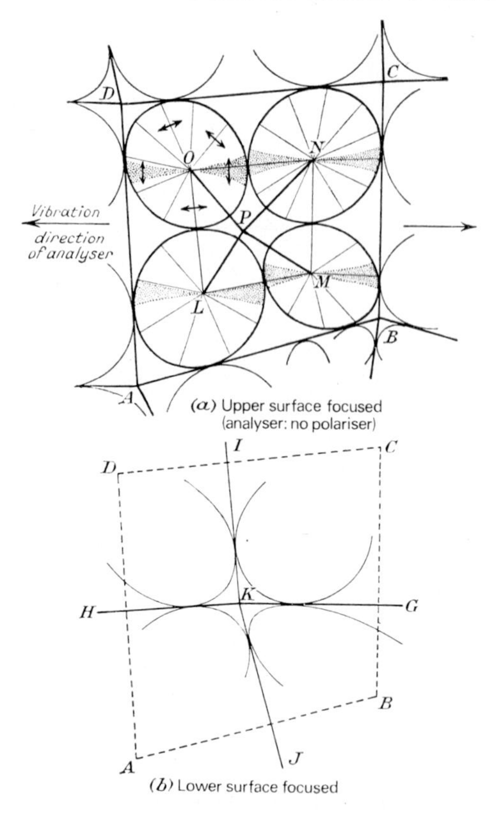

FIG. 2.9. Polygonal texture.

which is immediately below the point of intersection of the major axes of the ellipses (P in Fig. 2.9(*a*)). This point of intersection of the hyperbolae is a common corner in a system of polygonal areas on the lower surface which has now come into view. Thus in Fig. 2.9(*b*), K is a corner common to the polygons AHKJ, BGKJ, etc.

The above observations can be explained as follows. The medium is divided into pyramids and tetrahedra, the bases of the pyramids being the polygons observed on the upper and lower surfaces. For example, the base of the inverted pyramid KABCD (Fig. 2.10) may be taken to correspond to the polygon ABCD (Fig. 2.9(*a*)) and the apex of the pyramid, K, to the point lettered similarly (Fig. 2.9(*b*)). Each pyramid contains a family of cones of revolution, each being a focal conic "domain" such as that contained within the cone ABC (Fig. 2.8(*a*)). Also, the tetrahedra, such as ABKJ (Fig. 2.10), are focal conic domains in which the edges of the tetrahedron lying on the upper and lower surfaces of the preparation are, respectively, portions of an ellipse and a hyperbola, or the reverse. The boundaries of the polygons are in fact slightly curved.

When light passes through a phase having this texture it is everywhere resolved into ordinary and extraordinary rays, the former vibrating at

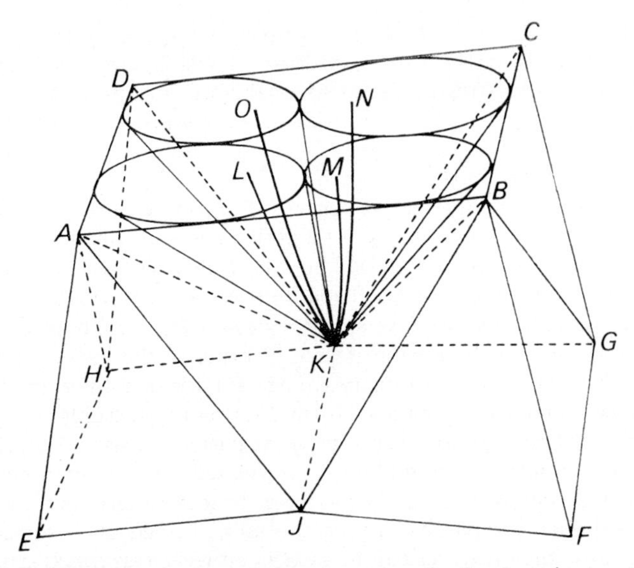

FIG. 2.10. Domains in the polygonal texture of a focal conic structure.

right angles and the latter parallel to the long axes of the molecules. Owing to the curvature of the strata, however, the direction of the axes changes continuously from point to point with the result that the extraordinary rays are deviated and become "lost". (Grandjean [8] calculated the path of an extraordinary ray in such a medium.) Little light corresponding to this component therefore reaches the objective of the microscope. The ordinary rays are not affected. Fig. 2.9(*a*) shows the vibration directions of the ordinary rays in different parts of the ellipses as short double-headed arrows. These are at right angles to the radial lines along which the long axes of the molecules lie, which in turn are at right angles to the cyclides. (Compare the figure with the ellipse and its included cyclides in Fig. 2.8(*b*).) If only the analyser is used, the light vibrating normally to its vibration direction is extinguished (see the figure), giving the dark brushes mentioned above. A similar explanation accounts for the brushes parallel to the vibration direction of the polarizer which are seen when only this polar is used.

Apart from the treatment in Friedel's paper [5], a very full account of the focal conic texture was given by Sir William Bragg in 1934 [9].

Smectic phases often display textures that are more difficult to interpret than the polygonal texture, and although they are all probably based on the focal conic arrangement the ellipses and hyperbolae cannot always be distinguished in the microscopic image. The chief examples are: (1) *b tonnets,* separate little objects having a characteristic elongation along an axis of rotational symmetry but otherwise a bewildering variety of shapes, are the form in which a smectic may separate on cooling the isotropic melt; (2) the *fan-like texture* (Friedel's "*plages en eventail*") consisting of irregularly disposed areas in which a fan-like pattern is revealed by using crossed polars; this may consist of very eccentric ellipses; (3) "*oily streaks*" (Friedel's "*stries huileuses*") which are long transversely striated bands consisting of chains of focal conic groups, and which sometimes appear when a well-defined focal conic texture is disturbed by shifting the cover slip. For further details about these textures Friedel's paper should be consulted.

The textures and optical properties of smectics other than Type A, and particularly of preparations of them that are optically inhomogeneous, have received less detailed study, but the following characteristics have been noted. Contorted layers of Type B phases are unstable and tend to be replaced by optically homogeneous regions ([1], second reference), no doubt as a result of the greater degree of order in the molecular arrangement in the layers. Stepped drops (p. 32) are formed by both B- and C-type phases, and the latter also form focal conic textures, though with characteristic breaks or interruptions in the pattern. A streaky ("schlieren") texture [49], somewhat resembling the nucleated domain texture shown by nematics (see below, Fig. 2.11) has been found in certain C- and other phases [44] when prepared between specially cleaned glass surfaces. It is not clear how such textures can be associated with a stratified structure.

Nematic Mesophases

These phases derive their name from the mobile thread-like lines observed in them when they have been prepared by rapid cooling of the melt. These lines are discontinuities in the structure, like the ellipses and hyperbolae in smectics, but unlike these they do not have to conform to any geometrical plan because there is no stratification in the medium.

A feature of most nematic phases when prepared directly from the crystalline phase confined between lightly etched glass surfaces is that the molecules in contact with the glass have a striking tendency to attach themselves sideways to it and govern the orientation of those in the bulk of the phase. This is shown by the following experiments. Molten *p*-azoxyanisole or *p*-azoxyphenetole is allowed to run between a slide and cover slip, previously treated with hydrofluoric acid. By cooling the preparation and inoculating the edge of the film with a grain of the crystalline substance, fairly large crystal plates are obtained. On heating the preparation, each plate becomes nematic without changing its outlines, and is optically uniform with an orientation presumably determined by that of the original crystal. Some areas of the preparation will probably be oriented so as to present approximately centred basal uniaxial interference figures (Fig. 2.6), and the effect of a compensator on such figures will show that the sign is *positive*, a consequence of the parallel arrangement of the molecules. (The nematic phases of some substances when confined between cleaned but unetched glass surfaces readily adopt a homeotropic orientation (p. 32), i.e. one in which the optic axis is precisely normal to the surfaces and thus gives a centred figure.)

If now the cover slip be slightly shifted sideways the borders of the uniform plates become doubled, because the molecules in the top layer are attached to the cover slip while those in the bottom layer remain stuck to the slide. If instead the cover slip is rotated a little, the medium acquires a kind of optical rotatory power, because it is now forced to adopt a twisted structure. This differs from true optical activity in that if plane-polarized light is passed into the medium along the axis of twist, it only remains plane-polarized if its vibration direction on entry is parallel to one or other of the vibration directions offered by the first layer. The vibration direction of the light then closely follows the twist and there is little or no wavelength dispersion of the rotation. (In true optical activity the rotation is approximately inversely proportional to the square of the wavelength. If this condition is not met, the light undergoes a more complicated rotatory effect in which it becomes elliptically polarized and only regains its plane polarization at regularly spaced depths of the material along its direction of transmission.

This behaviour is due to the fact that the pitch of the twist is very much greater than the wavelength of the light [9], [10], and is to be contrasted with that of a cholesteric (pp. 28, 44) in which the pitch and wavelength are of the same order.

It should be noted that it is not possible to twist a nematic in this way through more than 90°. If the rotation of the cover slip exceeds 90° by $\theta°$ and is, say, right-handed, the twist of the structure relaxes to a left-handed one of $(90 - \theta)°$, because this has a lower potential energy than that corresponding to a continuation of the right-handed rotation. This behaviour is the result of imposing an external twisting force on a system the molecules of which have no inherent tendency to adopt a twisted structure. By contrast, the twisted structures of cholesterics result directly from interactions between neighbouring molecules, and a twist of a given hand can continue through any angle whatever.

We must now consider the structure around the thread-like lines in nematics. Information about this structure has been obtained from certain textures studied by Lehmann, Grandjean and Friedel, and termed by the last "*plages à noyeaux*" (*nucleated domains*) (Fig. 2.11). Between crossed polars it shows an irregular network, of dark brushes branching out from a number of scattered points or nuclei, and passing continuously from one nucleus to another. Four brushes meet at some nuclei and two at others. The nuclei are the thread-like lines—attached at the ends to the glass surfaces—end on, as may be shown by slightly shifting the cover slip when the points become lines.

Friedel [5] termed the nuclei at which four brushes meet *integral nuclei*, and those with two only, *half nuclei*. Each of these is also of two kinds, *fixed* and *turning*, which differ as follows. When the stage of the microscope (S, Fig. 2.5) is rotated the brushes in the immediate neighbourhood of the nucleus remain unchanged, or practically unchanged, in position if it is a fixed one, but rotate in the same direction as the stage, and at twice the angle (or more in some cases), if it is of the turning type.

The interpretation is as follows. In making a complete circuit around a nucleus, the direction of the optic axis of the medium (i.e. the mean direction of the molecular long axes) changes by 360°, or 2π, if the nucleus

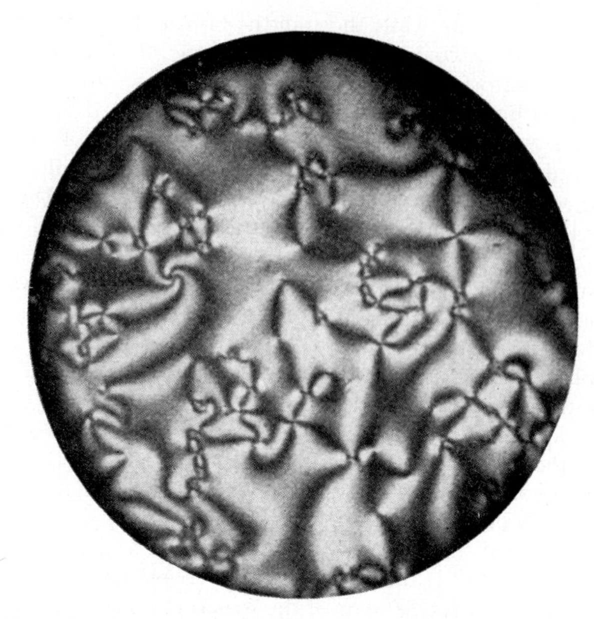

FIG. 2.11. Nematic phase of *p*-azoxyphenetole showing "plages à noyeaux". Both integral nuclei (4 brushes) and half-nuclei (2 brushes) can be seen [5].

is integral, and by 180°, or π, if it is a half-nucleus. The difference between fixed and turning nuclei is that around the former the direction of rotation of the optic axis is the same as that in which the circuit is made, whilst in the latter the directions are opposite. Four possibilities are illustrated in Figs. 2.12(*a*) to (*d*), in which are represented the nucleus as a central point, the directions of the optic axis by continuous lines, and the brushes by stippling. The positions of the analyser are denoted An_1 and An_2, and of the brushes B_1 and B_2. Those of the polarizer are omitted for clarity. In (*a*) (integral fixed) the optic axis lies along radial lines (it could also be along concentric circles around the nucleus), and the behaviour is similar to that of a spherulite. If An_1 is the initial analyser position and the stage is rotated in a clockwise direction (equivalent to rotating the polars anti-clockwise, so that the analyser moves, say, to An_2), the dark cross remains coincident with the cross-hairs of the eyepiece (E, Fig. 2.5), i.e. with the vibration directions of the polars. In (*b*) (integral turning type) the optic axis lies on a cross and on four sets of curves tangential to it as shown. If we start with the brushes coincident with this cross and rotate the stage clockwise (equivalent to rotating the analyser from An_1 to An_2, say), the brushes also move clockwise from B_1 to B_2, their position always marking the locus of points on the curves where these are at right angles or parallel to the analyser direction. The shape of the curves is such that angle B_1B_2 = angle An_1An_2, i.e. the brushes have moved through twice the angle of rotation of the stage. In (*c*) and (*d*) we have cases corresponding to (*a*) and (*b*) respectively, but for half-nuclei, around which the optic axis only undergoes a change of π for a complete circuit. The right-hand branch

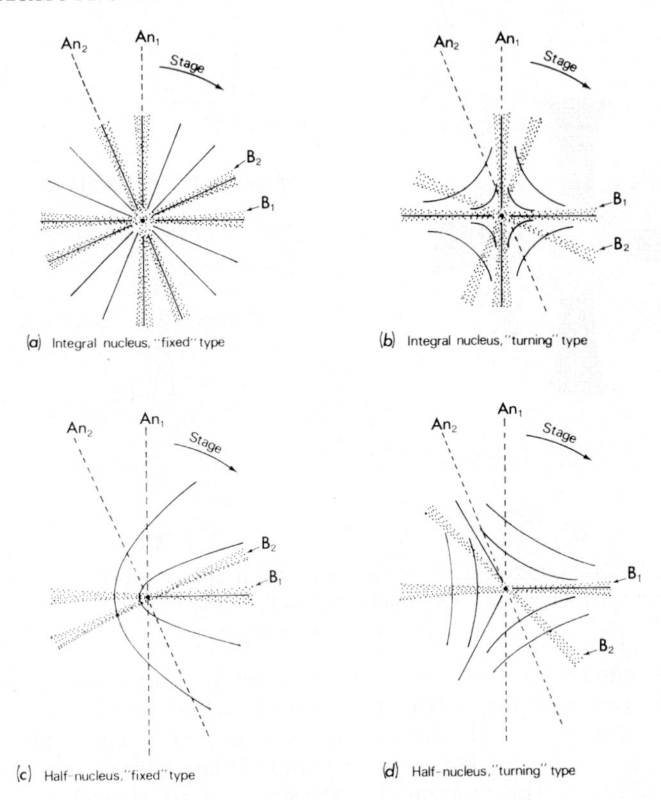

FIG. 2.12. Types of nuclei in "plages à noyeaux".

of the brush B_2 in (*c*) is only indicated approximately; for the structure shown it would be slightly curved. In (*d*) the angle B_1B_2 is about twice the angle An_1An_2, so that the rotation of the brush (i.e. the sum of these angles) is about three times that of the stage.

While the general shapes of the curves in Fig. 2.12 follow from the optical observations, they were also calculated by Oseen from a theory of the curvature-elasticity in liquid crystals [11]. A revision of the theory by Frank [12] did not affect the configurations represented by the curves (see also ref. [49]).

Cholesteric Mesophases

The optical properties of cholesterics are striking and differ significantly from those of other types of mesophases. The properties are best demonstrated by preparations between a slide and cover slip in which the optic axis is normal to the glass; this orientation often arises spontaneously or by imparting small movements to the cover slip. Conoscopic observations show that the sign is *negative* (Fig. 2.13). There is strong optical activity along the optic axis, and the rotation for a thickness of 1 mm may amount

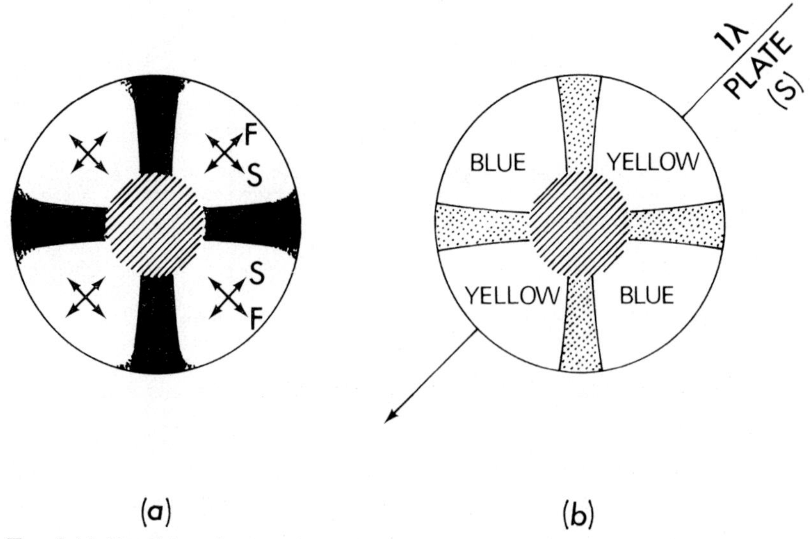

FIG. 2.13. Basal interference figure given by basal "plane" texture of cholesteric mesophase showing (*a*) optical activity, and (*b*) negative sign by insertion of unit retardation plate.

to some hundreds of complete turns. As a result of dispersion of the rotation the centre of the interference figure in white light is coloured, the colour being that of the light which emerges vibrating parallel to the vibration direction of the analyser. Thus if the analyser is rotated this colour changes. The rotation changes sign at a wavelength λ_0, which is characteristic of the substance and the temperature. Within a narrow band of wavelengths around λ_0 a circularly polarized component of the normally incident light is strongly reflected and a circularly polarized component of opposite hand is transmitted. When white light is directed obliquely at the preparation it is scattered and the wavelength of the rays making the same angle with the surface as the incident rays is always less than λ_0 at the same temperature, and declines with increasing obliquity of the light. When such preparations are viewed by diffused daylight the scattering of different wavelengths in different directions produces a striking display of vivid colours recalling the appearance of a peacock's feather.

The reflected component of normally incident light is right-handed for some cholesterics (*dextro*-type) and left-handed for others (*laevo*-type). A remarkable feature is that if right-handed circularly polarized light is incident on a *dextro*-mesophase, or left-handed light on a *laevo*-mesophase, it is reflected *without change of sense*. In all other cases in which circularly polarized light is reflected the sense of the rotation is reversed. If, however, the sense of the circular polarization of the incident light is opposite to that of the light normally reflected by the mesophase, the light is transmitted without change of sense.

The optical activity of cholesterics conforms to the following rules. In *dextro*-mesophases, the plane of vibration of the light is rotated to the right at wavelengths less than λ_0, and to the left at greater wavelengths. For a *laevo*-mesophase rotation to the left occurs at wavelengths less than λ_0, and to the right at greater wavelengths.*

(The meaning of "right" and "left" follows the usual convention, namely that the observer is supposed to be facing the oncoming light and a rotation which he sees as clockwise is called right-handed.)

Grandjean [14] discovered that when a little active amyl 4-*p*-cyano-benzylideneaminocinnamate is melted into a fresh cleavage crack in a sheet of mica, so that it takes the form of a thin wedge, the uniformly oriented mesophase is formed spontaneously on cooling, and under the microscope there can be seen (most clearly using crossed polars) a series of regularly spaced bands separated by sharp lines (Fig. 2.14). These lines

FIG. 2.14. Grandjean "planes" (schematic).

follow the contours of the wedge, and the separation between them corresponds to a constant change of thickness of the wedge. Friedel [5] observed the same structure with a number of other cholesterogens, confined not only in cleavages in mica but also in Madagascar orthoclase.

Grandjean and Friedel supposed that these lines were the edges of parallel reflecting planes in the medium, at which some kind of structural discontinuity existed. These hypothetical planes became known as the "*plans de Grandjean*" or *Grandjean planes*. On this basis Grandjean determined the distance between planes by reference to the Newton's colours shown at different depths of the cleavage crack before it was filled, and found it to be about 2000 Å (200 nm (SI)), and to increase with rise of temperature. Now by ordinary interference theory the wavelength λ of the light reflected by a transparent medium containing a series of equally spaced parallel reflecting planes is given by

$$\lambda = 2dn \sin \theta \qquad (2.1)$$

where d is the spacing between the planes, n is the refractive index of the medium, and θ the grazing angle of incidence of the light. For normal incidence the equation simplifies to

$$\lambda = 2dn \qquad (2.2)$$

* A recent theory by Chandrasekhar and Srinivasa Rao [13] predicts a second reversal of the sense of rotation at some wavelength above λ_0.

Friedel applied this equation to Grandjean's results (λ, i.e. λ_0, = ca. 6000 Å) to calculate n. The answer,

$$n = \lambda/2d = 6000/2 \times 2000 = 1{\cdot}5$$

was a plausible value for the mean refractive index of such material, and consistent with a value (1·68) obtained previously [15] for the ordinary refractive index of the mesophase.

The optical activity of cholesterics indicates that the structure has a twist about the optic axis, and the very high values of the activity can be accounted for in a general way if the pitch is of the same order as the wavelength of light. If too the long axes of the molecules are everywhere at right angles to the optic axis, this small pitch would account for the negative sign, since the structure would be optically equivalent to a system of long molecules arranged in parallel planes but inclined in different directions in these planes, this in turn being equivalent to a structure consisting of parallel plate-like molecules or ions. The refractive index for light vibrating parallel to the planes would be greater than for that vibrating normally to them. An apparent difficulty about this picture is that it does not account for the Grandjean planes, since there would seem no reason why the structure should develop surfaces of discontinuity separated by thousands of Å.

A theory by de Vries [16] disposes of this difficulty and accounts for the optical properties of cholesterics. The basis of the theory and the main conclusions are as follows. Accepting the continuously twisted, i.e. helicoidal, model, de Vries pointed out that it must consist of thin *birefringent* layers normal to the optic axis, each one turned through a small angle with respect to its neighbours. The characteristic reflection follows from the fact that because of this turn each linear component of the light experiences a change in refractive index in passing from one layer to the next. The changes will be an increase for the fast and a decrease for the slow component. According to the theory of reflection, the fast component will be reflected with phase reversal and the slow one without phase reversal. This explains why incident and reflected circularly polarized light have the same sense of rotation. From electro-magnetic considerations de Vries showed that for light at normal incidence the wavelength at which reflection at maximum intensity occurs (i.e. λ_0) is

$$\lambda_0 = pn \tag{2.3}$$

where p is as before the pitch and n the mean refractive index of the material. This equation is equivalent to equation (2.2), if we identify d in that equation with half the pitch, i.e. as being the distance between two successive levels at which the long molecular axes become parallel (though with an unsymmetrical molecule, pointing in diametrically opposite directions).

De Vries obtained the expression for the optical rotation of light travelling along the optic axis.

$$\frac{\partial \psi}{\partial z} = -\frac{\pi p}{4\lambda^2} \cdot \frac{(n_2 - n_1)^2}{(1 - (\lambda/\lambda_0)^2)} \tag{2.4}$$

where ψ is the rotation angle, z the thickness, and $(n_2 - n_1)$ the birefringence of a layer, n_1 and n_2 being its two refractive indices. The negative sign means that the sense of the rotation is opposite to that of the twist of the structure for wavelengths less than λ_0. The equation predicts a change in sense of the rotation at λ_0 as observed, for as λ approaches and passes λ_0, $\partial\psi/\partial z$ changes rapidly from very large negative to very large positive values. Another consequence is that since $\lambda_0 = pn$, a sufficient increase in p would put λ_0 in the infrared, so that no visible colours would be reflected. Also in that case $(\lambda/\lambda_0)^2$ would be very small in comparison with 1, so that the equation would simplify to

$$\frac{\partial\psi}{\partial z} = -\frac{\pi p}{4\lambda^2}\cdot(n_2 - n_1)^2 \qquad (2.5)$$

Equation (2.5) also follows from a treatment of the problem in which reflection is neglected, and has, for example, been deduced by Ward [17] by an extension of the method used by Mauguin [10] to derive the rotational effect of a twisted nematic (p. 39) to cases in which the pitch of the structure is of the same order as the wavelength of light. It also follows from the theory of Chandrasekhar and Srinivasa Rao (footnote, p. 43) as applied to wavelengths well below λ_0.

The appearance of the Grandjean "planes" is explained by de Vries as follows. The surfaces of the mica or other supporting material will impose a certain orientation on the molecules in contact with them, so that only when the thickness of the wedge is equivalent to a whole number of half turns of the helicoid will the structure "fit". Elsewhere there will be competition between the orientating influence of the surfaces and the tendency of the structure to continue the twist at its normal pitch, so that the structure will be disturbed and strained. The succession of disturbed and undisturbed sections of the structure as the wedge thickness changes will give the observed periodicity in the microscopic image.

These ideas have been developed in more detail by Cano [18] and Kassubek and Meier [19].

As far as is known, all compounds which form cholesterics have asymmetric molecules and are optically active. In general this affords an explanation of why the mesophases have twisted structures. It is to be expected that (*i*) if both optical enantiomers are available, one will give a *dextro*-mesophase and the other a *laevo*-mesophase, while (*ii*) if a racemic modification is used a mesophase with an untwisted structure and having nematic properties will result. With the single compounds so far investigated it seems that only (*ii*) has been confirmed, because only one of the enantiomers has been readily accessible (Gray [20]; Leclercq, Billard and Jacques [21]). Both (*i*) and (*ii*) have been confirmed with birefringent *solutions* of certain polypeptides (p. 50).

It is appropriate to recall that Friedel [5] found that when mixtures of amyl 4-*p*-cyanobenzylideneaminocinnamate (*dextro*-) and cholesteryl benzoate (*laevo*-) were heated, the characteristic cholesteric properties declined and at a certain temperature, depending on the composition of the mixture, gave nematic properties, there being no discontinuity in the change. From this and other evidence he became convinced that cholesterics and nematics are closely allied.

The Mesomorphism of Diisobutylsilanediol

The mesophase of this compound, (*iso*-$C_4H_9)_2Si(OH)_2$, shows unusual properties and is therefore considered separately. It was studied optically by Eaborn and Hartshorne [22], who found that although it showed fan-like and striated band textures (p. 37–38), suggesting that it was smectic, it was optically negative. There was, however, no sign of optical activity or of any of the characteristic properties of cholesterics.

It was suggested [22] that the structure of the mesophase might consist of parallel layers, deformable and able to glide over one another as in ordinary smectics, but with negative anisotropy, i.e. having a greater refractive index for light vibrating in the planes of the layers than at right angles to them. It was thought that the molecules in each layer might be held together by a statistically non-directional interaction between the hydroxyl groups. Optically negative lamellar mesophases can be formed in certain lyotropic systems.

LYOTROPIC MESOMORPHISM

In this section we shall be concerned with the optical properties of mesophases formed when solvents are added to certain types of compounds, namely polypeptides and amphiphiles. Such *lyotropic* systems are also thermotropic; thus a birefringent solution on heating becomes isotropic and regains its birefringence on cooling. Moreover, some amphiphiles show thermotropic mesomorphism without the addition of a solvent. However, the mesophases that they form by themselves are closely allied structurally to those that they form with solvents, and since it is on their lyotropic behaviour that most interest centres, we shall mainly confine our attention to that.

Solutions of Polypeptides [48]

Robinson *et al.* [23–6] studied the mesomorphism of various solutions of poly-γ-benzyl-L-glutamate (PBLG) and other polypeptides. When sufficiently concentrated these solutions are birefringent, and observations under the microscope also show that in them the solute has a twisted structure with an optically resolvable pitch. Increasing the concentration of polypeptide decreases the pitch, and in certain cases has led to solutions in which the pitch is no longer resolvable and which show typical cholesteric properties, e.g. the selective reflection of circularly polarized light. As this indicates, the work affords strong evidence for the continuous helicoidal model of the cholesteric state.

The formula of PBLG is

$$\left[\begin{array}{l} | \\ \mathrm{CO} \\ | \\ \mathrm{CH{\cdot}CH_2{\cdot}CH_2{\cdot}COOCH_2{\cdot}C_6H_5} \\ | \\ \mathrm{NH} \\ | \end{array}\right]_x$$

where the mean value of x may vary widely according to conditions of preparation. The specimens used had values ranging from 60 to 1255. X-ray and other evidence show that except in dilute solutions in certain solvents the polypeptide chain is wound into a tight helix in which adjacent coils are bound together by hydrogen bonds, with side chains $R = -CH_2{\cdot}CH_2{\cdot}COOCH_2{\cdot}C_6H_5$ projecting radially from the helix. (This is an example of the so-called α-helix structure first proposed by

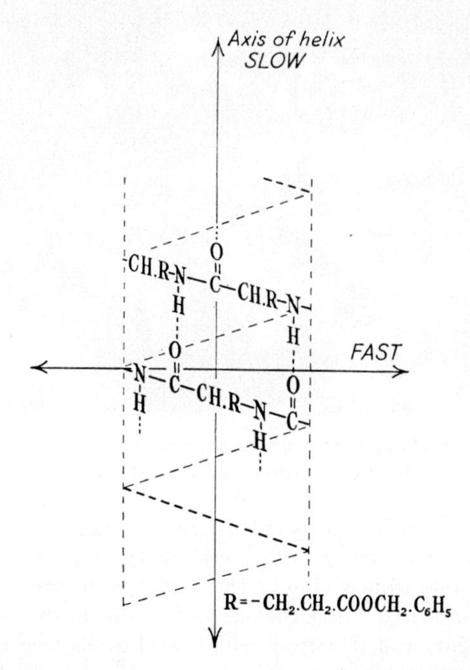

FIG. 2.15. α-Helix structure in poly-γ-benzyl-L-glutamate (diagrammatic).

Pauling, Corey and Branson [27] for proteins.) The structure is shown diagrammatically in Fig. 2.15. The helix is very rigid, and the polymer molecule is to be pictured as a straight stiff rod with projecting side chains.

Dilute solutions of PBLG are isotropic and optically active. In most solvents the optical activity is due to the molecular α-helix configuration. With increase in concentration a two-phase system is produced at a concentration A, which depends both on solvent and degree of polymerization

x. The phase richer in PBLG is birefringent. This two-phase system is stable over a concentration range A to B; B is also dependent on solvent and degree of polymerization, and above B only the birefringent phase exists. Within the range A to B the concentrations of PBLG in the two phases are constant; only their relative amounts change as the total concentration changes.

When the birefringent phase is observed in thin layers by transmitted light, parallel and equally spaced narrow bands, alternately dark and bright, are seen at quite moderate magnifications. Between crossed polars the bright bands show polarization colours in white light. When the birefringent phase has just been formed, or has just been introduced into its

FIG. 2.16. PBLG in dioxan: two-phase region. (Crossed polars, with vibration directions parallel to edges of figure.)

containing vessel, the bands usually show no regular pattern apart from maintaining their parallelism and constant spacing, but follow irregular sinuous curves and whorls (Fig. 2.16). After some time, however, a tendency for the bands to arrange themselves parallel to the walls of the vessel becomes apparent, and if a thin cell is used areas free from bands and showing a uniform colour between crossed polars gradually develop.

The spacing (S) between adjacent dark and bright bands increases with decreasing concentration of PBLG, and varies with the solvent. It is, however, independent of degree of polymerization, wavelength of light, and shape or thickness of specimen. In the solutions studied, S ranged from 10^{-4} to 10^{-2} cm.

The optical examination of the birefringent solutions yielded the following results:

(1) The uniform areas in the thin cells showed a very high optical

rotatory power, greater by several powers of ten than those of the dilute isotropic solutions, and of the order typically shown by cholesterics.

(2) By use of crossed polars and a graduated quartz wedge, a periodic change of path difference corresponding with the alternation of dark and bright bands was found. When the specimen was illuminated by light vibrating at right angles to the bands they disappeared, showing that for this direction of vibration the medium has a constant refractive index. Therefore there must be a periodic change of refractive index for light vibrating parallel to the bands.

These results can be accounted for by a helicoidal structure (Fig. 2.17),

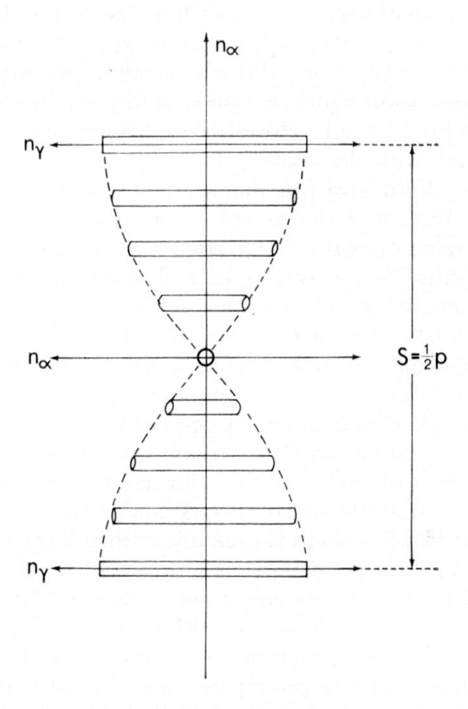

FIG. 2.17. Helicoidal structure.

in which the directions or mean directions of the molecular rods in successive layers are indicated by single rods. The refractive indices for light vibrating along and across the rods are denoted by n_γ and n_α respectively. The spacing of the bands (S) would be the distance along the axis of twist over which the molecules undergo a rotation of 180°, i.e. half the pitch. The middle of a dark band would occur at the "head-on" view of the rods, where the medium would present the single index n_α, as at the middle of the figure. The model accounts for the constant refractive index for light vibrating at right angles to the bands, i.e. parallel to the axis of twist, since at all points along this axis the index is determined solely by the cross-section of the rods, and is n_α. The uniform areas would be views along the axis of twist, and their colour the result of dispersion of the rotation. Note that n_γ and n_α correspond respectively to n_2 and n_1 in equation (2.5).

The optical rotatory powers of fourteen different solutions of PBLG having values of S ranging from 5×10^{-4} to 10^{-2} cm had rotations ranging from 20,000 to 140,000° per mm of thickness [23]. Application of equation (2.5), by inserting values for rotation, pitch (taken as 2S) and wavelength, and dividing by the volume fraction concentration to allow for the fact that the measurements were made on solutions in an inactive solvent, whereas the equation was derived for the case of an undiluted substance, gave values of the layer birefringence, $(n_2 - n_1)$, falling within 20% of a mean value of 0·0248; this constancy indicated that the equation was probably correct. The equation was so applied because it was not feasible to determine $(n_2 - n_1)$ directly, except approximately by comparing the compensation values obtained with a graduated quartz wedge for the dark and bright bands. However, such a determination gave 0·025, in good agreement with the above value.

Robinson and Ward also found that solutions containing equal concentrations of PBLG and the *dextro*- enantiomorph (PBDG) are birefringent, but have no optical activity, do not show bands and behave like an ordinary nematic. Similar results were obtained by mixing solutions of PBLG in methylene chloride and in dioxan. These solutions have rotations of opposite signs, and the solvents must therefore induce twists of opposite hands in the solute, showing that the asymmetry of the glutamic acid residue is not the sole factor determining the twist. Evidently in the mixed solutions the twist is annulled. Direct determinations of the birefringence on these inactive solutions (corrected to unit concentration as above) gave values within $\pm 20\%$ of 0·025, i.e. the same as found for the birefringence of the layers in the optically active phases. Such determinations were possible because the inactive solutions readily formed large uniform areas in which the optic axis was parallel to the windows of the containing cell.

The results give a completely consistent picture on the assumption that birefringent solutions of PBLG and PBDG are lyotropic analogues of ordinary cholesterics with properties as predicted by de Vries, but with much larger pitches. The larger pitches accounted for the fact that no cholesteric "peacock" colours were reflected. Taking the smallest value of S obtained, namely 10^{-4} cm, and 1·5 for the mean refractive index n, λ_0 $(= pn = 2Sn)$ works out to *ca.* 30,000 Å, i.e. well into the infrared. Attempts were made to prepare solutions of PBLG with much smaller values of S by increasing the concentration, but these proved to be opaque and did not give iridescent reflections. However, it was found that concentrated solutions of poly-γ-ethyl-L-glutamate (PELG) in ethyl acetate did reflect brilliant iridescent colours. Moreover, the wavelength of the reflected light declined with increasing angle of reflection measured from the normal, and the reflected and transmitted beams were circularly polarized and in opposite senses, thus completing the analogy with non-lyotropic cholesterics.

As seen in Fig. 2.16 part of the birefringent phase in a two-phase system may appear as spherulites. It has been shown [24] that this is to be expected when the twisted structure in a small portion of the phase has to accommodate itself to the spherical shape imposed by the interfacial tension between itself and the isotropic phase. The spherulite may be pictured as a series of shells like the layers of an onion but of molecular thickness, to the surfaces of which the molecular rods are parallel. Along any radius of the sphere the directions of the rods follow a twist. This arrangement gives for general directions of observation a spiral pattern of alternating dark and bright bands when the microscope is focused on an equatorial plane of the spherulite, but also requires one radius along which the bands show a discontinuity. This discontinuity can be clearly seen on suitably oriented spherulites; there are no examples of these in Fig. 2.16.

Following a suggestion [24] that other materials having optically active rod-like molecules might form cholesteric-type solutions, Robinson examined solutions of deoxyribonucleic acid (DNA) in aqueous NaCl [28]. These showed equidistant dark and light bands with a spacing of about 10^{-4} cm like the solutions of PBLG. Uniform areas like those in the latter solutions were also observed but only in preparations that were so thin that no optical activity could be detected. Spherulites like those of PBLG solutions have been observed in solutions of transfer ribonucleic acid (RNA) [29].

Solutions containing Amphiphilic Compounds

The molecules of these compounds possess a *polar group* which tends to be soluble in water (*hydrophilic*) and insoluble in hydrocarbons (*lipophobic*), while the rest of the molecule tends to be insoluble in water (*hydrophobic*) and soluble in hydrocarbons (*lipophilic*). Compounds in which these two opposing tendencies are strong and fairly equally balanced form mesophases with water within ranges of concentration and temperature which depend on the compound. Amphiphiles are of two main types: (1) those in which the polar group is at the end of a long lipophilic chain; (2) those in which it is linked with two such chains. In the latter case the molecules commonly adopt a conformation with the two chains side by side or at an acute angle to one another, i.e. the molecules become peg-shaped. Examples of (1) are the soaps, for example the sodium and potassium salts of stearic, palmitic and oleic acids, the polar groups being —COONa and —COOK. Examples of (2) are the phospholipids and compounds such as sodium di-2-ethylhexyl sulphosuccinate ("Aerosol OT"), i.e.

$$\begin{array}{l} CH_3\cdot(CH_2)_3\cdot CH(C_2H_5)\cdot CH_2\cdot OOC\cdot CH_2 \\ \qquad\qquad\qquad\qquad\qquad\qquad\quad | \\ CH_3\cdot(CH_2)_3\cdot CH(C_2H_5)\cdot CH_2\cdot OOC\cdot CH\cdot SO_3Na \end{array}$$

the polar group being —SO_3Na in this last case. The two types are represented very formally in Fig. 2.18; the lipophilic chains will actually be of many different forms and will be flexible.

Mesophases also occur in ternary systems consisting of an amphiphile (A), water (W), and a compound (C) normally insoluble in water such as a hydrocarbon or a long-chain fatty acid or alcohol. In these systems W and

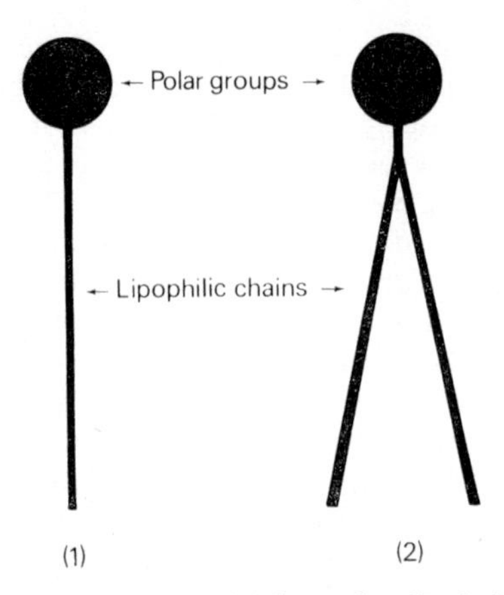

FIG. 2.18. Types of amphiphile molecules (schematic).

C are said to be "solubilized" in one another by A which acts as a "cosolvent". The study of these and other multi-component systems containing amphiphiles owes much to the separate researches of Lawrence, Luzzati and Ekwall and their collaborators. References to this work are given in a comprehensive review by Winsor [30]. The mesophases in these multi-component systems are in the main structurally and microscopically similar to those in binary systems of amphiphiles and water; we shall deal only with the latter, which will suffice to illustrate the contributions made by the study of optical properties.

MESOPHASES IN BINARY SYSTEMS OF AMPHIPHILES AND WATER

In these systems the amphiphile molecules aggregate into *micelles* owing to the tendency of the polar groups to associate with one another and with the water. In whatever phase is formed the pattern of the structure is believed to be defined mainly by the polar groups while the lipophilic parts of the molecules form liquid-like associations. Fig. 2.19 shows in (*a*) the spherical micelles in dilute (isotropic) solutions (there seems to be no doubt that in such solutions they must be spherical), and in (*b*), (*c*) and (*d*) the structures proposed for the main types of mesophases in the birefringent regions of the systems, the evidence for which will be discussed below. In the figure the liquid-like association of the lipophilic chains is indicated by their irregular arrangement, while the slightly irregular arrangement of the polar groups serves to indicate thermal displacements and to emphasize that the micelles are readily deformable [48].

(*i*) *The "neat" phase* (*or "G" phase*)

There is general agreement that this phase is lamellar, i.e. smectic, with the amphiphile forming double layers (Fig. 2.19*b*), separated by layers of water. The evidence comes from a number of X-ray diffraction studies, e.g. by Luzzati and Husson [31]. Except in certain cases discussed later, this phase is optically *positive*, like the smectics of single compounds.

(ii) The "middle" (or "M_1") phase

This does not seem to be formed if the amphiphile molecules are peg-like. It is stable at higher water concentrations than the G phase in cases in which both phases are formed. Luzzati *et al.* [31] have found from X-ray diffraction studies that the amphiphile molecules are grouped into rod-like micelles of indefinite length, arranged parallel to one another in a two-dimensional hexagonal pattern as shown in Fig. 2.19. The water forms the

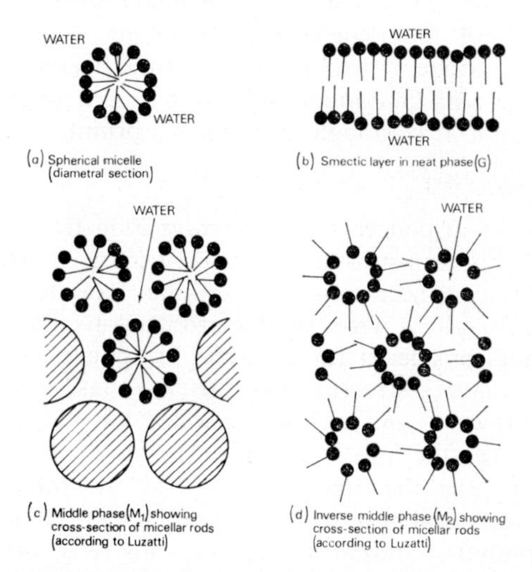

FIG. 2.19. Lyotropic mesophases. (Only molecules with single lipophilic chains are shown, but peg-shaped ones would normally be arranged similarly.)

intervening medium between the rods. This structure is in agreement with the optical sign found by Gilchrist, Rogers, Steel, Vaal and Winsor [32], which is *negative*. Clunie, Corkhill and Goodman [33], however, find from calculations based on X-ray and density data that the average diameter of the rods is less than corresponds to this structure, and suggested instead that they are linear chains of spherical micelles (Fig. 2.19*a*). But this would make the phase nearly isotropic (owing to the isotropic character of the spherical units in which the molecules are pointing in all directions) with a small positive form birefringence owing to the parallel-rod arrangement, and is therefore inconsistent with the observed negative sign.

(iii) The "viscous isotropic" (or "V_1" phase)

This appears in some systems at concentrations intermediate between those within which the G and M_1 phases are stable. Its structure is a matter of some doubt and is not one on which ordinary optical observations can give any information beyond showing that it is isotropic. X-ray diffraction studies by Luzzati *et al.* and by Clunie and others were first taken to indicate that it is based on a face-centred cubic lattice, but more recent work [34] favours a body-centred cubic lattice [47]. An interesting observation mentioned by Gilchrist *et al.* [32] is that if a preparation of a

V_1 phase between crossed polars is pressed with a glass rod it shows marked shear birefringence which rapidly relaxes on removal of the rod. This suggests that the building units of the phase are not necessarily inherently isotropic, and perhaps therefore not spherical micelles.

(iv) The isotropic "S_{1c}" phase

This mesophase has been found at higher water concentrations than those at which the M_1 phase is stable in the binary systems of water with (*a*) dodecalethylenglycol monolauryl ether (Fontell, Mandell and Ekwall [35] and (*b*) decyl-, dodecyl- and tetradecyl-trimethylammonium chlorides (Balmbra, Clunie and Goodman [36]). It has also appeared in certain ternary systems [36]. It is thought to possess a primitive lattice.

(v) "Inverse" phases ("V_2" and "M_2")

In some systems at concentrations *greater* than those at which the G phase is stable, another isotropic phase, V_2, occurs, to be succeeded with further increase of concentration by another middle phase, M_2. The structure of the V_2 phase poses similar problems to that of the V_1 phase and we shall not consider it further. The M_2 phase has been found by Luzatti *et al.* to have a structure like the M_1 phase, but with the polar groups directed inwards and enclosing a water core (Fig. 2.19(*d*)), so that the medium between the rods is lipophilic. From a consideration of interfacial energies between the aqueous, polar and lipophilic regions, Winsor [30] found that there will be a tendency for such an inversion to occur with increasing amphiphile concentration. The transverse orientation of the long axes of the molecules with respect to the axes of the rods is supported by conoscopic observations [32] showing the M_2 phase is optically *negative* like M_1, except in certain cases mentioned below.

If all the above mesophases occurred at a given temperature in a binary system, the sequence of appearance with increasing concentration of amphiphile would be:

$$S_{1c} \rightarrow M_1 \rightarrow V_1 \rightarrow G \rightarrow V_2 \rightarrow M_2$$

The complete series does not appear to have been found in any single system; often the only ones observed are M_1, V_1 and G, or G, V_2 and M_2, but whichever phases occur do so in the above order.

The textures of G and M phases

In 1954 Rosevear [37] published a detailed account of the textures shown by neat (G) and middle (M_1) phases of soaps and synthetic detergents. The M_2 phase had not at that date been recognized, but since the evidence is that it is structurally similar to the M_1 phase, apart from the "inverse" orientation, it may be expected to show similar textures. The paper is illustrated by many photomicrographs and should be studied by anyone concerned with microscopy of these phases. At that time it was generally thought that both neat and middle phases were smectic, and although the evidence above shows that this is not so, the two types do give textures with many superficial resemblances. This must be attributed to the parallel rod structure of M phases, which although very different from the lamellar G structure, has a very much greater degree of order than an ordinary nematic, and may in a sense be said to be built up of layers of rods (cf. [38]).

Although the G phase would be expected to show focal conic textures, Rosevear did not observe any obvious ellipses or hyperbolae. On the other hand both G and M phases commonly show textures with spherulitic and fan-like units (not only in the systems studied by Rosevear, but in other amphiphile–water systems; see Fig. 2.20), and in some cases at least these

FIG. 2.20. Spherulitic and fan-like textures of M_1 phase in Triton X 100–water system. Crossed polars with vibration directions parallel to edges of figure, ×100. (*After P. A. Winsor, Thornton Research Centre.*)

may be focal domains with the hyperbola lying in a vertical plane. Studies by Rogers and Winsor on the spherical droplets of the G phase separating from isotropic solutions, and showing an extinction cross typical of a spherulitic structure, indicate, however, that the smectic layers are arranged concentrically like the layers of an onion.

Rosevear [37] detected a number of differences in detail between the textures of his G and M phases, but a more general difference was that his G phases had a greater birefringence than the M phases. Obviously care must be taken in applying this criterion as a practical test; the polarization colours shown by sections of similar thickness and similar orientation of the optic axis to the plane of the preparation (as shown by their interference figures) must be compared.

One of the most useful ways of studying lyotropic mesophases stable at room temperature is to place a drop of the dilute isotropic solution of the amphiphile between a slide and cover slip and observe the sequence of

FIG. 2.21. Result of peripheral evaporation of isotropic solution in sodium 3-undecyl sulphate–water system, showing M_1, V_1 and G phases. Crossed polars with vibration directions parallel to edges of figure; ×65. (*After P. A. Winsor, Thornton Research Centre.*)

FIG. 2.22. Bâtonnet of M_1 phase of sodium caprylate–water system, formed by slow evaporation of isotropic solution. Crossed polars with vibration directions parallel to edges of figure; ×200. (*After P. A. Winsor, Thornton Research Centre.*)

mesophases formed as the water evaporates giving a concentration gradient across the preparation. An example of applying it to the system sodium 3-undecyl sulphate–water is shown in Fig. 2.21. The phases formed are M_1, V_1 and G. The striated texture of the M_1 phase is typical of this type of phase produced in this way. The striations correspond to a zig-zag alternation of extinction directions, and therefore of orientations of the structure, across the mesophase. The texture has been discussed by Rogers and Winsor [39].

Winsor and Rogers [40] have studied optically the mesophase *particles* which separate from isotropic solutions of amphiphiles on cooling or on

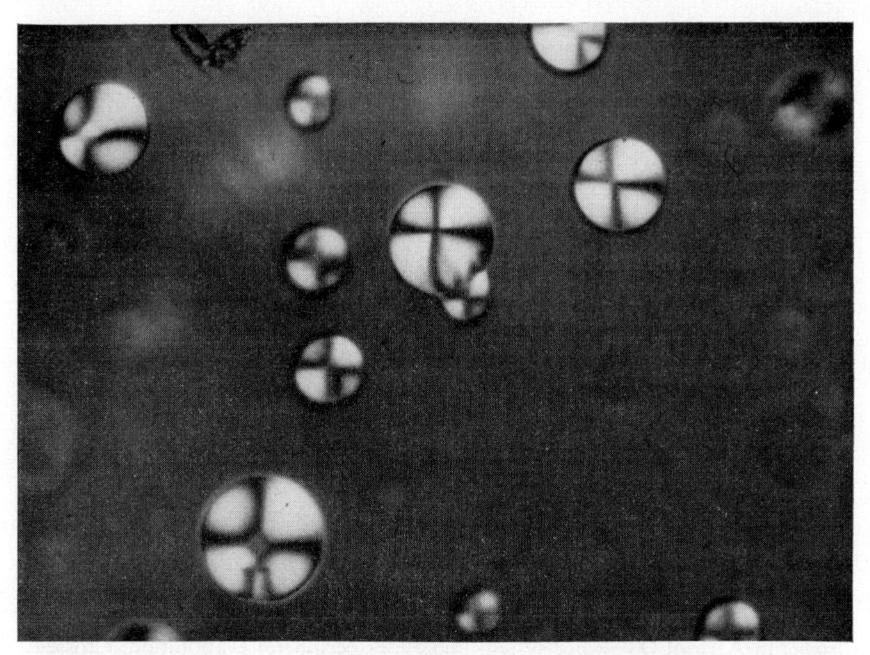

FIG. 2.23. Droplets of G phase in Aerosol OT–water system, formed by slow evaporation of isotropic solution. Crossed polars with vibration directions parallel to edges of figure; ×400. (*After P. A. Winsor, Thornton Research Centre.*)

isothermal evaporation, and which provide further points of difference between G and M phases. Both types can appear as bâtonnets. G phases can also appear as spherical droplets with the radial direction "slow", while M phases can deposit as platelets. The study has indicated that the structures of these are as follows.

Bâtonnets. These are figures of revolution about a central axis (Fig. 2.22). In G bâtonnets the smectic lamellae are normal to this axis, while in M bâtonnets the rods of the structure encircle the axis.

G droplets (Fig. 2.23). The smectic lamellae form concentric shells like the layers of an onion.

M platelets. The rods are normal to the platelets.

The optical sign of G and M phases

As stated earlier, the optical sign of most G phases is positive, and that of most M phases is negative. We shall now consider some cases of departure from these general rules.

In the G phase of aqueous solutions of Aerosol OT (p. 51) the area occupied per polar group in the smectic layers increases with dilution of the amphiphile. Winsor and Rogers [41] found that at room temperature at a concentration of 30% by weight of amphiphile the sign is negative, but if this solution be evaporated by the method above, the negative birefringence gradually approaches zero, and then becomes increasingly

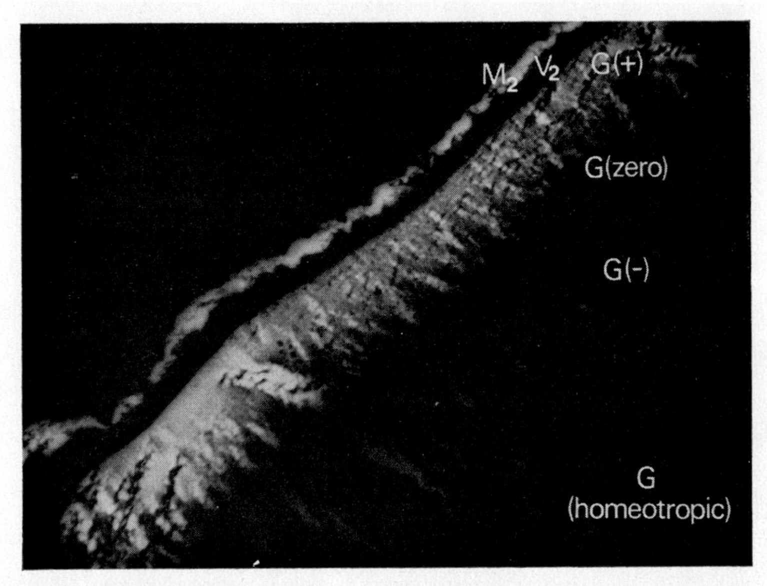

FIG. 2.24. Phase succession formed when optically negative G phase (*bottom right*) in Aerosol OT–water system undergoes peripheral evaporation. Crossed polars with vibration directions parallel to edges of figure; ×30 [41].

positive. The G phase is succeeded by a V_2, and then an M_2, phase, as shown in Fig. 2.24. The explanation first given for the change in sign of the G phase, which is in accord with the increase in area per polar group with dilution, is that at lower concentrations the two lipophilic chains in the peg-shaped molecules are splayed out and lie more nearly parallel to the layers than across them, so giving a negative sign, and that as the concentration increases the angle between the chains progressively decreases and the normal smectic positive structure develops. Similar results were obtained with the corresponding potassium salt. With Aerosol MA,

$$\begin{array}{l}(CH_3)_2{\cdot}CH{\cdot}CH_2{\cdot}CH(CH_3){\cdot}OOC{\cdot}CH_2\\ \qquad\qquad\qquad\qquad\qquad\qquad\quad |\\ (CH_3)_2{\cdot}CH{\cdot}CH_2{\cdot}CH(CH_3){\cdot}OOC{\cdot}CH{\cdot}SO_3Na\end{array}$$

although the positive birefringence of the G phase decreased with decreasing concentration, the phase broke down at room temperature into an isotropic solution before any change in sign took place.

Subsequently [42] Winsor and Rogers discussed the possible contributions which the negative form birefringence resulting from the layer structure of the G phase, and the anisotropic distribution of bonds other than C–C could have in determining the optical sign of the Aerosol OT system. They also report other cases of unusual signs recently discovered. The sodium and potassium salts of di-(2-ethylhexyl)acetic acid

$$\begin{matrix} CH_3{\cdot}CH_2{\cdot}CH_2{\cdot}CH_2{\cdot}CH(C_2H_5){\cdot}CH_2\diagdown \\ \qquad\qquad\qquad\qquad\qquad\qquad CH{\cdot}CO_2H \\ CH_3{\cdot}CH_2{\cdot}CH_2{\cdot}CH_2{\cdot}CH(C_2H_5){\cdot}CH_2\diagup \end{matrix}$$

have strongly negative G phases and positive M_2 phases, while the corresponding salts of di-(2-ethylhexyl)ethyl sulphuric acid

$$\begin{matrix} CH_3{\cdot}CH_2{\cdot}CH_2{\cdot}CH_2{\cdot}CH(C_2H_5){\cdot}CH_2\diagdown \\ \qquad\qquad\qquad\qquad\qquad\qquad CH{\cdot}CH_2{\cdot}O{\cdot}SO_2{\cdot}OH \\ CH_3{\cdot}CH_2{\cdot}CH_2{\cdot}CH_2{\cdot}CH(C_2H_5){\cdot}CH_2\diagup \end{matrix}$$

have weakly negative G phases and also weakly negative M_2 phases. All these compounds have peg-shaped molecules, and although at this stage any explanation of these unusual signs must be advanced with considerable caution, the change in the area per polar group for Aerosol OT seems to tip the balance somewhat in favour of the original explanation, i.e. the splaying out of the lipophilic chains.

Pelzl and Sackmann [43] found that the thallous salts of the normal fatty acids from C_5 up to *ca.* C_{11} have smectic (neat) phases with negative birefringence. Although the investigation has been concerned only with the anhydrous soaps, it is appropriate to deal with it, since the compounds are amphiphiles and the smectics concerned have the G structure (without the water layers) shown in Fig. 2.19(*b*). The negative sign is attributed to the negative influence of the layers of polar groups overriding the positive influence of the lipophilic chains because these are comparatively short. Only in the higher members of the series with their longer chains does the positive effect of the chains predominate. The change from negative to positive occurs in the soaps having 10 to 12 carbon atoms, the exact point of change depending on the temperature and wavelength.

As shown above, e.g. in Figs. 2.20 and 2.23, lyotropic mesophases can appear as spherulitic textures. It is customary to refer to the spherulites as being positive if the radial direction is "slow" and negative if it is "fast". However, the *sign of a spherulite* gives no necessary indication of the intrinsic optical sign of the material. It depends first on whether the optic axes of the structural elements composing the spherulite are oriented tangentially or radially, and secondly on the intrinsic optical sign of these elements. Thus positive elements with the axis oriented tangentially give a negative spherulite and negative ones with it oriented tangentially a positive one. The intrinsic optical sign of a mesophase must be determined from observations of the effect of a compensator on the interference figure given by an optically homogeneous region of the phase, as shown for example in Fig. 2.6 for the case in which the optic axis is parallel to the microscope axis. For methods of determining the sign from observations on interference figures for other orientations see ref. [7].

REFERENCES

[1] SACKMANN, H. and DEMUS, D. *Molec. Crystals* **2,** 81 (1966); *Fortschr. Chem. Forsch.* **12,** 349 (1969).
[2] CHISTYAKOV, I. G., SCHABISCHEV, L. S., JARENOV, R. I., and GUSAKOVA, L. A. *Molec. Crystals and Liqu. Crystals* **7,** 279 (1969).
[3] WIENER, O. *Abh. Sachs. Ges. (Akad.) Wiss.* **32,** 507 (1912).
[4] LEHMANN, O. *Z. physikal. Chem.* **4,** 468 (1889).
[5] FRIEDEL, G. *Annls. Phy.* **18,** 273 (1922).
[6] HALLIMOND, A. F. *The Polarizing Microscope*, 3rd edn, Vickers Instruments, York (1970).
[7] HARTSHORNE, N. H. and STUART, A. *Crystals and the Polarising Microscope*, 4th edn, Arnold, London (1970). This work has chapters on hot stages and on mesomorphism.
[8] GRANDJEAN, F. *Bull. Soc. fr. Minér. Cristallogr.* **42,** 42 (1919).
[9] BRAGG, SIR WILLIAM H., *Nature* **133,** 445 (1934).
[10] MAUGUIN, C. *Bull. Soc. fr. Minér. Cristallogr.* **34**, 6, 71 (1911).
[11] OSEEN, C. W. *Trans. Faraday Soc.* **29,** 881 (1933).
[12] FRANK, F. C. *Discuss. Faraday Soc.* No. 25 (1958).
[13] CHANDRASEKHAR, S. and SRINIVASA RAO, K. N. *Acta Crystallogr.* **A24,** 445 (1968).
[14] GRANDJEAN, F. *C. r. hebd. Séanc. Acad. Sci., Paris* **172,** 71 (1921).
[15] STUMPF, F. *Ann. Physik* **37,** 351 (1912).
[16] DE VRIES, H. *Acta Crystallogr* **4,** 219 (1951).
[17] WARD, J. C. *Report F.R.L. 213, Courtaulds Ltd*, available from Courtaulds Ltd, Patent Department, Coventry, CV6 5AE, England.
[18] CANO, R. *Bull. Soc. fr. Minér. Cristallogr.* **91,** 20 (1968).
[19] KASSUBEK, P. and MEIER, G. *Molec. Crystals and Liqu. Crystals* **8,** 305 (1969).
[20] EABORN, C. and HARTSHORNE, N. H. *J. Chem. Soc.* 549 (1955).
[21] GRAY, G. W. *Molec. Crystals and Liqu. Crystals* **7,** 127 (1969).
[22] LECLERCQ, M., BILLARD, J. and JACQUES, J. *Molec. Crystals and Liqu. Crystals* **8,** 367 (1969).
[23] ROBINSON, C. *Trans. Faraday Soc.* **52,** 571 (1956).
[24] ROBINSON, C., WARD, J. C. and BEEVERS, R. B. *Discuss. Faraday Soc.* No. 25, 29 (1958).
[25] ROBINSON, C. and WARD, J. C. *Nature* **180,** 1183 (1957).
[26] ROBINSON, C. *Molec. Crystals* **1,** 467 (1966).
[27] PAULING, L., COREY, R. B. and BRANSON, H. R. *Proc. Nat. Acad. Sci.* **37,** 205 (1951).
[28] ROBINSON, C. *Tetrahedron* **13,** 219 (1961).

[29] Spencer, M., Fuller, W., Wilkins, M. H. F. and Brown, G. L. *Nature* **194,** 1014 (1962).
[30] Winsor, P. A. *Chem. Rev.* **68,** 1 (1968).
[31] Luzzati, V. and Husson, F. *J. Cell Biol.* **12,** 207 (1962). This paper gives references to other work on lyotropic mesophases by Luzzati and Husson and their associates.
[32] Gilchrist, C. A., Rogers, J., Steel, G., Vaal, E. G. and Winsor, P. A. *J. Colloid Interface Sci.* **25,** 409 (1967).
[33] Clunie, J. S., Corkhill, J. M. and Goodman, J. F. *Proc. R. Soc.* **A285,** 520 (1965).
[34] Luzzati, V., Tardieu, A., Gulik-Krzywicki, T., Rivas, E. and Reiss-Husson, F. *Nature* **220,** 485 (1968).
[35] Fontell, K., Mandell, L. and Ekwall, P. *Acta Chem. Scand.* **22,** 3209 (1968).
[36] Balmbra, R. R., Clunie, J. S. and Goodman, J. F. *Nature* **222,** 1159 (1969).
[37] Rosevear, F. B. *J. Am. Oil Chem. Soc.* **31,** 628 (1954).
[38] Bucknell, D. A. B., Clunie, J. S. and Goodman, J. F. *Molec. Crystals and Liqu. Crystals* **7,** 215 (1969).
[39] Rogers, J. and Winsor, P. A. *J. Colloid Interface Sci.* **30,** 500 (1969).
[40] Winsor, P. A. and Rogers, J. *Paper presented at the Vth International Congress on Surface Active Substances, Barcelona*, September 1968.
[41] Winsor, P. A. and Rogers, J. *Nature* **216,** 477 (1967).
[42] Winsor, P. A. and Rogers, J. *J. Colloid and Interface Science* **30,** 247 (1969).
[43] Pelzl, G. and Sackmann, H. *Molec. Crystals and Liqu. Crystals* **15,** 75 (1971).
[44] Demus, D., Diele, S., Klapperstück, M., Link, V. and Zaschke, H. *Molec. Crystals and Liqu. Crystals* **15,** 161 (1971).
[45] Vorländer, D. *Trans. Faraday Soc.* **29,** 913 (1933).
[46] Taylor, T. R., Fergason, J. L. and Arora, S. L. *Phys. Rev. Lett.* **24,** 359 (1970).
[47] *Liquid Crystals and Plastic Crystals, Vol. 1* (edited by G. W. Gray and P. A. Winsor), Ellis Horwood, Ltd., Chichester, 1974, Chap. 5.
[48] See also *Liquid Crystals and Plastic Crystals, Vol. 2* (edited by G. W. Gray and P. A. Winsor), Ellis Horwood, Ltd., Chichester, 1974, Chap. 4.4.
[49] Nehring, J. and Saupe, A. *J. chem. Soc. Faraday II* **68,** 1 (1972).
[50] Elliot, G. *Chemistry in Britain* **9,** 213 (1973).

3

X-Ray Diffraction by Liquid Crystals—Non-amphiphilic Systems

J. FALGUEIRETTES and P. DELORD

The first attempts to study liquid crystals by X-ray analysis followed immediately the discovery of X-ray diffraction [1, 2]; they indicated that the X-ray diffraction patterns for mesomorphic phases and the corresponding ordinary liquids (amorphous) were similar. This result made it possible to state that there was no three-dimensional periodicity in structure of these phases [3].

Afterwards, more accurate experiments detected some differences between the X-ray patterns of amorphous isotropic liquids and mesophases [4, 5]. In particular, experiments with a nematic oriented by electric or magnetic fields gave X-ray patterns in which the main diffuse ring split into two crescents [6].

Detailed examinations of experiments made earlier than 1940 are available in Gray's book [7] and the review by Brown and Shaw [8]. We shall be concerned with developments since 1950. Improvements in X-ray diffraction techniques and a better knowledge of the nematic state, have allowed more critical experiments to be made. Chatelain's work on nematic phase orientation by rubbing glass or mica sheets has been of great importance, because this made it possible to obtain oriented specimens in a reproducible manner [9]. By the combined action of rubbing and applying a magnetic field parallel to the direction of rubbing, we may obtain "single liquid crystals" or "liquid monocrystals" of highly uniform orientation.

Falgueirettes made the first quantitative study of X-ray scattering by a single nematic liquid crystal of *p*-azoxyanisole [10]. By considering the scattering from an isolated molecule and then that from molecules arranged parallel to one another in a cluster, he proved that intermolecular interferences were possible only when the scattering vector was perpendicular to the major axis of the cluster. Such a cluster of molecules has no real physical existence, but by considering such bundles, it is possible to describe the short-range order in the nematic and to interpret the X-ray diffraction data. This result, which is of fundamental importance, made it possible to correlate the intensity along the crescent with the number of molecules taking part in the intermolecular scattering and to deduce the molecular angular distribution function $f(\alpha)$; $f(\alpha)$ is the number of molecules per unit volume having their long axes at angles within the limits α and $\alpha + d\alpha$ to the optic axis of the preparation. For *p*-azoxyanisole, distribution functions obtained in this way lead to values of the degree of order (S), defined as $1 - \frac{3}{2}\overline{\sin^2 \alpha}$ which agree, within experimental error, with values obtained by other methods (Fig. 3.1). We should

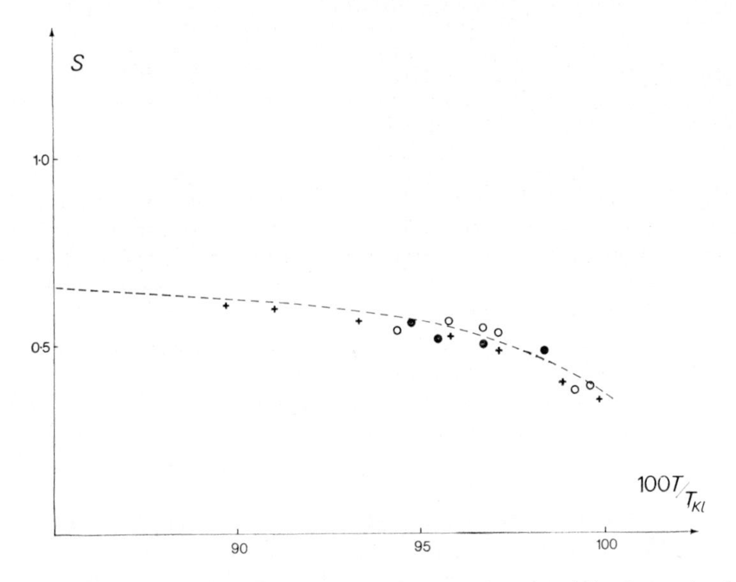

FIG. 3.1. Effect of temperature on the degree of order (S), determined by different methods, for the nematic mesophase of *p*-azoxyanisole. Experimental points were determined as follows:

(+) optical studies [11];
(●) X-ray diffraction [10];
(○) nuclear magnetic resonance spectroscopy [12].
T_{Kl} is the temperature of transition from the nematic phase to the amorphous liquid.

remember that the degree of order (or orientation factor) is expressed as an average deviation of orientation of the molecular axes, since allowance must be made for oscillations of the long axes caused by thermal effects.

Since that time, Delord has studied other nematic compounds such as *p*-azoxyphenetole and *p*-anisaldazine but with a diffractometer [13]. His results corroborate those of Falgueirettes.

Methods arising from work with long-chain molecules by Vainshtein [14] provided a quantitative interpretation of equatorial diffraction by an oriented nematic or smectic by use of cylindrical distribution functions. If one supposes that the system has cylindrical symmetry, one may use the cylindrical Patterson function linking the intensity scattered in the equatorial plane of reciprocal space to an interatomic distance function in projection on the equatorial plane of the real space. Chistyakov, Kosterin and Chaikovski studied many mesomorphic compounds in this way, e.g. the structures of α-benzeneazo(anisylidene-α'-naphthylamine)—(structure I)

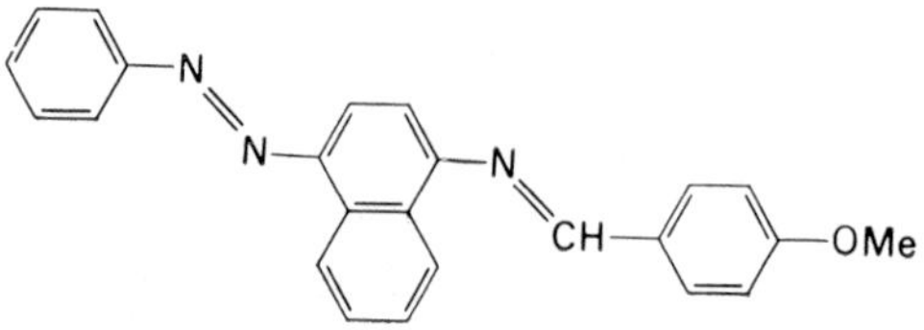

I

in the vitrified liquid crystal state [15], of nematic *p*-azoxyanisole in electric fields [16] and magnetic fields [17], and of 4-*p*-n-nonyloxybenzylideneaminotoluene and cholesteryl caprate in the liquid crystal states [18].

In this section, we shall examine in greater detail in what circumstances scattering of X-rays by nematics or smectics permits us to determine the "short-range order". By "short-range order" is meant the liquid-like order which exists in the mesophases; it is well known now that mesophases may be distinguished by a long-range order which can be studied by light scattering but which does not give rise to X-ray diffraction, except for small angle diffraction. Classical X-ray methods allow us to obtain short-range order only.

We will not deal here with experimental techniques; details will be found in the following publications: [10, 13, 18, 19].

X-ray Diffraction by a Nematic Phase

According to Vainshtein, one may express the symmetry of a classical nematic:

$$\infty \tau_{\infty}(z)\ W(x, y)$$

The first symbol ∞ presupposes a random orientation of the molecules about their long axes, i.e. it is assumed that each molecule may adopt all orientations around its long axis, with the same probability. The long axes of the molecules which lie parallel to one another are oriented along the principal z axis. The second symbol $\tau_{\infty}(z)$ allows for an infinitely small statistical translation of the molecules in the z direction, for it is supposed that the centres of molecules are arranged randomly in the z direction. Consequently there is no correlation of the projections of the centres of gravity of the molecules on the principal optic axis. Finally, one may describe the short-range order in the xy plane in a nematic by a distribution function $W(x, y)$ or $W(r)$ in cylindrical co-ordinates.

It is well known that from the X-ray diffraction of a liquid, only a statistical average of all interatomic distances may be obtained. This average function, a distribution function, is defined as the probability of finding two atoms at a given distance one from the other.

For the crystal, the distribution function shows discrete peaks because of the three-dimensional periodicity of the lattice. However, for the amorphous liquid, the lack of periodicity gives a continuous function. We must not however assume that a crystal is an absolutely rigid assembly of atoms in a periodic pattern. Because of thermal energy, the atoms vibrate about their lattice centres. These vibrations, which can be assumed to be completely uncorrelated (the relative shift in position of any two atoms about their nodes is independent of their distance of separation), do not destroy the long-range order of the lattice.

In a liquid, a second kind of lattice distortion arises. There is no long range order as in a crystal—only a short-range order remains. The law governing the relative shift in position of two atoms now depends on their distance of separation and there is no rigid specification of the equilibrium position of each atom.

The crystal may therefore be characterized by a distribution function with distortions of the first kind, and the distribution function for the amorphous liquid requires distortions of the second kind [14].

The nematic phase will therefore be characterized by a liquid-like distribution function $W(x, y)$ in projection along the optic axis. However, we know that, in a nematic, there are correlations between the orientations of the molecular axes over distances of, at a minimum, $0{\cdot}2\ \mu$; in a "single nematic liquid crystal", the correlations will extend throughout the sample. Thus the short-range order will be much better defined than in an ordinary liquid.

This description of nematic structure on a molecular scale, i.e. over distances of some tens of Å at the most, does not take into account the angular distribution of the molecular axes called $f(\alpha)$, but it is nevertheless adequate for the X-ray study of nematics considered as having cylindrical symmetry.

Let us now consider X-ray scattering by such a system. The intensity of X-rays scattered by an amorphous liquid is a time and space averaged intensity [20] and the distribution function, obtained by means of a Fourier integral, describes in a statistical manner, the surroundings of each molecule. Consider a nematic obtained without special care to obtain uniform orientation. This nematic is constituted by a multitude of domains, each possessing its own direction of preferred molecular orientation. There is no preferred orientation for the specimen as a whole; the symmetry of the system is cylindrical on a molecular scale, and becomes spherical on a macroscopic scale (over distances of say 1 mm) because the molecular axes for the domains may assume all possible orientations.

The X-ray diffraction pattern of this system has the same symmetry as that for an amorphous isotropic liquid, i.e. a symmetry of revolution around the direction of the X-ray beam. For such a system atomic radial distribution functions may be computed; it suffices to suppose that the scattering elements are the atoms without taking into account the molecular structure of the mesophase. Effectively these functions will be very difficult to interpret because the distances separating atoms of a given molecule are of the same order of magnitude as the distances separating atoms belonging to neighbouring molecules.

Moreover, the use of molecular radial distribution functions cannot be justified, for each molecular axis is unable to take up all orientations with respect to the axes of neighbouring molecules. This is the difficulty when studying scattering by a cholesteric, for here there is an infinite axis of symmetry (the twist axis) which gives an equal probability for all orientations of the molecular axes in planes perpendicular to this axis. Only one study of a cholesteric has been made using atomic radial distribution functions, namely that of cholesteryl caprate [18].

Only scattering experiments, with oriented smectics or nematics, which lead to diagrams with cylindrical symmetry, give the cylindrical distribution function $W(r)$. Following these remarks, we may now examine the X-ray scattering given by oriented nematics or smectics.

X-ray Diffraction Patterns for Oriented Nematic Mesophases

In Fig. 3.2(*a*), we see a diagram of the X-ray diffraction pattern for an oriented nematic mounted as a film perpendicular to the incident X-ray beam; the main optic axis is also perpendicular to the X-ray beam. The molecules are represented as lines, implying that they may be considered as possessing rotational symmetry; the arrow C_∞ shows the preferred orientation of the molecular axes. In other words, thermal fluctuations in orientation of the molecules are neglected in the drawing of the specimen but have their effect on the X-ray diagram; if there were no thermal fluctuations in orientation of the molecular axes, intermolecular scattering would be concentrated in the equatorial plane.

The X-ray pattern shows two symmetry directions. One is parallel to the optic axis (C_∞) and can be described as the meridional section of reciprocal space in the plane of the film. The other direction is perpendicular to the main optic axis; this is the equatorial section of reciprocal space in the plane of the film.

The principal feature of the X-ray pattern is that the main ring splits into two crescents for each of which the intensity is a maximum in the equatorial section; these crescents are mainly due to intermolecular scattering [10].

There are also other rings for which the intensity has a maximum in the meridional section; these are due to intramolecular scattering. These rings, which are less intense than the crescents due to intermolecular scattering, appear distinctly only on patterns obtained after long exposure times. Such an X-ray diagram easily allows us to locate the main optic axis of an oriented nematic phase and to determine all variations of this direction in different experiments.

Sometimes, additional rings, of maximum intensity in the equatorial section, are observed in experiments employing very strong magnetic

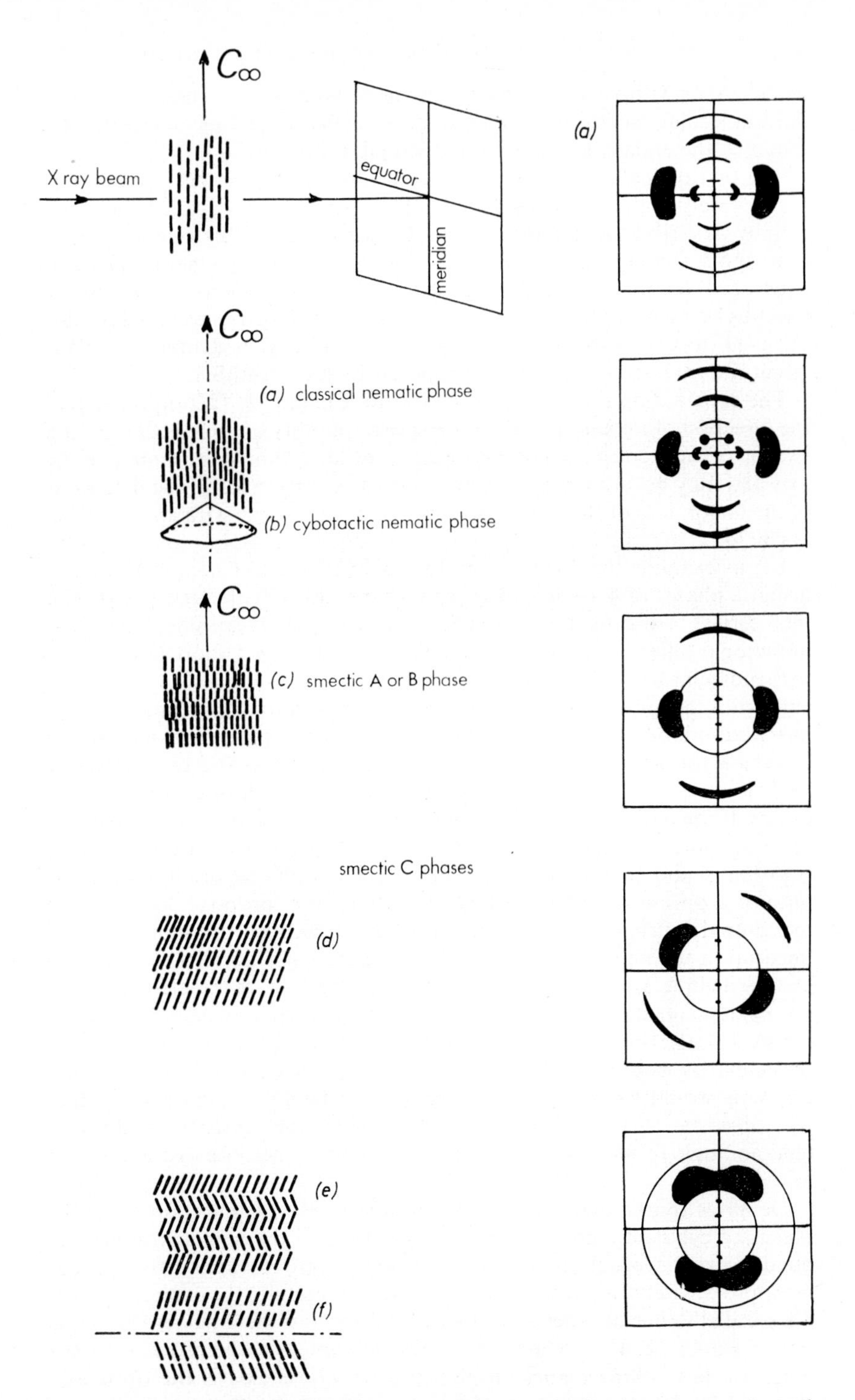

FIG. 3.2(*a*)–(*f*). X-Ray patterns of the main types of oriented nematic and smectic mesophases.

fields [16] or with very sensitive detection apparatus [21]. These are due to intramolecular scattering—interferences between waves scattered by atoms in the equatorial plane of individual molecules.

The first of the meridional rings has great importance and corresponds quite closely with the molecular length (this ring is also called the inner ring by de Vries). Sometimes, for an aligned nematic, this ring splits into four spots symmetrically arranged with respect to the meridional and equatorial sections (Fig. 3.2(*b*)). This has been observed with X-ray patterns of nematics of some 4,4′-di-n-alkoxyazoxybenzenes [22] and also of bis-(4′-n-octyloxybenzylidene)-2-chloro-1,4-phenylenediamine [23]. Different interpretations have been proposed by these authors.

The explanation given by Chistyakov and Chaikovski [22] supposes that the direction of maximum diamagnetic susceptibility of a molecule (or of a swarm of molecules) is not necessarily the direction of the main optic axis and may be at a definite angle δ to this. In this case, all the directions of the major axes of the molecules would envelop a cone with an azimuthal angle 2δ.

Let us examine the results given by Chistyakov and Chaikovski for the nematic phases of 4,4′-di-n-alkoxyazoxybenzenes. For the first member of this series, 4,4′-dimethoxyazoxybenzene (or *p*-azoxyanisole), the first meridional reflection appears on the flat X-ray film as a short straight line perpendicular to the optic axis. For the highest homologues, i.e. the molecules with the greatest length, the reflection splits into definite maxima and has a dumb-bell shape. In terms of reciprocal space, the meridional reflection for *p*-azoxyanisole corresponds to flat discs, and for the higher homologues, the discs are no longer flat, but become thinner towards their centre. If the temperature is now increased, the discs first of all change into discs of uniform thickness, and then at the transition to the amorphous isotropic liquid they become crescent-shaped, with the resulting appearance of a diffuse ring. Chistyakov and Chaikovski propose that this first meridional reflection divides into four spots because the molecules are lined up in the magnetic field with the long axes not parallel to, but at a certain definite angle to the lines of force of the field, the angle depending on the diamagnetic anisotropy of the particle (which may be the molecule, the swarm, the domain . . .). However, as de Vries [23] points out, such an arrangement would affect the main equatorial reflection such that their crescents would be considerably longer (by a factor of almost two) than was observed. For 4,4′-diethoxyazoxybenzene, the crescent would subtend an angle of 86° about the centre of the film; the observed angle is at most 50°.

De Vries thinks that these spots are due to Bragg reflection from the planes of cybotactic groups. He supposes that, in the nematic, molecular bundles exist in which the molecular centres of gravity are correlated and lie approximately in parallel planes; in such a bundle, the molecular axes are parallel to one another and tilted with respect to these planes (the angle of tilt would be 45°). When the molecules are oriented parallel to the magnetic field, these planes envelop a cone whose axis is the optic axis (Fig. 3.2(*b*)). There are four orientations of the planes that satisfy Bragg's law and give rise to diffuse spots. De Vries therefore proposes the existence of a "cybotactic" nematic mesophase in which, compared with a classical nematic phase, there is an additional order, i.e. a substantial proportion of the molecules is arranged in "cybotactic groups" and the molecular centres are in well-defined planes. However, in no case has it been possible to confirm this definitely on the basis of X-ray results. De Vries cites the optical properties of these nematics as supporting evidence [23].

In addition to the classical nematic phase, de Vries in fact defines two types of cybotactic nematic mesophase, one termed a "normal cybotactic mesophase" with the molecular axes perpendicular to the cybotactic planes, and the other a "skewed cybotactic mesophase", with the molecular axes tilted with respect to these planes. He has noticed that the smectic phase preceding a "skewed cybotactic nematic mesophase" is one with the molecular axes tilted relative to the smectic planes (the angle of tilt is about 45°); hence, the nematic and the preceding smectic are very similar in these cases.

A further point is that the skewed smectic, which is identical to the phase classified as smectic C by Sackmann and his co-workers, has, in several compounds been found to be followed on heating by a smectic A (Fig. 3.2(*c*)), and one would expect that such a normal smectic would be followed on further heating by a normal cybotactic or a classical nematic. This led de Vries to examine the optical properties of some nematic textures and to conclude that pseudo-isotropy may be characteristic of classical and particularly of normal cybotactic mesophases, whereas failure to give pseudo-isotropy may be characteristic of a skewed nematic which generally follows direct from a smectic C.

Quantitative Interpretation of X-ray Diffraction Patterns of Nematic Mesophases by Cylindrical Distribution Functions

Chistyakov's review [24] gives a detailed description of methods which permit the calculation of cylindrical distribution functions. We simply note that the Bessel function of zero order, $J_0(2\pi Rr)$, for systems with cylindrical symmetry, plays the part of the function $(\sin 2\pi sr)/2\pi sr$ for systems with spherical symmetry.

s is the modulus of the scattering vector in reciprocal space in the case of spherical symmetry. Only the modulus of the scattering vector arises in the expression of distribution functions; $s = 2 \sin \theta/\lambda$, where 2θ is the scattering angle. R is the scattering vector in the equatorial plane of reciprocal space for cylindrical symmetry. r is the distance between two atoms or molecules in the case of spherical symmetry and between two atoms or molecules in projection on the equatorial plane of real space in the case of cylindrical symmetry.

One can, by application of a Fourier–Bessel transform, calculate an atomic or molecular cylindrical distribution function from an interference function $i(R)$ obtained from the intensity in the equatorial plane of the diffraction pattern:

$$2\pi r Z(r) = 2\pi r Z_{\mathrm{m}} + 4\pi^2 r \int_0^\infty i(R)\, J_0(2\pi R r) R\, \mathrm{d}R \tag{3.1}$$

$Z(r)$ is a normalized distribution function; Z_{m} is either the average number of atoms per Å^2 or the number of molecular centres of gravity per Å^2.

Atomic Cylindrical Distribution Functions

The system is considered to be constituted of atoms, and its molecular structure is neglected; thus the method is insensitive to the fact that two atoms may or may not belong to the same molecule.

The interference function $i(R)$ has the form:

$$i(R) = \frac{I(R) - [\mathrm{f}^2(R) + C(R)]}{\mathrm{f}^2(R)} \tag{3.2}$$

where $I(R)$ is the experimentally determined equatorial density, $\mathrm{f}(R)$ is an averaged atomic scattering factor and $C(R)$ is the Compton incoherent scattering.

The atomic distribution function gives the probability of finding two atoms at a distance r, in projection on the equatorial plane of real space, the atoms belonging to either the same or two different molecules. However by comparison with close packed molecular models, one may say that the shape of this function is mainly determined by intermolecular statistics [16]. This is because, in a bundle of parallel molecules, the greatest distances between atoms belonging to the same molecule in projection on a plane perpendicular to the bundle axis, are smaller than the greatest distances between atoms belonging to neighbouring molecules.

More recently Delord has established that, for *p*-azoxyanisole, these functions may be considered as being distributions of "non-point molecules" and are understood by postulating a hexagonal disposition of the centres of gravity of the molecules in projection on a plane perpendicular to the main optic axis.

Consider results obtained for *p*-azoxyanisole. Fig. 3.3 illustrates the atomic cylindrical distribution functions for a single liquid crystal at various temperatures. The highest temperature (132°C) is slightly lower than the clearing temperature of the nematic (135°C). The middle temperature (115°C) is near to that of the transition from crystal to nematic (118°C). The lowest temperature (100°C) corresponds to a high degree of supercooling. The greatest distance between two atoms of one molecule in projection in the plane perpendicular to the molecular axis, is about 3·5 Å, but most of the interatomic distances are about 2·5 Å. It is thus assumed that the peak at 4·6 Å is mainly associated with distances between neighbouring molecules. The multiples of 4·6 Å, i.e. 9·2 Å, 13·8 Å, 18·4 Å, agree well with the theoretical positions of the centres of gravity of successive peaks for a liquid-like distribution function and give the second, third and fourth co-ordination cylinders. It may be shown that

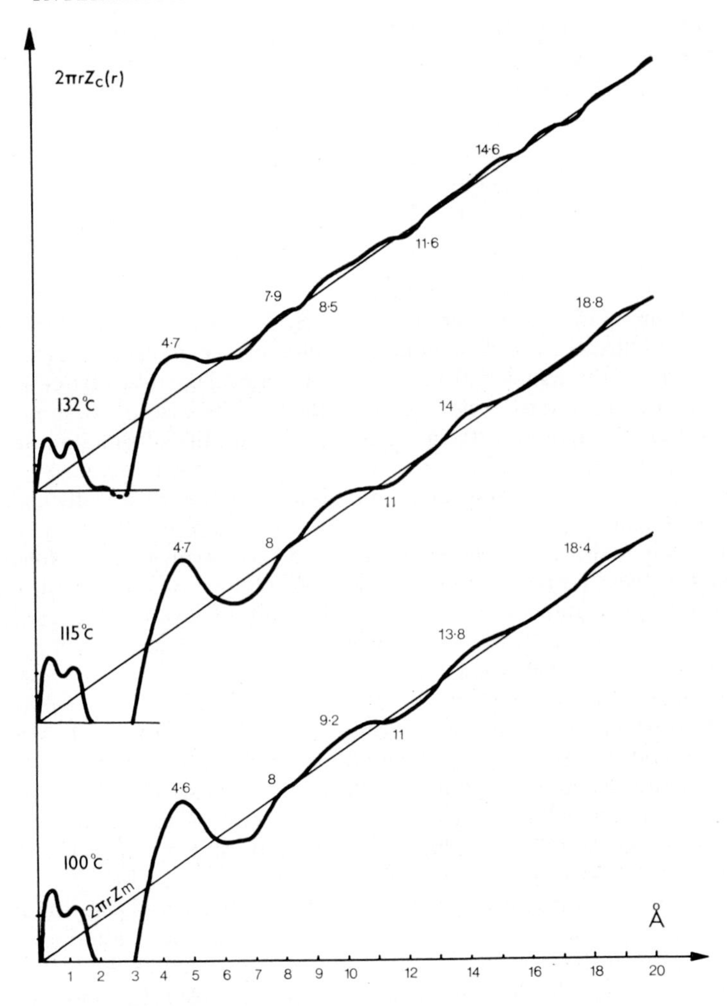

FIG. 3.3. Atomic cylindrical distribution functions for a single nematic liquid crystal of *p*-azoxyanisole at various temperatures.

the peak at 8 Å agrees with an hexagonal disposition of the projections of the molecular centres. The distribution functions obtained in this way allow us to define the short-range order and to follow its variations as a function of temperature.

Cylindrical Distribution Function for the Molecular Axes

To obtain a distribution for the molecular axes, one may take an interference function of the form:

$$i(R) = \frac{I(R) - [\overline{F_M^2}(R) + C(R)]}{\bar{F}_M^2(R)} \tag{3.3}$$

and substitute in Equation (3.1) in which Z_m is now the number of molecular centres of gravity per Å^2.

$\overline{F^2_M}(R)$ and $\bar{F}^2_M(R)$ are respectively the square mean and mean square averaged molecular scattering factors, in the equatorial plane of reciprocal space, for a molecule which may adopt all possible orientations about its major axis; these have the form:

$$\overline{F^2_M}(R) = \sum_j \sum_k f_j f_k J_0(2\pi R r_{jk})$$

$$[\bar{F}^2_M](R) = \left[\sum_j f_j J_0(2\pi R r_j)\right]^2$$

The main interest of a molecular cylindrical distribution function would be to give directly the distribution pattern of the molecular centres in projection on the equatorial plane of real space. Theoretically, computation of these functions is not quite justified in the case of mesomorphic compounds by analogy with the conditions of validity of radial molecular distribution functions [25]. For systems with cylindrical symmetry, computation of cylindrical molecular distribution functions is fully justified if there is statistical independence of the orientations of the molecules about their major axes; these conditions do not seem to be fulfilled for small r values, corresponding to the first cylinder of co-ordination. In addition, the distances between neighbouring molecules are generally about 4·8 Å, and we may query whether there is room for rotation of molecules around their long axes.

In concluding this part, we may say that only the atomic cylindrical distribution function allows us to obtain short-range order in a nematic. This method is applicable to all cybotactic or classical nematics.

We finally dwell on conditions which allow us to calculate accurate distribution functions; these conditions are now well established from studies of the liquid state [20]. In the case of systems with cylindrical symmetry one may state them as follows.

The equatorial intensity must be known for high values of the modulus of the scattering vector, $R = 2\sin\theta/\lambda$, where 2θ is the scattering angle. $R \approx 2$ seems to be a minimum value; this value, which leads to $2\theta = 90°$ when $\lambda = 0{\cdot}707$ (MoK$_\alpha$ radiation), is never fulfilled by photographic methods with a flat plate normal to the beam. Only diffractometer techniques or a cylindrical film allow us to record intensity measurements at such scattering angles. Consequently $I(R)$, the experimental intensity measurement, is restricted to a region $R \leqslant R_M$ of reciprocal space, R_M being determined by geometrical conditions of the experiment and the wavelength used. For calculating cylindrical distribution functions, one must insert R_M as the upper limit of the Fourier–Bessel integral (see equation 3.1); it is equivalent to multiplying $I(R)$ by a shape function [26]. The distribution function obtained in this way is the product of convolution of the true distribution function by the shape amplitude (the Fourier–Bessel transform of the shape function). The resulting effects on calculated distribution functions are ghost peaks without physical significance; these vanish only if R_M is great enough and if $i(R)$ is multiplied by a convergence factor.

The introduction of a sharpening function, i.e. division of $i(R)$ by f^2 or F^2, increases resolution of distribution functions, but also increases the importance of measuring the intensity at high angles of scattering; these intensities are generally the least accurate because of their very small values. Introduction of a convergence factor may allow us to palliate these effects.

At present, there is only one result complying with the above conditions, namely that in Fig. 3.3. In our opinion, it is mainly because the above conditions have been neglected that the distribution functions for all nematics so far studied are not specific functions of the second kind characteristic of a paracrystalline structure [27–29].

Atomic Linear Distribution Function

There is another way to obtain quantitative results from the X-ray pattern. When a given system has a structure characterized by a preferred orientation, for instance a cylindrical axis of symmetry, one may study such a structure in projection on this axis; this gives information about the periodicity in the atomic structure of the scattering elements as well as about the periodicity in disposition of the scattering elements themselves.

In the particular case of an oriented nematic, if one supposes that the statistical distribution of the molecular centres in projection on the main optic axis is uniform, then the only information contained in the meridional section of reciprocal space relates to interferences between waves scattered by atoms within individual molecules, i.e. to intramolecular scattering) [14]. It is possible to compute, from the meridional intensities, a one-dimensional Fourier transform which leads to a linear distribution of atoms in projection on the main optic axis.

For *p*-azoxyanisole, the functions obtained in this way can be fully interpreted in terms of the internal molecular structure providing that one

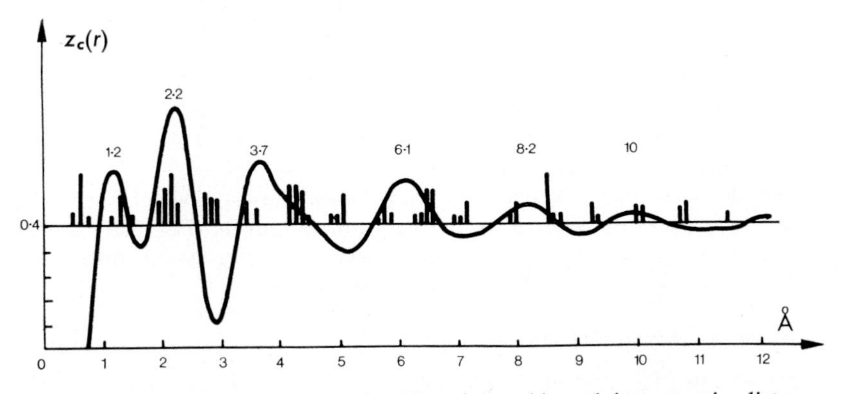

Fig. 3.4. Atomic linear distribution function $z_c(r)$ and interatomic distance function in projection on the major axis of the molecule of *p*-azoxyanisole. Results relate to the supercooled oriented nematic phase at a temperature of 100°C [21].

takes into account the angular distribution function $f(\alpha)$ of the molecular axis about the optic axis [21]. In Fig. 3.4, we see the linear distribution function of atoms in the nematic phase of *p*-azoxyanisole at 100°C and the corresponding interatomic distance function for the molecule.

These functions are rather insensitive to all possible correlations between molecular centres because the Fourier transform is little affected by large variations in intensity in low angle scattering. They will be typical of nematic structure in projection on the optic axis only if all precautions relating to the computation of a Fourier transform are taken.

X-ray Diffraction by Smectic Phases

Since Herrmann's work on thallium oleate and stereate, there have been, up to the time of writing, few published results on X-ray diffraction by non-amphiphilic smectics. Nevertheless, such studies are most interesting because of the additional order in smectics; the molecular centres lie in well-defined planes and give a periodicity of smectic layers. We feel that many results on smectics, in particular on their polymesomorphism, are to be expected in the next few years.

We adopt the classification of Sackmann and Demus and discuss mainly three kinds of smectic: smectic A, B and C [30]. There are other kinds of smectic; for example, Sackmann has recently discussed smectic D and E, Demus has examined smectic F and G, and de Vries has proposed a smectic H. However, there is not enough information for us to discuss these modifications in this section.

The majority of known smectics are of type A, with the molecular centres lying in planes perpendicular to the mean direction of orientation of the long molecular axes (Fig. 3.2(*c*)). Probably smectic A phases are identical with those smectic phases which Friedel supposed to be the only ones. Smectic A, in cases of smectic polymorphism, is always the high-temperature form; if the molecules show high thermal mobility, the perpendicularity of their axes to the planes will be a statistical average position.

The smectic B form, observed only in substances showing at least a smectic A phase, differs from the previous type in having a higher degree of order within each layer; probably there is a hexagonal close packing of the molecular centres.

We shall designate by smectic C those mesophases in which the molecular axes are tilted relative to the planes of the layers (Fig. 3.2(*d*))—see however p. 75. We shall investigate successively the X-ray diffraction patterns of smectic A and B and then those for smectic C.

Smectic A and B Mesophases

According to Vainshtein [14], one may express the symmetry of the smectic A or B phase as follows:

$$\infty C\tau(z)\, W(x, y)/2.$$

As in the case of the nematic state, the first symbol ∞ denotes a random orientation of the parallel molecules with respect to rotation about their major axes. $C\tau(z)$ allows for a periodicity C due to the packing of the layers in the z direction; the probability of finding the projections of the molecular centres on the optic axis is periodic. Again as for the nematic state, one may describe the disposition of the long molecular axes within each layer by a statistical liquid-like distribution function $W(x, y)$.

In smectic B, the molecular arrangement is hexagonal. The shape of the molecules is considered to be the main cause of smectic layer formation because of the convenient way in which anti-parallel packing may occur. According to Vainshtein, therefore, there is almost always a statistical diad axis which denotes the anti-parallelism of neighbouring molecules in each layer; this axis lies in a plane perpendicular to the main optic axis.

As in the case of nematics, results obtained with oriented specimens are of most interest. The X-ray pattern (Fig. 3.2(*c*)) of oriented smectics A and B is similar to that for an oriented nematic. We see two crescents whose intensity is a maximum on the equatorial section of reciprocal space; the meridional direction is perpendicular to the layer planes. In addition, regularly spaced meridional spots indicate layer stratification and are superimposed upon intramolecular scattering. A higher degree of sharpness of the main rings and crescents allows us to distinguish a smectic B from a smectic A diagram; the X-ray diffraction patterns for the smectic phases of ethyl *p*-methoxybenzylideneaminocinnamate oriented by an electric field illustrate this property very well. This compound has recently been studied [31] both by X-ray diffraction and by differential thermal analysis. These studies show that one nematic and two smectic mesophases are formed. The comparison between the smectic A and the smectic B diagram shows effectively a higher sharpness of the main ring corresponding to intermolecular scattering in the case of smectic B. The nematic phase seems to be a classical or a normal cybotactic nematic mesophase.

Cylindrical distribution functions, computed from the intensity in the equatorial section of reciprocal space, should allow us to obtain the short-range order within the layers, but there are no such results available. The conditions required to obtain accurate results are the same as those stated for cylindrical distribution functions for the nematic state.

We must mention now recent work by Levelut and Lambert [32] on ethyl 4-*p*-ethoxybenzylidenaminocinnamate and terephthalylidene-bis-(4-n-butylaniline); both compounds, on heating, show a smectic B, this phase preceding a smectic A in the first case, and a smectic C in the second. In both cases, the smectic B obtained by heating keeps the external shape of the single crystal, thereby providing a favourable case for X-ray study.

Each smectic layer in each of the two smectic B mesophases has a periodic structure resulting from an hexagonal close packing of the long axes of the molecules. In smectic B of the ester, the average orientation of the long axes of the molecules is normal to the layer planes, whereas in that of the Schiff's base, the long axes are tilted at a mean angle of 58° to the layer planes. This tilt angle agrees well with that obtained by Taylor, Arora and Fergason [38] for the succeeding smectic C of the Schiff's base.

For each compound therefore, the transition from smectic B to the relevant higher temperature smectic involves little orientational change of the long axes of the molecules relative to the layer planes. By their elegant studies, the authors have also shown that on passing from one smectic B layer to another there is a displacement of the regular orientation which destroys any long-range order in a direction perpendicular to the smectic planes.

This work provides the first example of a smectic B with molecules tilted with respect to the smectic planes, and shows convincingly the great value of X-ray diffraction for the structural analysis of oriented smectics.

Smectic C Mesophases

One may consider other kinds of smectics (smectic C phases) with the molecules tilted relative to the smectic planes. Such a mesophase may precede the isotropic liquid, a nematic or a smectic A, and is easy to distinguish from a tilted smectic B. One may conceive of several smectic C types according to the relative disposition of the successive layers [33] (Fig. 3.2(*d*)–(*f*)).

A layer packing keeping the same direction of tilt in all the layers is one possibility (Fig. 3.2(*d*)). Amongst the compounds known to give this structure, are the bis-(4′-n-alkoxybenzylidene)-2-chloro-1,4-phenylenediamines with nine to eighteen carbon atoms in the chains [34]. A smectic C in which the molecules are tilted to the right and left in alternate layers (Fig. 3.2(*e*)) is another possibility, and Chistyakov *et al.* have proposed such a structure for the smectic phase of *p*-n-nonyloxybenzoic acid [31].

One may also have a twisted smectic C obtained by progressive rotation of the direction of the major molecular axes through a constant angle on passing from layer to layer [33]; such a smectic is optically active (see Chapter 2.1 of Vol. 1).

The preceding descriptions of smectic C phases are rather schematic, and do not take into account thermal motions which may cause undulations of the smectic planes or a statistical distribution of the direction of inclination of the molecular axes in the smectic planes [22]. Chistyakov [22] has proposed that the latter situation applies to the smectics formed by the 4,4′-di-n-alkoxyazoxybenzenes with seven to ten carbon atoms in the alkyl groups.

Consider now the X-ray diffraction patterns given by these different kinds of smectic C mesophase. A tilted smectic leads to a diagram very similar to the smectic A diffraction patterns. Only the meridional section is different and the intensity maxima due to the layer periodicity are on a line tilted with respect to the meridional section (Fig. 3.2(*d*)).

A smectic C in which the molecules are tilted to the same extent but in opposite directions in alternate layers leads to a very characteristic diffraction pattern with four crescents with maxima in the equatorial planes of the two directions of the long molecular axes (Fig. 3.2(*e*)).

A twisted smectic C or a smectic C with a statistical distribution of the direction of inclination of the molecular axes in the smectic planes should lead to very similar X-ray patterns which would also be difficult to distinguish from a smectic A diagram except perhaps by a greater angular width of the two main crescents.

By X-ray study of any oriented smectic C, it is easy to obtain the layer periodicity and hence to calculate a mean tilt angle of the molecular axes relative to the smectic planes provided that one assumes that the molecules remain rigid and uncrumpled. It would be more difficult to obtain data relating to the detailed distribution of the molecular centres within smectic planes. Few results of any kind on smectic C phases exist, and to our knowledge these are restricted to the examples cited above and discussed in greater detail below.

Compounds in the series of 4,4′-di-n-alkoxyazoxybenzenes may be denoted by the number of carbon atoms in each alkyl group; the C_7, C_8, C_9 and C_{10} compounds give one smectic and one nematic phase. For these compounds, when the transition from the nematic to the smectic occurs, the dumb-bell-shaped reflection, peculiar to the nematic, divides into two spots whose intensities are often different [22]. Chistyakov and Chaikovski do not think that this is explained by a predominant tendency of the molecules to adopt one particular inclination of their long axes with respect to the principal axis of the preparation, for this would involve the splitting of each of the two main equatorial reflections into four (see Fig. 3.2(*c*)). They conclude rather that there is a statistical distribution of the directions of inclination of the molecular axes in the smectic planes.

Another interesting smectic has been observed by Chistyakov with *p*-n-nonyloxybenzoic acid [31] which gives both a nematic and a smectic. The X-ray diagram of the smectic shows four crescents and indicates that the molecules may assume alternating directions of inclination in successive smectic planes. Chistyakov supports this by interpreting optical data relating to the phase on a model of the smectic structure analogous to the Fig. 3.2(*e*), but an alternative structural model is that shown in Fig. 3.2(*f*); this would lead to the same X-ray patterns.

Finally, de Vries has shown that the bis-(4′-n-alkoxybenzylidene)-2 chloro-1,4-phenylenediamines with nine to eighteen carbon atoms in each of the alkyl chains have a smectic as well as a nematic. Their smectics could be identified by their X-ray diagrams as tilted smectics (Fig. 3.2(*d*)). The molecules are inclined to the planes at an angle of about 45°'

We must stress here that these three different kinds of compound, the bis-(4′-n-alkoxybenzylidene)-2-chloro-1,4-phenylenediamines, 4,4′-di-n-heptyloxyazoxybenzene and *p*-n-nonyloxybenzoic acid give closely analogous X-ray patterns for the nematics and that the structural interpretations given by the different authors are quite distinct. If we follow de Vries, all these patterns are however characteristic of skewed cybotactic nematics. Moreover the X-ray patterns of the smectics to which these nematics give rise on cooling and the structures proposed for them are similar, except for *p*-n-nonyloxybenzoic acid; we note particularly that we do not know if the smectic structure proposed for this compound could give rise to a cybotactic nematic mesophase. There are as yet insufficient results to make it possible to reach definite conclusions.

In conclusion, we may say that X-ray diffraction remains a very valuable and precise method for investigating mesophases, especially if one can use oriented specimens. In combination with other methods such as optical studies of the mesophases or differential thermal analysis, X-ray diffraction is an essential technique for the investigation of the polymorphism of mesophases [35]. The discovery of other smectics such as smectic D, for which a cubic structure is presumed, and smectic E, the X-ray pattern of which points to a higher order compared with the A, B and C phases, must give a high importance to X-ray studies [36].

Finally, we note that there is only one report of experimental work involving small angle scattering; this relates to work by Gravatt and Brady, on the nematic and amorphous liquid states of *p*-azoxyanisole [37].

REFERENCES

[1] LINGEN, J. S. V. D. *Ber. dt. chem. Ges.* **15,** 913 (1913).
[2] FRIEDEL, G. *Annls Phys.* **18,** 273 (1922)—see p. 376.
[3] HUCKEL, E. *Phys. Z.* **22,** 561 (1921).
[4] STEWART, G. W. *Phys. Rev.* **38,** 931 (1931).
[5] STEWART, G. W. *Trans. Faraday Soc.* **29,** 982 (1933).
[6] GLAMANN, P. W., HERRMANN, K. and KRUMMACHER, A. H. *Z. Kristallogr. Kristallgeom.* **74,** 73 (1930).
[7] GRAY, G. W. *Molecular Structure and the Properties of Liquid Crystals*, Academic Press, London and New York (1962).
[8] BROWN, G. H. and SHAW, W. G. *Chem. Rev.* **57,** 1049 (1957).
[9] CHATELAIN, P. *Bull. Soc. fr. Minér. Cristallogr.* **66,** 105 (1943).
[10] FALGUEIRETTES, J. *Diffusion des Rayons X par un Monocristal Liquide du Type Nématique*, Thèse, University of Montpellier, France, C.N.R.S. No. 160 (1958).
[11] CHATELAIN, P *Bull. Soc. fr. Minér. Cristallogr.* **78,** 262 (1955).
[12] SAUPE, A. *Angew. Chem. Internat. Edn.* **7,** 97 (1968).
[13] DELORD, P. and FALGUEIRETTES, J. *C. r. hebd. Séanc. Acad. Sci., Paris* **260,** 2468 (1965).
[14] VAINSHTEIN, B. K. *Diffraction of X-Rays by Chain Molecules*, Elsevier, Amsterdam (1966).
[15] CHISTYAKOV, I. G. and VAINSHTEIN, B. K. *Soviet Phys. Crystallogr.* **8,** 458 (1964).
[16] VAINSHTEIN, B. K., CHISTYAKOV, I. G. and CHAIKOVSKI, V. M. *Soviet Phys. Dokl.* **12,** 405 (1967).
[17] CHISTYAKOV, I. G. *Soviet Phys. Crystallogr.* **12,** 770 (1968).
[18] CHISTYAKOV, I. G. *Soviet Phys. Crystallogr.* **8,** 691 (1964).
[19] GULRICH, L. W. and BROWN, G. H. *Molec. Crystals Liqu. Crystals* **3,** 493 (1963).
[20] PINGS, C. J. *Structure of Simple Liquids by X-Ray Diffraction*, North Holland Publishing Company, Amsterdam (1968).
[21] DELORD, P. *Diffusion des Rayons X par une Phase Nématique Orientée*, Thèse, University of Montpellier, C.N.R.S. No. 4372 (1970).

[22] CHISTYAKOV, I. G. and CHAIKOVSKI, V. M. *Liquid Crystals 2* (edited by G. H. Brown), Gordon & Breach, New York (1968), Part II, p. 803.

[23] DE VRIES, A. *Molec. Crystals Liqu. Crystals* **10,** 31 and 219 (1970).

[24] CHISTYAKOV, I. G. *Soviet Phys. Usp.* **9,** 551 (1967).

[25] FOURNET, G. *Handb. Phys.* **32,** 1238 (1957).

[26] DELORD, P. and MALET, G. *C. r. hebd. Séanc. Acad. Sci., Paris* **270,** 1107 (1970).

[27] HOSEMAN, R. and MULLER, B. *Liquid Crystals 2* (edited by G. H. Brown), Gordon & Breach, New York (1968), Part II, p. 139.

[28] LEMM, K. *Molec. Crystals Liqu. Crystals* **10,** 259 (1970).

[29] HOSEMAN, R. and MULLER, B. *Molec. Crystals Liqu. Crystals* **10,** 273, (1970).

[30] SACKMANN, H. and DEMUS, D. *Liquid Crystals* (edited by G. H. Brown, G. J. Dienes and M. M. Labes), Gordon & Breach, New York (1967), p. 341.

[31] CHISTYAKOV, I. G., SCHABISCHEV, L. S., JARENOV, R. I. and GUSAKOVA, L. A. *Liquid Crystals 2* (edited by G. H. Brown), Gordon & Breach, New York (1968), Part II, p. 813.

[32] LEVELUT, A. M. and LAMBERT, M. *C. r. hebd. Séanc. Acad. Sci., Paris* **272,** 1018 (1971).

[33] SAUPE, A. *Liquid Crystals 2* (edited by G. H. Brown), Gordon & Breach, New York (1968), Part I, p. 59; HELFRICH, W. and OH, C. S. *Molec. Crystals Liqu. Crystals* **14,** 289 (1971).

[34] DE VRIES, A. *Acta Crystallogr.* **A25,** S135 (1969).

[35] SACKMANN, H., DIELE, S. and BRAND, P. *Acta Crystallogr.* **A25,** S537 (1969).

[36] DIELE, S., BRAND, P. and SACKMANN, H. *Molec. Crystals Liqu. Crystals* **16,** 105 (1972).

[37] GRAVATT, G. C. and BRADY, A. W. *Liquid Crystals 2* (edited by G. H. Brown), Gordon & Breach, New York (1968), Part II, p. 819.

[38] TAYLOR, T. R., ARORA, S. L. and FERGASON, J. L. *Phys. Rev. Lett.* **25,** 722 (1970).

Additional references

DE VRIES, A. *J. chem. Phys.* **56,** 4489 (1972).

DE VRIES, A. *Acta Crystallogr.* **A28,** 659 (1972).

DELORD, P. and MALET, G. *Molec. Crystals Liqu. Crystals* (1973) to be published.

4

X-ray Diffraction by Liquid Crystals—Amphiphilic Systems

K. FONTELL

INTRODUCTION

A characteristic of the liquid crystal structures in amphiphilic systems is that, while they have no short-range crystalline order (<5 Å), there is a superior order that is crystalline in one, two or three dimensions. In a phase diagram, the liquid crystalline phases can be bounded on one side by isotropic solutions with no long-range order, and on the other side by phases that possess both crystalline long- and short-range order.

An X-ray diffraction examination of a liquid crystal specimen provides information not only on the state of organization of the hydrocarbon chains but also on the one-, two- or three-dimensional crystallographic lattices of the secondary structure. To obtain an impression of the structure of the unit aggregates of a phase the X-ray findings must be combined with other data, such as the composition and density of the sample and the size, shape and chemical properties of the molecules. This information does not suffice for a unique definition of the structure; it is also necessary to assess the likelihood that a particular structure can exist in the light of what else is known about the system as a whole.

All amphiphilic mesophases were formerly thought to have a lamellar structure. One of the reasons for this was the paucity of the X-ray diffraction patterns for many such systems.

THE EXPERIMENTAL PATTERN

The diffraction pattern obtained for an amphiphilic liquid crystal is characterized by a series of sharp reflections corresponding to interplanar spacings ranging from 10 to above 100 Å, and a wide diffuse reflection with a position corresponding to a spacing of 4·5 Å. For a system containing water there is further a diffuse reflection corresponding to 3·2 Å.

The position and appearance of the 4·5 Å reflection resemble those obtained for liquid paraffin. The reflections corresponding to the large distances ("long Bragg spacings") are often as sharp as those obtained from well-crystallized substances. The number of such reflections depends both on the amphiphilic substance, the liquid crystal phase and the composition. Between 2 and 7 reflections are usually obtained, but a larger number may sometimes be observed.

When the reflections are few, the interpretation of the diffraction pattern may present difficulty, but when they are more numerous it is possible to divide the interference patterns into categories and hence to classify the phases according to their one-, two- or three-dimensional periodicity. However, as the reflections are located close together and in many cases are of low intensity, such a classification may be unreliable.

EXPERIMENTAL EQUIPMENT

Because of the consistency of amphiphilic liquid crystals usually only X-ray diffraction methods similar to those used for polycrystalline systems (powder methods) can be applied. In the choice of apparatus, account should be taken of the fact that part of the information is obtained from an angular range near the primary beam, within what is usually referred to as the low-angle (small-angle) region. (For the most commonly used radiation, copper K_α ($\lambda = 1{\cdot}54$ Å), these diffraction angles, 2θ, lie between 10° and below 1°). Moreover, the reflections that occur are of weak intensity. For these reasons two separate cameras are often used, one of a more conventional design for the wide-angle region corresponding to crystal spacings of around 4·5 Å (an ordinary powder camera), and the other for the long spacings corresponding to the low-angle region. In order to shorten the exposure time and reduce the background scattering the cameras should be evacuable (to 10^{-1} mm Hg). It is also advisable to use monochromatic radiation. This may be obtained by filtering (e.g. through nickel foil for copper K_α-radiation) or by means of a focusing crystal monochromator. The low-angle cameras can be furnished with point or slit collimation; point collimation has the advantage that an impression of orientation effects within the specimen can be obtained—though at the expense of the intensity.

Because of the vacuum, the specimens must be enclosed in thin-walled (0·01 mm) glass capillaries or cuvettes. The latter can be designed for examination of the specimen under the polarizing microscope.

Geiger counter diffractometer techniques may, of course, be used but the low intensities of the reflections in the vicinity of the primary beam, their spottiness and the orientation effects render the examinations cumbersome. Such a technique is, however, indispensable for the determination of absolute intensities.

INTERPRETATION OF EXPERIMENTAL X-RAY DIFFRACTION PATTERNS

The following account of the interpretation of X-ray diffraction patterns is concerned mainly with the structures which occur in aqueous systems containing the fatty acid soaps and some other similar amphiphiles of more or less polar nature. The phase structures existing in anhydrous soap systems at elevated temperatures are outlined only briefly. By virtue of the simple molecular structure of the soaps, their systems serve as excellent models, and conclusions reached with them can be generalized for application to more complex systems.

In the analysis of the experimental results, one may distinguish between the short-range order which is dependent on the state of organization of the hydrocarbon chains (and of the water molecules), and the long-range order of the secondary structure which is composed of the structural aggregates within the system.

The interpretation is based upon treating the actual X-ray diffraction photographs as powder patterns. The position of the reflections are converted to interplanar spacings by means of the familiar expression (Bragg equation)

$$n\lambda = 2d \sin \theta$$

where λ is the wavelength used, d the interplanar spacing between successive identical crystal planes, 2θ the diffraction angle and n is an integer—"the order of reflection".

The symmetry of the lattice is determined by finding an equation with which all the observed spacings agree and from this the dimensions of the unit cell can be calculated. One disadvantage inherent in the powder technique is that it usually is not possible to index all the reflections unequivocally. However, in combination with other available data and especially when the lattices are of high symmetry it may be possible to propose a "crystalline" structure for the phase.

The verification of the correctness of a proposed structure lies in the degree of agreement between observed and calculated intensities of reflections. A comparison is difficult to make in the case of amphiphilic liquid crystals, owing to the small number of reflections observed. Nevertheless, it has been possible to do this in some cases when the diffraction pattern was studied as a function of the water content. When a structure is formed by structural units of constant dimensions separated by variable amounts of water, the intensity of the reflections is proportional to the Fourier transform of these structural units sampled at the lattice points. The ratio of the observed intensities must thus match the amplitudes of the Fourier transform.

The Short-range Order of the Hydrocarbon Chains

The hydrocarbon chains in an amphiphilic liquid crystal are usually considered to be in the liquid state. This state is dependent on the temperature and is favoured by the presence of water or organic solvent.

The identical position and appearance of the 4·5 Å reflection found for liquid paraffin [29, 34] provides strong, but not conclusive, evidence for the liquid state of the hydrocarbon chains. The inconclusiveness arises because hydrocarbon chains pack in a variety of ways, with the result that crystalline structures, too, can give rise to a reflection at 4·5 Å; if the regions with long-range order are small enough (less than 10^{-5} cm), this reflection can be diffuse.

Further evidence for the liquid state of the hydrocarbon chains, derived from studies of the long spacings, is the negative coefficient of thermal expansion of these spacings observed with liquid crystals containing one- or two-dimensional lattices. This phenomenon has been interpreted by Luzzati by analogy with the conditions in a rubber thread stretched by a constant force, where equilibrium exists between the effects of the external force tending to orient the molecular chains, and the thermal movement of the molecules, tending to produce greater disorder [29, 34]. An increase in temperature displaces this equilibrium towards a more disordered state which is manifested in a linear contraction of the specimen that is visible to the naked eye. The value obtained experimentally for the expansion coefficient in amphiphilic liquid crystals is in close agreement with the theoretical [34]. Note that there is no volumetric contraction; it is only the one- or two-dimensional structure that would be expected to contract. However, Lawson *et al.* have reported a reduction in the long spacings—albeit only a small one—on raising the temperature of their three-dimensional, optically isotropic viscous phase in the dodecylamineoxide–water system [26].

Other physico-chemical studies also point to the liquid state of the hydrocarbon chains (IR, NMR, DTA, and density [5, 27, 43, 45] and miscibility).

By virtue of their liquid state, the paraffin moieties of various molecules are able to form a homogeneous mixture. This is why the structures of multi-component systems often can be interpreted analogously to those of simple systems. Some authors, however, have criticized the concept of the liquid state of the hydrocarbon chains in amphiphilic liquid crystals [50, 61].

The Long-range Order

As mentioned above, within the low-angle region the relative positions of the reflections vary from one type of liquid crystal structure to another. The pattern may be so distinctive that it is possible to recognize the phase type solely on a visual inspection of the X-ray diffraction pattern.

When the number of reflections is small it is not always possible to categorize phases according to their one-, two- or three-dimensional long-range periodicity, and in such cases guidance can be obtained from observations with the polarizing microscope and knowledge of the position of the phase in the phase diagram in relation to temperature and concentration. For example, phases with one- or two-dimensional periodicity are usually optically birefringent, whereas those with three-dimensional periodicity can be optically isotropic.

Though opinions differ as to whether an amphiphilic phase exhibiting three-dimensional periodicity and whose hydrocarbon chains are in the liquid state should be defined as liquid crystalline or truly crystalline [18, 29], in this article all the amphiphilic phases having the diffuse 4·5 Å reflection and sharp reflections in the low-angle region will be considered as liquid crystals. These phases have been called liquid-paraffin phases by Luzzati [29, 31]. The structures of the "gel" and "coagel" will not be dealt with as they have crystalline "short" reflections in place of the diffuse 4·5 Å reflection.

In the evaluation of the geometric parameters of a particular structure (such as the interfacial area per polar group and the dimensions of the unit aggregates composing the liquid crystalline structure) one basic concept is a division into separate polar and non-polar regions. For systems of two or more components the partial molar volumes of the individual components can be estimated and it is then assumed that they do not differ significantly from the true volumes within the system itself. If experimental data are unavailable, it is possible to estimate the volume fractions of the regions proceeding from tabulated values of volume for different molecules and molecular parts [21, 47, 49]. As the specific volumes of the amphiphiles and the systems containing them are often quite close to unity, many workers have approximated the volume fractions by using the weight fractions [33, 47]; relative values of the structural parameters are then obtained, and hence an impression of the extent to which they are dependent on the composition.

Perhaps the most important parameter in judging the plausibility of a proposed structure is the derived interfacial area per polar group. This area will include for systems containing several amphiphilic compounds all their polar groups unless otherwise stated. The procedure for the calculation of this area will be explained in its appropriate context. This parameter represents the molecular crowding at the interface within a particular structure of the amphiphile system and the calculation of its value will be affected by any mutual solubility of the components of the system. In structures possessing one-dimensional periodicity (lamellar structures) the value obtained is independent of where one places the dividing line between the polar and non-polar parts of the molecule but this assignment may be of immense importance in two- and three-dimensional structures. The effect of the assignment is especially noticeable for compounds possessing large polar groups, such as cation-active and non-ionic amphiphiles.

Structures with One-dimensional Periodicity

Here we are concerned with phases having a lamellar structure. They are usually spontaneously birefringent and besides the diffuse reflection with spacing corresponding to 4·5 Å they usually show sharp reflections in the low-angle region with d-values in the ratio $1 : 1/2 : 1/3 : 1/4$. The point collimation diffraction patterns indicate a tendency for preferred orientation of the lamellae within the sample parallel to the container walls. The intensity falls off steeply for the higher orders, and often with aqueous systems only the first reflection is visible. However, anhydrous systems may display a large number of reflections in the low-angle region.

The most common phase possessing a lamellar structure in aqueous amphiphilic systems is the neat phase (phase D in Ekwall's terminology [8, 40, 41], phase LL in Luzzati's [29, 34] and phase G in Winsor's [65]. The structure is assumed to be composed of an alternation of indefinitely extended continuous polar and non-polar layers (Fig. 4.1). The former are made up of the polar groups of the amphiphile and the water present in the system. The non-polar layers contain the hydrocarbon moieties of the amphiphilic molecules.

From the repeat distance derived from the diffraction pattern the structural parameters can be obtained. For a two-component system, if V is the total volume of the sample, v_a and v_w the partial molar volumes, and n_a and n_w the numbers of moles in the sample then $V = n_a v_a + n_w v_w$ (where the subscripts a and w refer to the amphiphilic and aqueous components respectively), and the volume fractions will be

$$\phi_a = \frac{n_a v_a}{V} \quad \text{and} \quad \phi_w = \frac{n_w v_w}{V}$$

respectively. If d is the observed interlayer spacing, the thicknesses of the amphiphilic and aqueous layers are given by

$$\begin{aligned} d_a &= \phi_a d \\ d_w &= d - d_a \; (= \phi_w d) \end{aligned} \tag{4.1}$$

If the hydrophilic groups are assumed to be evenly distributed on both sides of the amphiphilic layer, the interfacial area per mole of polar groups can be expressed by

$$S_{\text{molar}} = 2\,\frac{v_a}{d_a} \quad \left(= 2\frac{V}{dn_a}\right) \tag{4.2}$$

The interfacial area per individual polar group is then obtained by dividing the above expression by the Avogadro number N.

Other versions of the formulae have been used [16, 17, 29, 34, 44 and 51].

In the neat phase the repeat distance can change with the water content in various ways, depending on the type of amphiphile and the composition of the system. The logarithmic version of equation (4.1)

$$\log d = \log d_a - \log \phi_a \tag{4.3}$$

implies that, if the thickness of the amphiphilic layer is unaffected by the water content and if the amphiphile (or mixture of amphiphiles) is insoluble in water, for one-dimensional swelling the slope of a plot of $\log d$ versus $(-\log \phi_a)$ should be unity.

If the thickness of the amphiphilic layer remains constant but the amphiphile (or mixture of amphiphiles) is partly soluble in the water lamellae

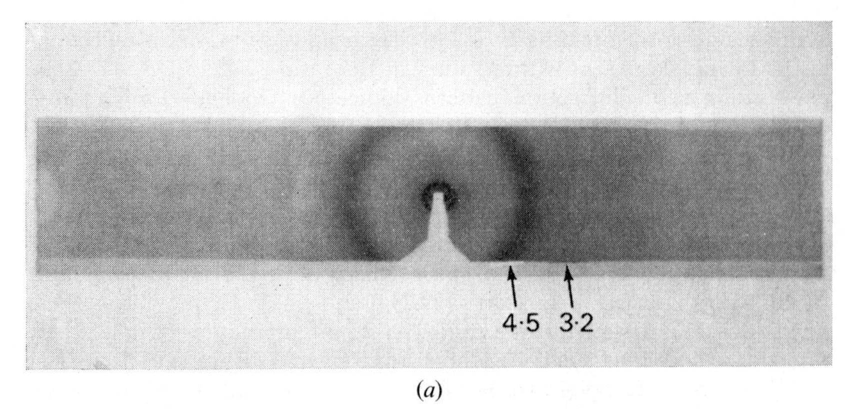

(*a*)

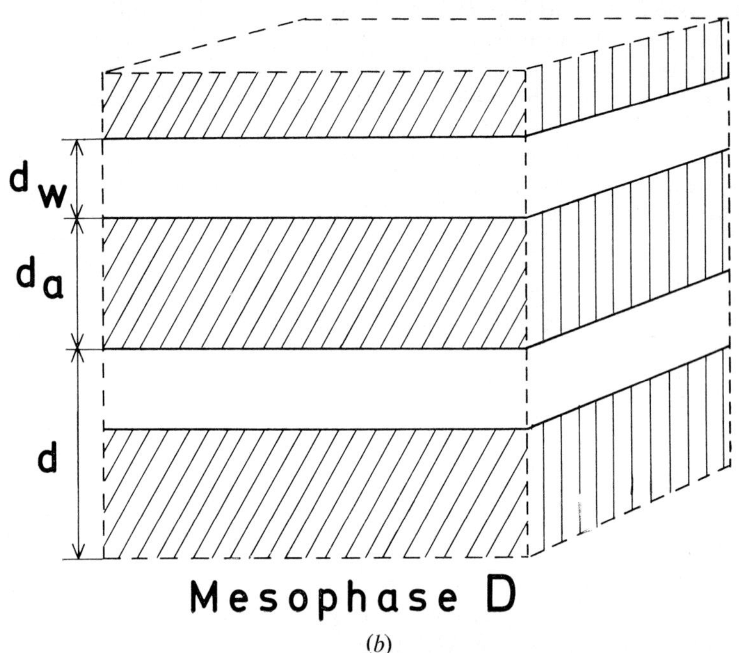

(*b*)

FIG. 4.1. (*a*) Debye–Scherrer pattern obtained for a neat phase specimen in the system sodium caprylate–decanol-1–water (33·7–36·9–29·4 by weight, 20°). The radiation was nickel-filtered copper K_α. Note the 3 low-angle reflections in the ratio 1 : 2 : 3 close to the primary beam and the diffuse 4·5 and 3·2 Å reflections due to the paraffin chains in "liquid" state and to water, respectively.

(*b*) Schematic representation of the structure of the lamellar neat phase. The structure consists of coherent double layers of amphiphilic molecules separated by water layers. The amphiphilic molecules are anchored at the interfaces of the structure by their polar groups.

d, the repeat distance: d_a, d_w thickness of amphiphile and water layers, respectively.

the slope will exceed unity. If there is any interaction of the polar groups with themselves and with the water molecules which causes an increase in area per polar group with dilution the slope will be decreased. Such interaction is sometimes so strong that the spacings remain constant or even diminish when the water content is increased. McBain *et al.* observed this many years ago and distinguished between expanding and non-expanding lamellar structures [42].

A unit slope for the plot of log d versus ($-\log \phi_a$) is obtained for the neat phase chiefly with binary systems containing non-ionogenic association colloids of industrial or natural origin such as polyethanoxy derivatives, mono- or di-glycerides, lecithin, etc. [10, 23]. Unit slope is also observed

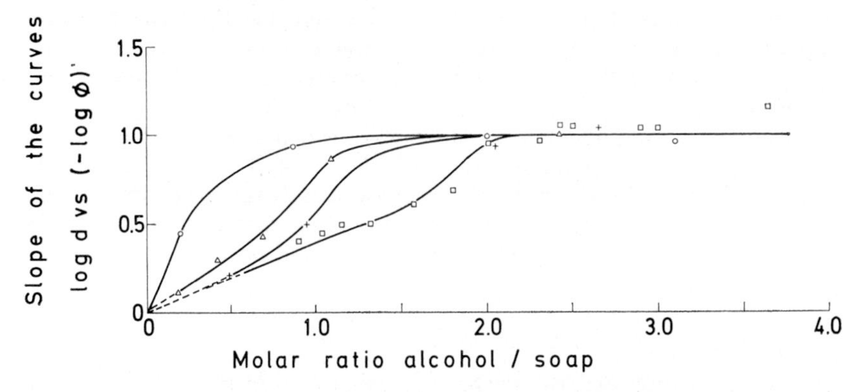

FIG. 4.2. The slope of the curves log d vs (-log ϕ) for the lamellar neat phase D in ternary systems of an alkali soap, long-chain alcohol and water as a function of the molar ratio alcohol to soap. d, the repeat distance; ϕ volume fraction of mixed amphiphiles.

(○) potassium oleate–decanol–1–water [13]
(△) potassium caprate–octanol-1–water [13]
(+) potassium caprylate–decanol–1–water [13]
(□) sodium caprylate–decanol-1–water [10]

for part of phase D in aqueous ternary systems containing alkali soaps (or some other similar ionic amphiphile) and a long-chain alcohol (or some other similar long-chain compound), (10, 12, 16). The slope depends on the ratio of soap to alcohol. When the slope is unity the capacity of the phase to incorporate water before separation of a second phase occurs increases [12, 16]. When the amount of the weakly polar long-chain compounds falls below a certain value, the slope decreases (Fig. 4.2) and there is a gradual transformation from an expanding to a non-expanding lamellar structure.

In amphiphilic systems containing short-chain alcohols, the great water solubility of the alcohol affects the results and the slope will be above unity. Allowance for the solubility of alcohol in the aqueous zone of the liquid crystal structure suffices to explain the data [39].

Aqueous amphiphilic systems may form other phases with a lamellar structure. One of these is Ekwall's phase B, which occurs in several three-component systems of amphiphiles and water [8, 12, 16, 40, 41]. The X-ray reflections are weaker and more diffuse and correspond to greater spacings than for phase D with the same water content. The difference between these two lamellar structures was ascribed to differences in the configurations of the molecular chains [16].

Ekwall's phases C and K also give X-ray reflections with d-values in the ratio $1:1/2:1/3$ and might thus have lamellar structures [8, 12 16, 40, 41]. Other considerations, however, have pointed to two-dimensional periodicity in these cases (see below).

In systems of anhydrous fatty acid soaps yet another lamellar phase may be formed. This often occurs at elevated temperature. Since the soaps crystallize in layer lattices, it might be expected that the basic structure would be preserved as the temperature is raised. This is, however, apparently, not the case. First a succession of different two- and three-dimensional structures appears, and then, just before the transformation to an isotropic melt, appears a "neat soap" phase, which gives five sharp reflections with d-values in the ratio $1:1/2:1/3:1/4:1/5$ in the low-angle region [29, 52]. The structure of this phase closely resembles that of the aqueous neat phase; it contains a continuous succession of double layers of amphiphile molecules, with the polar groups located in the interfaces (cf. Fig. 4.1).

Structures with Two-dimensional Periodicity

These structures are considered to be composed of parallel rod-shaped aggregates of indefinite length arranged in a two-dimensional lattice, which may be square, hexagonal, rectangular or oblique. The rods may be built up around a hydrocarbon core consisting of the amphiphile with the polar groups facing the surrounding water or there may be the reverse structure with a polar core, comprising the polar groups and any water, in a non-polar hydrocarbon environment. More complex rod-shaped aggregates have also been proposed. All these structures are spontaneously birefringent and, under the polarizing microscope, their characteristic appearance distinguishes them from lamellar structures. Ideally, for an aqueous phase with a two-dimensional structure, whose individual unit aggregates remain unaltered by changes in water content, the plot of $\log d$ against $(-\log \phi_a)$ should have a slope of $1/2$.

Phases with Hexagonal Periodicity

The X-ray pattern for phases with hexagonal periodicity is characterized by a ratio of $1:1/\sqrt{3}:1/\sqrt{4}:1/\sqrt{7}:1/\sqrt{12}$ for the long spacings. There is also the diffuse 4·5 Å spacing. The pinhole patterns, even when the sample is rotated, may show discrete spots with a 30° displacement of the spots between the different rings [40] (Fig. 4.3).

The structure in this phase is presumed to be one of indefinitely extended long cylindrical aggregates, the axis of each cylinder coinciding with a six-fold axis of symmetry. In the typical case, the common middle soap

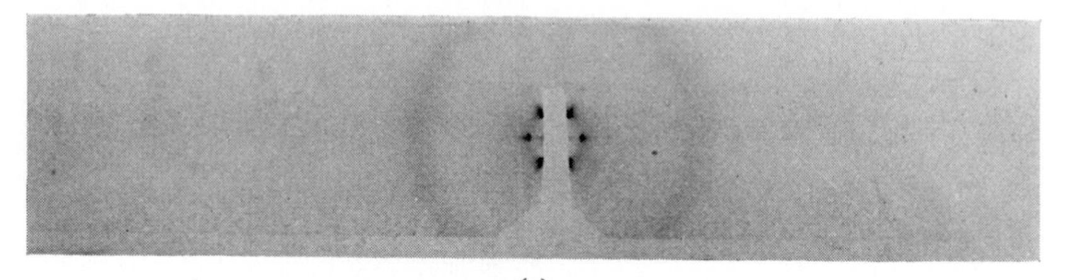

(*a*)

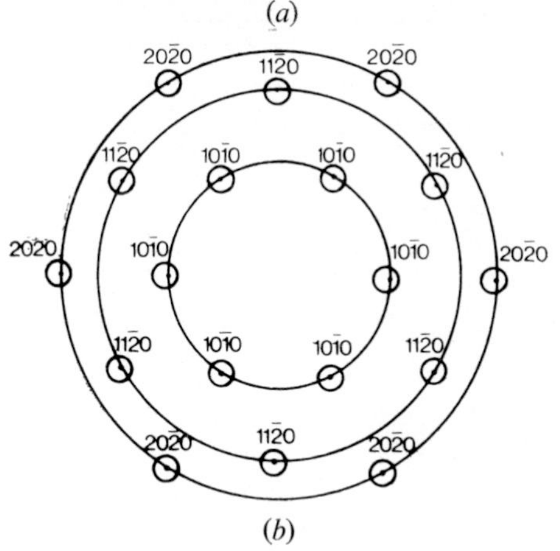

(*b*)

FIG. 4.3. (*a*) The Debye–Scherrer pattern from a middle phase specimen in the system sodium caprylate–caprylic acid–water (48·2–5·3–46·5% by weight, 20°C). Note the assorted set of spots close to the primary beam in hexagonal array, the sets being mutually displaced by 30° of arc. The diffuse reflection with the position 4·5 Å is that of the paraffin chains in the liquid state.
(*b*) Idealized picture of the central portion of the pattern.

phase, the hydrocarbon groups of the amphiphile are considered to fill the cylindrical aggregate, the polar groups lying in the interface with the water continuum (Fig. 4.4). The hexagonal lattice parameter, a, is obtained from the innermost reflection with the spacing d in accordance with the expression

$$a = \frac{2d}{\sqrt{3}} \tag{4.4}$$

The radius r_a, for the rod-shaped aggregates, if cylindrical, is obtained from the expression

$$r_a = d\left(\frac{\sqrt{3}}{2\pi}\phi_a\right)^{1/2} \tag{4.5}$$

and the interfacial area per mole of polar groups from the expression

$$S_{\text{molar}} = 2\frac{v_a}{r_a} \quad \left(= 2\frac{V}{r_a n_a}\phi_a\right) \tag{4.6}$$

The interfacial area, S, per individual polar group is obtained by dividing by the Avogadro number N. As in the case of the lamellar structure other versions of these formulae have been used.

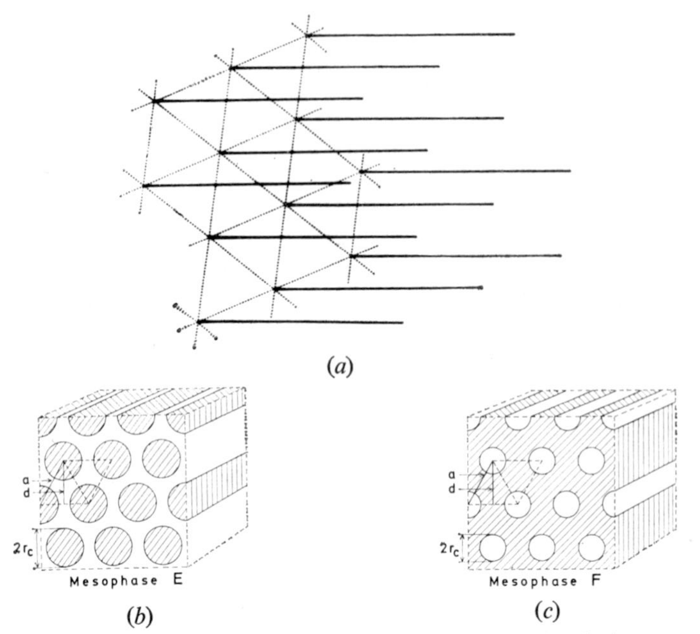

FIG. 4.4. The structure of the two-dimensional hexagonal phases.
(*a*) Two-dimensional hexagonal array of indefinitely extended long parallel rods. The heavy lines represent the rod axes (Space group p6 [66])
(*b*) Phase E, presumed structure of the "normal" hexagonal phase (middle phase). Rod-like unit aggregates with a hydrocarbon core in aqueous environment.
(*c*) Phase F, presumed structure of the "reversed" hexagonal phase. Rod-like unit aggregates with a polar core in hydrocarbon environment.
d, the spacing corresponding to the innermost reflection; *a*, unit cell parameter of the two-dimensional network; $2r_c$, diameter of the rod-like aggregates.

The above structure, with rod-shaped amphiphilic aggregates containing amphiphile in a water continuum, is the generally accepted one for a middle phase. Clunie and co-workers, however, proposed a different interpretation of the structure of the middle phase encountered in a system of trimethylaminododecanoimide and water [6], of also [20], namely one in which the amphiphile forms spherical unit aggregates, which, arranged as a "string of beads", constitute the long hexagonally arrayed rod-shaped aggregates. Their interpretation was based mainly on the anomalous jump in the area per polar group, when calculated on the basis of the rod-shaped model, that apparently occurred on the transformation from the hexagonal phase of the adjacent viscous isotropic phase, which is located on the higher concentration (neat phase) side of the middle phase. In the absence of evidence of any separating two-phase region, they regarded the transformation as a second-order one and having observed no discontinuities in other physical parameters, they inferred that the primary units composing the two structures are the same. Their calculations of the area per polar group for the viscous isotropic phase were, however, based on the assumption that the structure here is one of spherical aggregates in cubic closest packing. As is evident from the section on viscous isotropic phases, this postulate no longer appears acceptable. Small [53] presented a similar interpretation of the structure of the hexagonal phase found in a system comprising a bile-acid salt, lecithin and water.

The fact that liquid crystal structures with hexagonal symmetry may occur at several regions in one and the same phase diagram separated by regions showing quite different structures shows that there are a number of variants. Besides the structure in which the cylindrical aggregates are considered to be filled with amphiphile and surrounded by water, with the polar groups located in the interface between the two regions, there is the reverse structure of polar cylinders in a non-polar lipophilic environment (Fig. 4.4) [14, 33]. (For the calculation of the structural parameters, equations (4.5) and (4.6) have to be suitably modified.) It is impossible to decide which structure is the correct one solely on the basis of the X-ray diffraction findings; the pointers include the lattice parameter, the molecular dimensions of the amphiphiles, the composition of the specimen, the position of the phase in relation to other phases in the phase diagram, and the calculated values of the interfacial area per polar group. In a phase diagram the two variants occur on opposite sides of the lamellar neat phase.

In these two phases the rod-shaped aggregates should not always be regarded as having a circular cross-section. In the case of the reverse structure they can be regarded as hexagonal prisms with a polar core—and likewise for the normal variant with non-polar core, provided that all the water is bound to the polar groups.

Two related phases of "middle" character have been reported in the composition region intermediate between the middle and lamellar phases. One of them is the "phase mediane déformé" [29, 34]. The region of existence of this phase is stated to be small and the experimental results (only two reflections) and interpretation are uncertain. The structure might arise through an orthorhombic distortion of the hexagonal arrangement of the middle phase. The circular cross-section of the rod-shaped "normal" aggregates would then presumably be deformed into an ellipse [34].

Another type of hexagonal phase is found in systems of certain sodium and potassium soaps and referred to as the "phase hexagonale complexe" [2, 3, 7, 10, 29, 33, 34]. This phase displays up to six reflections, which satisfy the two-dimensional hexagonal criteria, and its microscopic appearance is furthermore typical of a hexagonal phase [34]. Its large cell-parameter (almost twice that of the ordinary middle phase in the same system) points to a complex structure. The model favoured by the Luzzati group consists of tubes composed of amphiphilic double layers, which contain water and are located in a water continuum [34]. Balmbra, Bucknall and Clunie, on the basis of electron microscope studies, propose the reverse structure [2, 3] (cf. Chapter 1).

Other Phases with Two-dimensional Periodicity

In addition to the liquid crystal phases with hexagonal symmetry there are other phases with two-dimensional periodicity in anhydrous and aqueous amphiphilic systems.

Luzzati reported the existence in aqueous amphiphilic systems of a "phase rectangulaire" characterized by two independent series of interplanar spacings—$a : a/2 : a/3$; and $b : b/2 : b/3$ [29, 34], the ratio between the two repeat distances a and b, being dependent on the type of amphiphile. This indicates a two-dimensional network with two independent parameters. The pattern may be accounted for by assuming a rhombic lattice of rod-shaped aggregates of "normal" type with a rectangular cross-section (phase R, Fig. 4.5). From the values of a and b and the composition of the specimen it is possible to calculate the dimensions of the rectangular cross-section and the mean area per polar group at the interface between the polar and non-polar regions. However, on the basis of electron microscopy Eins considers a rhombic face-centred structure for this phase [7]).

The rectangular phase has also been encountered by Ekwall *et al.* in

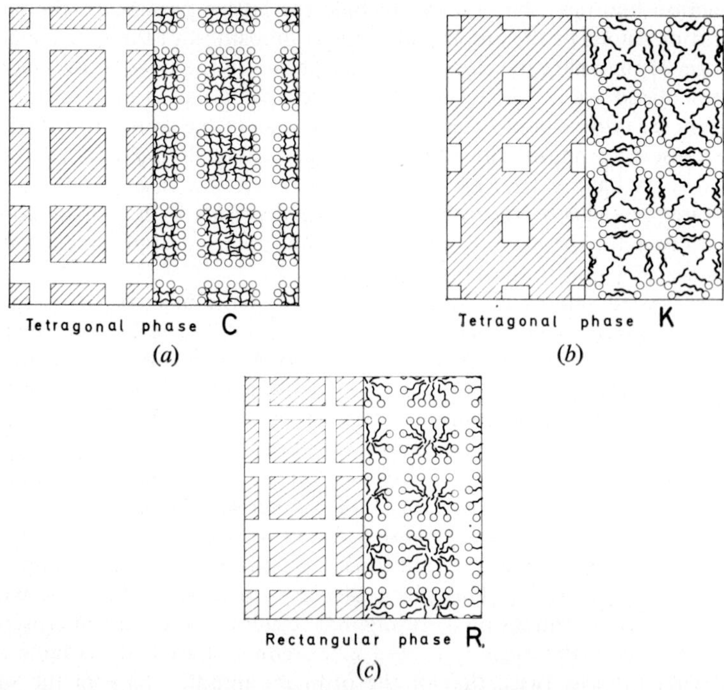

FIG. 4.5. The structure of tetragonal and orthorhombic liquid crystalline phases [11].

(*a*) Phase C, presumed structure for the normal two-dimensional tetragonal type. Rod-like aggregates with hydrocarbon core in aqueous environment; rods with predominantly square cross-section in tetragonal array.

(*b*) Phase K, presumed structure for the reversed two-dimensional tetragonal type. Rod-like aggregates with polar core in hydrocarbon environment; rods with predominantly square cross-section in tetragonal array.

(*c*) Phase R, presumed structure for the normal two-dimensional rectangular type. Rod-like aggregates with hydrocarbon core in aqueous environment; rods with rectangular cross-section in orthorhombic array.

potassium soap systems [12]. The phase is located between the middle and neat phases.

The special case where $a = b$ is, of course, conceivable. Such a structure may be possessed by the phases C and K (Fig. 4.5), observed by Ekwall and co-workers in a number of amphiphilic systems [8, 12, 40, 41], which display a series of sharp reflections with d-values in the ratio $1:1/2:1/3$. The curve of log d versus $(-\log \phi_a)$ for the phase C has a slope of about 1/2. A prerequisite for the existence of this phase is an extremely narrow range of soap-to-alcohol ratios, and Ekwall accordingly ascribed the occurrence of this phase to a strong attraction between these components. The phase K is located in the phase diagram on the other side of the lamellar phase D. It has been observed only in the potassium caprate–octanol–1–water system and its region of existence is extremely limited. The locations of these phases indicate that their structures are complementary. Phase C would have the "normal" structure, whilst phase K would be of the "reversed" type. The nature of such a structure would, however, imply the presence of a reflection, corresponding to the diagonal of the square lattice, but this has not been observed [12, 16].

Luzzati has reported the existence of a two-dimensional "square" phase with spacings in the ratio of $1:1/\sqrt{2}:1/\sqrt{4}:1/\sqrt{5}$ in the system lysozyme and phosphatidyl inositol with water contents ranging from 0 to 27% [22]. This phase obviously has a "reverse" structure.

In anhydrous systems of alkali metal soaps, Luzzati *et al.* found structures in which they considered the polar groups (still retaining some degree of the arrangement present in the solid crystal), to be arranged in ribbon-shaped aggregates, which were dispersed in the surrounding liquid hydrocarbon moiety in a two-dimensional lattice, either rectangular or oblique [29, 52]. Fig. 4.6 shows the structure of an anhydrous soap phase containing such ribbon-like elements. When the cell is rectangular the ribbons are

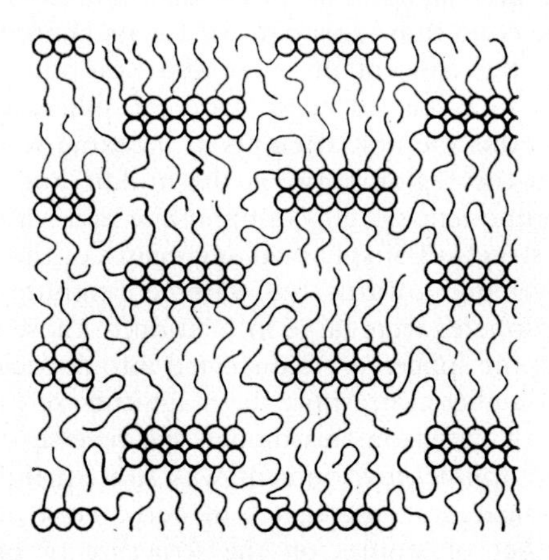

FIG. 4.6. Structure of the two dimensional rectangular-centred liquid crystalline phase displayed by anhydrous soaps at elevated temperature; cross-section showing the structure of the ribbon-like elements [29].

highly symmetric (two mirror planes); if the cell is oblique the symmetry is lower (one two-fold axis). This structural interpretation was based on a large number of reflections in the low-angle area. Phases with these structures were in earlier studies referred to as sub-waxy, waxy, super-waxy and sub-neat [4, 63, 64]. The differences between their structures lie essentially in the angle between the crystallographic axes and in the lattice parameters [29, 52].

Structures with Three-dimensional Periodicity

Phases whose inner structures display a three-dimensional arrangement in spite of the liquid state of the hydrocarbon moiety may be either optically isotropic or birefringent. In a number of cases the structure probably consists of short rod-shaped aggregates which, joined in threes or fours at each end, form the crystallographic structure. However, not all the liquid crystal phases belonging to this group are necessarily built up according to this principle.

Optically Isotropic Phases

Gel-like, optically isotropic, highly viscous phases have been encountered in numerous systems. They can occupy a number of different positions in the phase diagrams. Some are anhydrous or contain very little water, while others lie between the "reversed" hexagonal and the neat phases, between the neat and the middle phases or between the middle phase and the isotropic micellar aqueous solutions. Besides the diffuse reflection at 4·5 Å, the phases give a series of sharp reflections at low diffraction angles. They are often few in number, but in certain cases as many as 17 independent reflections have been detected. The positions of the reflections for a particular phase vary very little with the water content and the temperature. The range of the water content of the phase and its temperature range are furthermore often very restricted, which indicates that highly specific conditions are required for its characteristic structure.

On the basis of the concept of polar and non-polar regions and the optical isotropy, a face-centred cubic array of spherical aggregates was postulated, and on this the first attempts at indexing were based. For the phase occurring in the region between the middle and neat phases in a number of two-component systems (amphiphile and water), Luzzati *et al.* at first assumed a structure of spherical aggregates of the amphiphile, with the polar groups facing towards the water continuum [33, 34]. As the dimensions of the spheres were large in relation to those of the molecules, it was inferred that the spheres were distorted into dodecahedra. However, on the basis of anomalies concerning the magnitude of the interfacial area per polar group calculated on this model, the reverse structure of water spheres in an amphiphilic environment was subsequently proposed [35]. This idea of spherical unit aggregates in cubic close packing has been applied in a number of studies on the structure of optically isotropic viscous phases [6, 15, 24, 25, 28, 29, 33, 34, 46].

When in 1964 Spegt reported the existence of a body-centred cubic structure on the basis of fifteen or so reflections from anhydrous strontium myristate at elevated temperature (223°C [54]), Segerman commented that the sub-set of the four innermost strong reflections from a body-centred structure, which would have been observed in a system which gave fewer reflections, might easily be misconstrued as those from a face-centred structure. He accordingly expressed doubt as to the existence of a face-centred structure in amphiphilic systems [50].

However, there are in addition other optically isotropic viscous phases where the diffraction patterns point to a primitive or body-centred structure [15, 30, 36, 54, 57, 59]. Furthermore, sometimes it has been necessary to consider the presence of complex structures of concentric spheres in attempts to account for the experimental findings [15]. A further factor that has been considered is the rôle of water [15, 26]. At least some of the water must be regarded as structurally included in the unit aggregate, as water of hydration or in some other way.

A structure for the optically isotropic strontium myristate phase was subsequently proposed by Luzzati and Spegt [36]. Space group Ia3d (No. 230 [66]) accounted for the observed absences without conflicting with any of the observed reflections. Moreover, only the special set 48 g of co-ordinates on the two-fold axis was possible, and gave an excellent fit if the locus of the strontium ions was presumed to coincide with these positions. It was concluded that the strontium ions are grouped in short rod-shaped aggregates and that each end of a rod adjoins two others; the three rods of each group are co-planar and are related by a three-fold axis. The unit cell contains twenty-four such rods which define two infinite three-dimensional networks, mutually interwoven but otherwise unconnected (Fig. 4.7). The spaces between the rods are occupied by liquid paraffin chains anchored by—COO^- groups to the strontium ions; the disordered chains form a coherent matrix. As the region with a high electron density along the two-fold axis is of finite extent, the strontium ions may be located at some distance from the two-fold axis and thus leave space for the carboxylate groups. This structure gives close agreement between the theoretical and observed interference patterns as regards both the position and intensity of the reflections [36, 37]. The phase is found throughout the series of anhydrous strontium soaps from C_{12} to C_{22}. Though X-ray patterns do not in every case possess so many details, they show such close resemblance that the essential structure must be the same throughout. There are usually up to seven reflections in the same sequence as regards both position and intensity ratios [37].

The same X-ray pattern is found for the isotropic viscous phase in a number of two-component systems with a range of water contents (Table 4.1). A fresh analysis of the findings for the isotropic phase which lies between the lamellar and the reversed hexagonal phases (phases D and F) in the system decaethylene glycol monolauryl ether, oleic acid and water [15] has also shown that its X-ray diffraction pattern may be interpreted in the same way.

Like the two-dimensional structures, the cubic structures have two variants, "normal" and "reversed", or, as Luzzati prefers, type I if the paraffin chains are inside the rods and the water outside, and type II if the

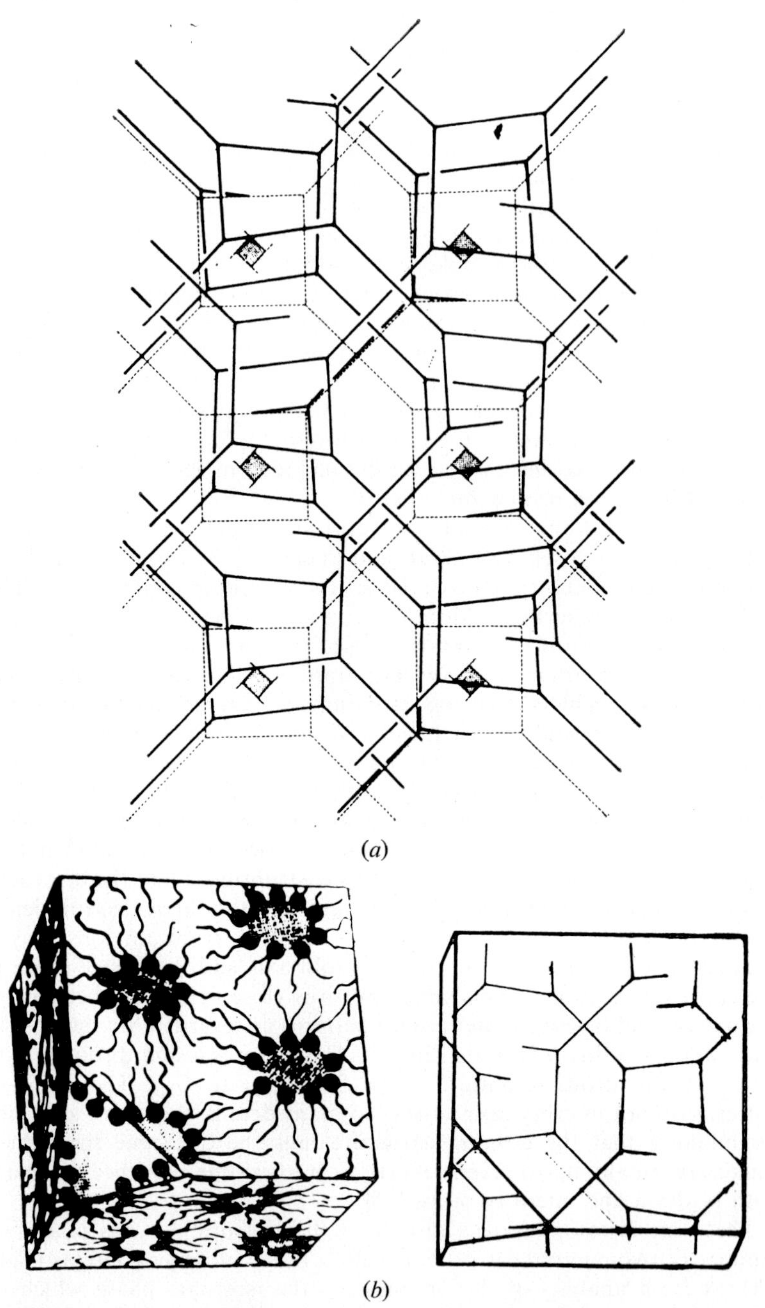

FIG. 4.7. The representation by Luzzati [36] of the structure in isotropic viscous phases belonging to the space group Ia3d [66]. Rods of finite length joined in threes at each end to form two three-dimensional networks, mutually interwoven and unconnected.

(*a*) Perspective view. The dotted lines are the projection of the heavy lines on the basal plane. The thin lines show the limits of one unit cell. Note the right-hand and left-hand screws penetrating the structure.

(*b*) Attempt to depict the actual arrangement in a system of galactolipid and water [49].

TABLE 4.1. *Dimensions of the body-centred cubic structure 1a3d in aqueous amphiphilic systems*

	Type I Paraffin chains inside the core								Type II Polar groups inside the rods			
Amphiphile	KC_8	KC_{12}	KC_{14}	KC_{16}	$C_{12}TACl$	$C_{12}TAB$	$C_{16}TAB$	$TAC_{12}I$	Aerosol OT	Lec.	Gal.	SrC_{14}
Ref.	[15]	[37]	[37]	[37]	[1]	[37]	[37]	[37]	[13]	[37]	[37]	[37]
a (Å)	60·5	78·6	90·7	99·5	79	76·9	98·7	82·9	69	96·2	110·2	62·4
t° (C)	20	100	100	100	22	70	70	20	20	82	64	235
v ($cm^3\ g^{-1}$)	(∼1)	0·99	1·01	1·03	(∼1)	$0{\cdot}99_7$	$1{\cdot}02_7$	1·08	(∼1)	1·025	0·986	1·165
ϕ	0·66	0·64	0·62	0·62	0·85	0·82	0·80	0·70	0·75	0·97	0·88	1
l (Å)	21·3	27·8	32·1	35·2	27·9	27·2	34·9	29·3	24·4	34·0	39·0	22·2
r (Å)	11·2	14·0	15·9	17·5	17·5	15·9	20·2	15·7	8·9	11·1	15·6	—
S ($Å^2$)	46·0	46·8	47·5	48·8	41·1	51·0	49·6	47·5	65	67·0	73·3	—

Notations in table 4.1

KC_n: $CH_3(CH_2)_{n-2}COOK$

SrC_{14}: $(CH_3(CH_2)_{12}COO)_2Sr$

$C_{12}TACl$: $CH_3(CH_2)_{11}N(CH_3)_3Cl$

C_nTAB: $CH_3(CH_2)_{n-1}N(CH_3)_3Br$

$TAC_{12}I$: $CH_3(CH_2)_{10}CO\overset{-}{N}\overset{+}{N}(CH_3)_3$

Aerosol OT: Sodium di-2-ethylhexyl sulphosuccinate

Lec.: hen egg lecithin

Gal.: galactolipids from maize chloroplasts

a: parameter of the cubic cell

t: temperature

v: specific volume of the sample

ϕ: volume fraction of the amphiphile

l: length of the rods

r: radius of the rods (containing non-polar or polar material, respectively)

S: area per polar group at the interface between polar and non-polar material opposite arrangement obtains [29, 37]. The chief criteria for selecting the correct alternative are the position of the phase in the phase diagram and the types of phases with which it is in equilibrium. Further, an abnormal upward or downward jump in the calculated value for the interfacial area per polar group is regarded as unlikely on transformation from one phase to another. It has been often found, however, that the values calculated for the interfacial area per polar group are compatible with either alternative (Fig. 4.8). Luzzati considers that the structure of the

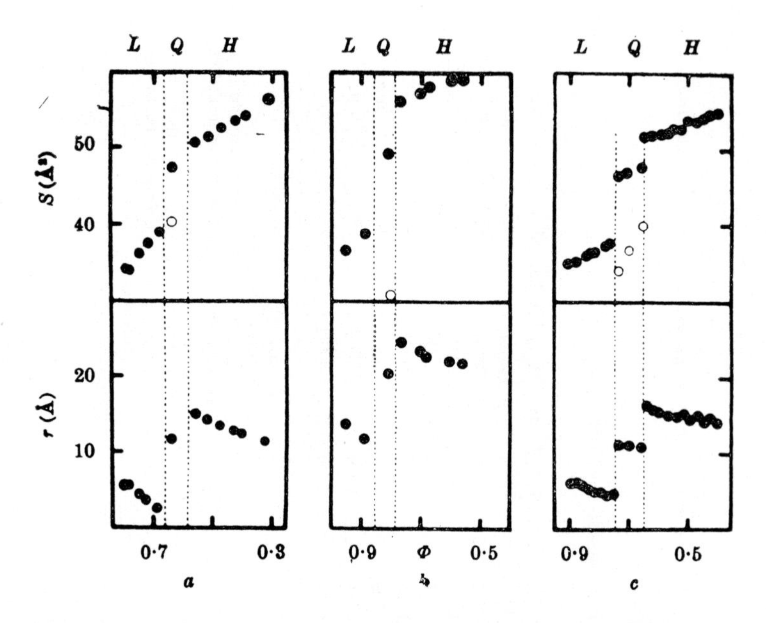

Fig. 4.8. Comparison by Luzzati of correlating principles for the lamellar cubic and hexagonal structures of some lipid–water systems [37].

(*a*) sodium myristate; (*b*) cetyltrimethylammonium bromide; (*c*) trimethylamino-dodecanoimide.

L, lamellar; Q, cubic; H, hexagonal structure.

S, interfacial area per polar group (Å²); ϕ, volume fraction of lipid; *r*, radius of the rods in the cubic and hexagonal structures, half-thickness of the lipid leaflet in the lamellar structure, respectively. The open circles refer to calculations based on type II, full circles to calculations based on type I structures of Luzzati.

isotropic phases in the lecithin and galactolipid systems is the same as for the non-aqueous strontium soaps (i.e. the reversed variant, type II of 1a3d), whereas potassium soaps, hexadecyl- and dodecyltrimethylammonium bromides display type I [37]. In the dodecyltrimethylammonium chloride system there are two cubic phases on either side of the middle phase [1]. That between the middle and the neat phases would be of type I of 1a3d; the structure of the other is discussed below.

The isotropic viscous phase in the sodium di-2-ethylhexyl sulphosuccinate (Aerosol OT)–water system, and presumably also the above-mentioned phase in the decaethylene glycol monolauryl ether–oleic acid–water system would be the reverse variant, type II of this 1a3d structure.

Figure 4.9 shows that, with this interpretation, in the Aerosol OT/water system the correlating principle of the decrease in interfacial area

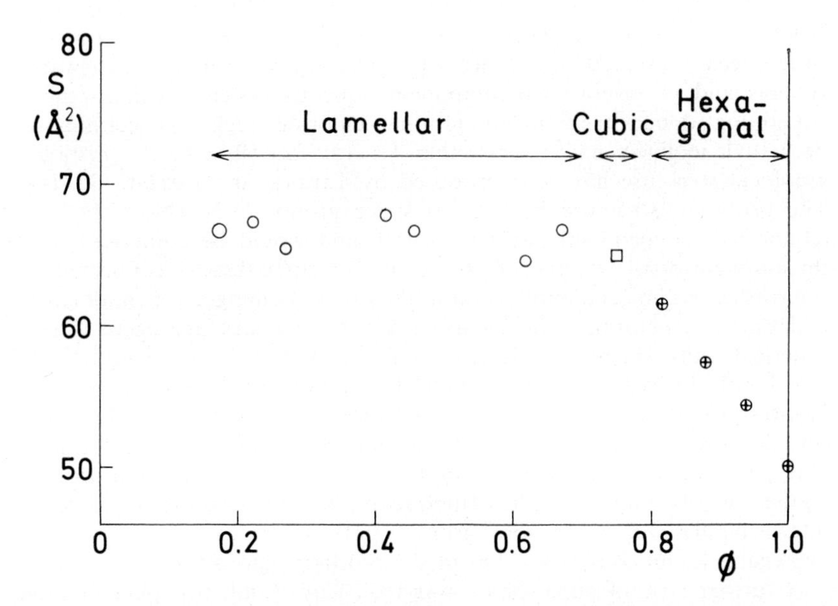

FIG. 4.9. Comparison of the correlating principle of the calculated interfacial area per polar group for the lamellar, cubic and hexagonal structures in the Aerosol OT–water system at 20°C [13]. The cubic structure is assumed to belong to space group 1a3d [66] and be of type II of Luzzati [37]. The hexagonal structure is also of type II, "reversed structure". For the two- and three-dimensional structures, the polar rod aggregates are assumed to be cylindrical in structure and the area per polar group is taken as the cross-section at the interface between the polar part (in which the water, sodium ion and sulphonate group are included) and the rest of the Aerosol OT molecule.

S, Interfacial area per polar group ($Å^2$);
ϕ, volume fraction of Aerosol OT.

per polar group with increase in concentration accords well with the values for the adjacent phases D and F (lamellar and reversed hexagonal).

The occurrence of such a cubic phase may depend on the division of the indefinitely extended rod-shaped aggregates of the adjoining hexagonal phase into smaller units in a manner which maintains a balance between the area per polar group and the volume of the hydrocarbon part of the amphiphile. The rod-shaped aggregates are found to be short, with the length often nearly equal to the diameter [see Table 4.1 and ref. 37].

Not all cubic phases have this structure, whether of the "normal" or "reversed" variant. There are at least three different places in the phase diagram where such phases may occur, and the X-ray diffraction patterns of some of these phases are inconsistent with the above interpretation. For instance, it was found that the viscous isotropic phase occurring in the three-component systems of sodium caprylate, hydrocarbon (n-octane or *p*-xylene) and water gave a pattern that could be indexed as a primitive cubic lattice giving reflections with *d*-values in the ratio $1/\sqrt{4}:1/\sqrt{5}:1/\sqrt{6}:1/\sqrt{8}:1/\sqrt{10}:1/\sqrt{18}:1/\sqrt{22}$ [15]. The same pattern has since been observed in other hydrocarbon-containing soap–water systems and in several two-component aqueous systems such as those containing dodecyltrimethylammonium chloride, egg lysolecithin or decaethylene glycol monolauryl ether [1, 15, 32, 59, 60]. A crystallographical structure has been proposed by Tardieu and Luzzati [32, 60]. The proposed structure belongs to space group Pm3n (No. 223 [66]) (confirmed independently by Clunie [1]), and would be composed of a three-dimensional network of short rods of equal length connected to three others at one end and to four at the other, forming a system of cages (chlathrate structure), each enclosing a spherical unit aggregate. These spherical units are in a body-centred array. The structure would be of type I with the hydrocarbon chains in the interior of the rods and spheres. Figure 4.10 attempts to depict this structure. The credibility of the structure is, however, impaired by the high number of independently adjustable parameters required by it for the calculation of the intensities. A further feature that could be criticized is that the assumption of equal interfacial area values per polar group results in a volume of the spherical aggregates about four times that of the rod-like aggregates [60].

A further type of cubic phase was found by Rand in a phosphatidyl ethanolamine–water system of unspecified water content [59]. It gave the reflection pattern $\sqrt{2}:\sqrt{3}:\sqrt{4}:\sqrt{6}:\sqrt{8}:\sqrt{9}$. Luzzati *et al.* concluded that the space group is Pn3m (No. 224 [66]) and that the structure is composed of short rods of definite length connected in fours at each end to give tetrahedrons, and forming two interwoven three-dimensional networks each of the diamond-type structure, the two networks being independent of each other [59].

It should be stressed that several isotropic viscous phases display so few reflections that it is not possible to index the diffraction patterns unequivocally and to determine the crystallographic structures. The many cases when two reflections with *d*-values in the ratio $1/\sqrt{3}:1/\sqrt{4}$ have been obtained do not constitute proof that the structure is of the face-centred cubic type; this ratio fits a body-centred structure equally well.

The reason for dealing in detail with the optically isotropic viscous phases lies in the two contradictory concepts of their structures—an arrangement of spherical aggregates in face-centred cubic packing† (a

† For a third concept see Gray and Winsor, *Liquid Crystals and Plastic Crystals, Vol. 1*, Ellis Horwood Ltd., Chichester (1974); also *Molec. Crystals Liqu. Crystals* (1974).

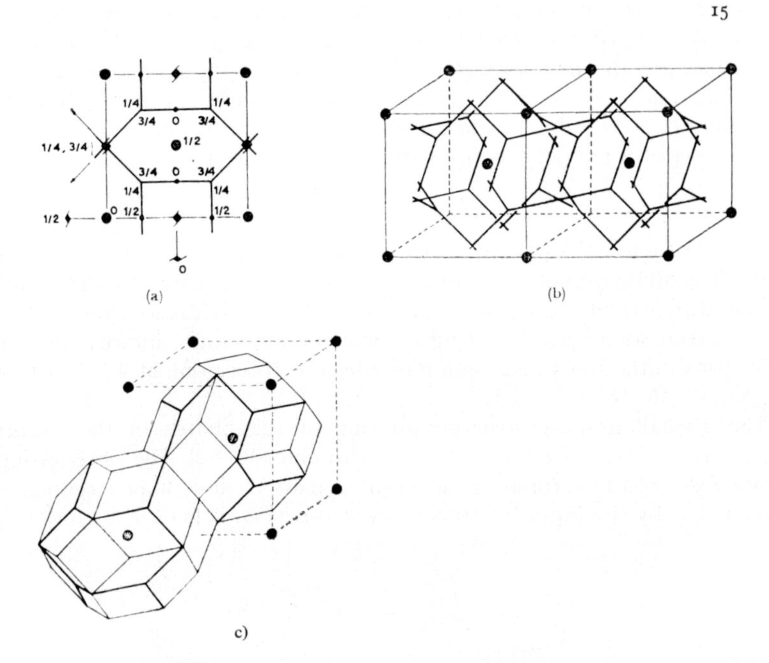

FIG. 4.10. The representation by Luzzati of the structure in isotropic viscous phases belonging to the space group Pm3n [66].

In (a) and (b) the axes of the rods are represented by heavy bars, the limits of the unit cells by thin lines, the centres of the spheres by black circles. The rods are of finite length, and lie along two-fold axes. The rods are joined three by three at one end, four by four at the other. The interior of the rods and spheres is occupied by the paraffin chains, the external volume by water. The rods and spheres are quite bulky, almost in contact with each other.

(a) Projection on the plane *ab*, with the position of some of the structural elements. The intersections of four rods are shown by dots. The fractional figures are the *z* co-ordinates of the intersections of the rods. (b) Perspective view of two unit cells. Note that the joining rods build up a three-dimensional network. (c) Front view of two cage-like truncated octahedra, each enclosing a spherical micelle. The rods (heavy lines) sit on the hexagonal faces. Note that the packing of the polyhedra is space-filling body-centred; the position of the rods indicates that the unit cell is primitive with two polyhedra per cell [32, 60].

concept that was generally held until quite recently), and experimentally adduced primitive or body-centred structures presumably composed of rod-shaped aggregates.

The body-centred structure 1a3d appears to be established for the cubic phases that exist with little or no water (i.e. on the side of the neat phase with little water) and for those found in the region between the neat and the middle phases; for the other phases their structures remain to be verified.

As regards the unit aggregates, in many cases the amount of water is so small that it must be regarded as chemically bound by the polar groups and consequently the aggregates will be of such a shape that they fill the available space. Only in the phase in the region between the isotropic micellar aqueous solutions and the middle phase is the water content high enough for part of it conceivably to be present in the free state.

Birefringent Phases

In addition to the optically isotropic viscous phases, other mesophases with three-dimensional structures have been recognized. In their high-temperature studies Luzzati *et al.* found that anhydrous uni- and di-valent metal soaps give birefringent phases with a high number of sharp reflections in the low-angle region besides the diffuse one at 4·5 Å [19, 38, 48, 52, 55, 56, 58].

The crystallographic structure of one of the phases in the sodium laurate system was interpreted as orthorhombic [48, 52]. The polar groups were considered to form an orthorhombic array of disc-shaped aggregates surrounded by the liquid hydrocarbon chains (Fig. 4.11).

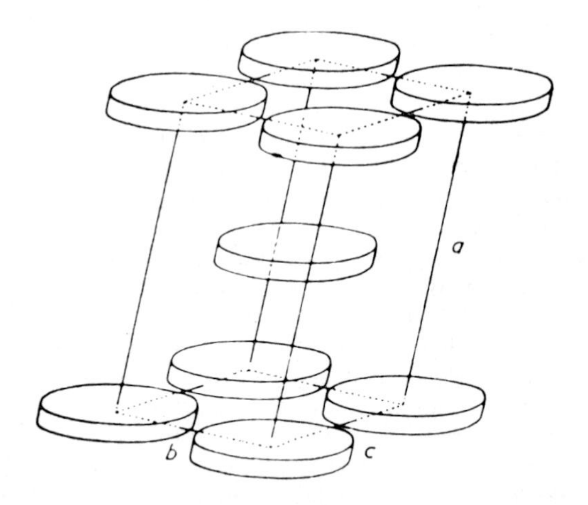

FIG. 4.11. Structure of the three-dimensional orthorhombic body-centred mesophase of anhydrous sodium laurate at elevated temperature. The discs are the locus of the polar groups [29].

For the divalent soaps they found structures [55, 56, 58] in which the polar groups were considered [38] to form short rod-shaped aggregates. Grouped in threes or fours, these make up two-dimensional networks which form a hexagonal or square pattern, respectively (Figs. 4.12 and 4.13). These structures would thus be composed of rod-shaped aggregates similar to those postulated for some of the optically isotropic viscous phases.

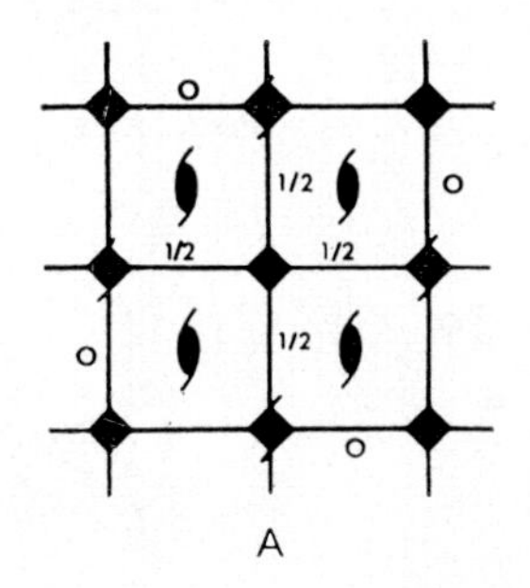

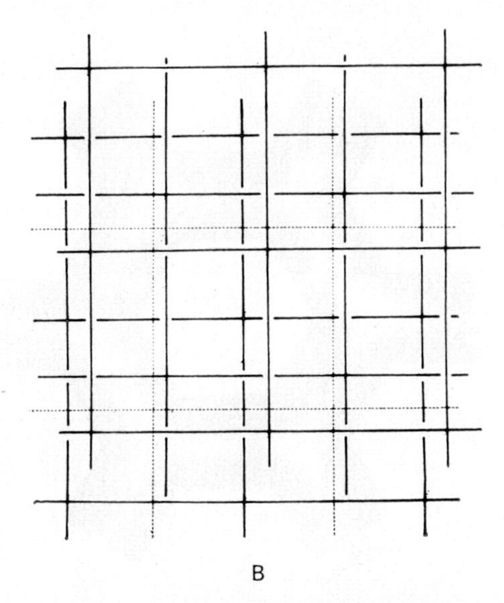

FIG. 4.12. The body-centred tetragonal phase structure of a liquid crystalline anhydrous amphiphile, space group 1422 [66, 38].

The loci of the polar groups, represented by heavy lines, are rods of indefinitely extended length that sit on two-fold axes.

(A) Projection of one unit cell along the tetragonal axis c; the rods are parallel to the plane *ab*; the fractional figures are the *z* co-ordinates of the rods.

(B) Perspective view. Note that the structure is formed by identical rod-like elements, joined four by four at each end, forming two-dimensional planar square networks. The networks are stacked on each other but otherwise unconnected [29].

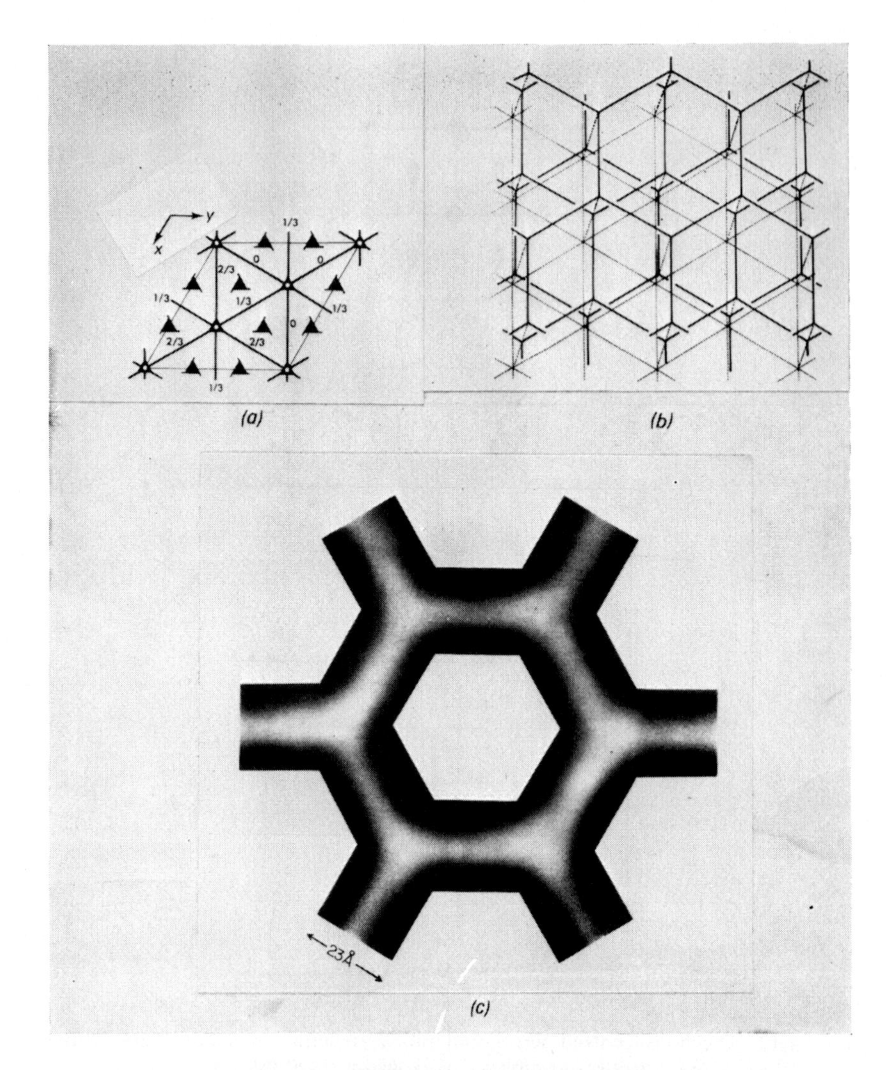

FIG. 4.13. The rhombohedral structure of a liquid crystalline anhydrous amphiphile, space group R3m [66, 30, 38].

The loci of the polar groups (heavy lines) are rods of finite length that sit on mirror planes. The rods are joined by threes at each end and form planar two-dimensional networks, stacked on each other but otherwise unconnected.

(*a*) Projection of one unit cell along the hexagonal axis, with the position of some of the symmetry elements. The rods are perpendicular to the *c*-axis, the fractional figures are the *z* co-ordinates of the rods.

(*b*) Perspective view. Note the two-dimensional hexagonal networks, stacked but mutually unconnected.

c) Attempt to depict the two-dimensional network in the rhombohedral phase of a system of egg lecithin and water; the size of the rods drawn to scale [29].

CONCLUDING REMARKS

The X-ray crystallographic information concerning a particular amphiphilic mesophase does not usually furnish an unequivocal structure. Only in rare instances has a structure (even tentative) been arrived at through conventional crystallographic techniques [29, 36, 54, 60, 62]. It is generally necessary to rely on a more inductive approach. To obtain a reliable concept of the structures, phase diagrams should be constructed. It is easy to misconstrue the structure of a particular phase if X-ray diffraction data are considered in isolation. Within a sequence of phases the existence of each phase and the model adopted for its internal structure should be clearly discernible as logical consequences of its composition. Further its structural parameters, calculated from the X-ray diffraction data on the basis of the model adopted, should bear realistic relationships to each other and to values of the parameters in neighbouring phases.

Even in quite simple amphiphilic systems a large number of phases may occur, each with its own internal structure. When the system contains several amphiphiles, especially those of complex molecular constitution, the numbers of different phases may be considerably increased. However, in aqueous systems containing mixtures of amphiphiles and in which both the hydrocarbon parts and the polar groups are in the liquid state, the various hydrocarbon and polar groups respectively are able to intermingle to form homogeneous hydrocarbon and polar zones; this explains why such multi-component systems can often be formally treated as simpler systems.

Factors which govern the formation of a particular structure are the nature, size and physico-chemical state (at the temperature prevailing) both of the polar groups and of the hydrocarbon part. The polar group may be anionic, cationic or non-ionic or a mixture of these types, and the hydrocarbon group may be open-chain, branched-chain, saturated or unsaturated, or may contain benzene rings.

The relative sizes of the polar and non-polar parts of the amphiphile are important in regulating the types of phase produced. Thus in binary aqueous systems where the polar groups are large in relation to the hydrocarbon portion, the formation of the "normal" hexagonal phase (middle phase) is favoured, while if the hydrocarbon part is relatively bulky the "reversed" hexagonal phase is formed.

Attempts have been made by Winsor [65] and Ekwall [11] to compile logical schemes, for the series of phase transformations which occurs in aqueous amphiphilic systems with progressive changes in composition, e.g. in water content or in the proportions and chemical nature of the amphiphiles present, or on progressive solubilization of hydrocarbons. In both schemes the liquid crystalline lamellar, "neat" phase occupies a central position. On either side of this phase, with its planar amphiphilic layers, lies series of complementary phases. The two schemes are broadly similar, but differ in detail.

The variation of the mean interfacial area per polar group (including all types of polar groups in systems containing mixtures of amphiphiles), as calculated from measured X-ray diffraction data on the basis of an assumed structural model, has often been regarded as a correlating principle. It has been assumed that this variation will display a regular trend from one phase to another with progressive changes in composition. Other things being equal, the area per polar group will be lowest for "reversed" structures of low water content and reach its maximum for "normal" structures of the highest water content. In calculating the interfacial area per polar group it must be remembered that not all of the amphiphilic molecules need be located in the interface between the polar and non-polar regions. If the amphiphile is appreciably soluble in water (or, if a mixture of amphiphiles is present, if part of the mixture is appreciably soluble in water) a significant proportion of the amphiphile will be contained within the aqueous region of the mesophase. If, on the other hand, an appreciably hydrocarbon-soluble amphiphile is present, a proportion of it will be included within the hydrocarbon region of the mesophase. If, in either case, these solubility effects are neglected in calculating the interfacial area per polar group, a low area will be obtained, e.g. if allowance is not made for partial exclusion of the decanol-1 from the interface between polar and non-polar regions in the reversed hexagonal phase of alkali soap–decanol-1–water systems, an unacceptably small area per polar group is obtained, i.e. the polar groups appear to be unreasonably crowded at the interface [12, 16]. Specific interaction between the polar groups of two amphiphiles can also result in the formation of a structure with an area per polar group different from that otherwise to be expected [Phase C, refs. 12 and 16.]

A further correlating principle in the selection of the most credible model is that, where a succession of phases is encountered as a consequence of a regular variation in composition or temperature, there must be a logical progression in the structures of the aggregates of the successive phases; a backward step in a sequence is improbable. For example, in earlier interpretations of the structure of the optically isotropic viscous phases on either side of the neat phase between this phase and either of the two hexagonal phases improbable steps were encountered when the unit aggregates were assumed to be spherical (either "normal" or "reversed"). A solution to this problem for the cubic phases may lie in Luzzati's more recent proposals discussed above. Difficulties of the same type concern the structure of the "complex hexagonal" phase for which two complementary structural alternatives have again been advocated. A proposal for a particular structural model must take account of all available information and any departure from a logical sequence of structures should be accepted only on the strongest of experimental evidence.

ACKNOWLEDGEMENT

This work has been performed as part of a research programme sponsored by the Swedish Board for Technical Development.

REFERENCES

[1] Balmbra, R. R., Clunie, J. S. and Goodman, J. F. *Nature* **222,** 1159 (1969).
[2] Balmbra, R. R., Bucknall, D. A. B. and Clunie, J. S. *Molec. Crystals Liqu. Crystals* **11,** 173 (1970).
[3] Balmbra, R. R. and Clunie, J. S. *Intern. Conf. Liquid Cryst.*, 3rd, Berlin (August 1970).
[4] Benton, D. P., Howe, P. G., Farnard, R. and Puddington, I. E. *Can. J. Chem.* **33,** 1798 (1955).
[5] Chapman, D. *J. chem. Soc.* **152,** 784 (1958).
[6] Clunie, J. S., Corkill, J. M. and Goodman, J. F. *Proc. R. Soc. A***285,** 520 (1968).
[7] Eins, S. *Molec. Crystals Liqu. Crystals* **11,** 119 (1970).
[8] Ekwall, P., Danielsson, I. and Mandell, L. *Kolloidzeitschrift* **169,** 113 (1960).
[9] Ekwall, P. and Mandell, L. *Acta Chem. Scand.* **22,** 699 (1968).
[10] Ekwall, P., Mandell, L. and Fontell, K. *Acta Chem. Scand.* **22,** 1543 (1968).
[11] Ekwall, P., Mandell, L. and Fontell, K. *Molec. Crystals Liqu. Crystals* **8,** 157 (1969).
[12] Ekwall, P., Mandell, L. and Fontell, K. *J. Colloid Interface Sci.* **31,** 508, 529 (1969).
[13] Ekwall, P., Mandell, L. and Fontell, K. *J. Colloid Interface Sci.* **33,** 215 (1970).
[14] Fontell, K., Ekwall, P., Mandell, L. and Danielsson, I. *Acta Chem. Scand.* **16,** 2294 (1962).
[15] Fontell, K., Mandell, L. and Ekwall, P. *Acta Chem. Scand.* **22,** 3209 (1968).
[16] Fontell, L., Mandell, L., Lehtinen, H. and Ekwall, P. *Acta Polytechn. Scand.*, Ch. 78, III (1968).
[17] François, J., Gilg, B., Spegt, P. A. and Skoulios, A. E. *J. Colloid Interface Sci.* **21,** 293 (1966).
[18] Friedel, G. *Annls. Phys.* **18,** 273 (1922).
[19] Gallot, B. and Skoulios, A. E. *Acta Crystallogr.* **15,** 826 (1962).
[20] Gilchrist, C. A., Rogers, J., Steel, G., Vaal, E. G. and Winsor, P. A. *J. Colloid Interface Sci.* **25,** 409 (1967).
[21] Gulik-Krzywicki, T., Rivas, E. and Luzzati, V. *J. molec. Biol.* **27,** 303 (1967).
[22] Gulik-Krzywicki, T., Shechter, E., Faure, M. and Luzzati, V. *Nature* **223,** 1116 (1969).
[23] Gulik-Krzywicki, T., Tardieu, A. and Luzzati, V. *Molec. Crystals Liqu. Crystals* **8,** 285 (1968).
[24] Husson, F., Mustacchi, H. and Luzzati, V. *Acta Crystallogr.* **13,** 668 (1960).
[25] Larsson, K. *Z. phys. Chem. (Frankfurt)* **56,** 173 (1967).
[26] Lawson, K. D., Mabis, A. J. and Flautt, T. J. *J. phys. Chem.* **72,** 2058 (1968).

[27] LAWSON, K. D. and FLAUTT, T. J. *Molec. Crystals* **1,** 241 (1966).
[28] LUTTON, E. S. *J. Am. Oil Chem. Soc.* **43,** 28 (1966).
[29] LUZZATI, V. *Biological Membranes* (edited by D. Chapman), p. 71, Academic Press (1968).
[30] LUZZATI, L., GULIK-KRZYWICKI, T. and TARDIEU, A. *Nature* **218,** 1031 (1968).
[31] LUZZATI, V., GULIK-KRZYWICKI, T., TARDIEU, A., RIVAS, E. and REISS-HUSSON, F. *The Molecular Basis of Membrane Function* (edited by D. C. Tosteson), p. 79, Prentice-Hall, New Jersey (1969).
[32] LUZZATI, V., GULIK-KRZYWICKI, T. and TARDIEU, A. *Intern. Conf. Liquid Crystals*, 3rd, Berlin (August 1970).
[33] LUZZATI, V. and HUSSON, F. *J. Cell Biol.* **12,** 207 (1962).
[34] LUZZATI, V., MUSTACCHI, H., SKOULIOS, A. and HUSSON, F. *Acta Crystallogr.* **13,** 660) (1960).
[35] LUZZATI, V. and REISS-HUSSON, F. *Nature* **210,** 1351 (1966).
[36] LUZZATI, V. and SPEGT, P. A. *Nature* **215**, 701 (1967).
[37] LUZZATI, V., TARDIEU, A., GULIK-KRZYWICKI, T., RIVAS, E. and REISS-HUSSON, F. *Nature* **220,** 485 (1968).
[38] LUZZATI, V., TARDIEU, A. and GULIK-KRZYWICKI, T. *Nature* **217,** 1028 (1968).
[39] MANDELL, L. Manuscript in preparation.
[40] MANDELL, L. and EKWALL, P. *Acta Polytechn. Scand.* Ch. 74, I (1968).
[41] MANDELL, L., FONTELL, K. and EKWALL, P., *Adv. Chem. Ser. Am. chem. Soc.* No. 63, p. 89 (1967).
[42] MARSDEN, S. S. and MCBAIN, J. W. *Acta Crystallogr.* **1,** 230 (1948).
[43] MOORE, R. J., GIBBS, P. and EYRING, H., *J. phys. Chem.* **57,** 172 (1953).
[44] PARK, D., ROGERS, J., TOFT, R. W. and WINSOR, P. A. *J. Colloid Interface Sci.* **32,** 81 (1970).
[45] PHILLIPS, M. C., WILLIAMS, R. M. and CHAPMAN, D. *Chem. Phys. Lipids* **3,** 234 (1969).
[46] RAND, R. P. and LUZZATI, V. *Biophys. J.* **8,** 125 (1968).
[47] REISS-HUSSON, F. and LUZZATI, V. *J. phys. Chem.* **68,** 3504 (1964).
[48] REISS-HUSSON, F. and LUZZATI, V. *Adv. Biol. Med. Phys.* **11,** 87 (1967).
[49] RIVAS, E. and LUZZATI, V. *J. molec. Biol.* **41,** 261 (1969).
[50] SEGERMAN, E. *Surface Chemistry* (edited by P. Ekwall, K. Groth and V. Runnström-Reio), p. 157, Munksgaard, Copenhagen (1965).
[51] SKOULIOS, A. *Adv. Colloid Interf. Sci.* **1,** 79 (1967).
[52] SKOULIOS, A. and LUZZATI, V. *Acta Crystallogr.* **14,** 278 (1961).
[53] SMALL, D. M. *J. Am. Oil Chem. Soc.* **45,** 108 (1968).
[54] SPEGT, P. A. Thesis, Univ. of Strasbourg (1964).
[55] SPEGT, P. A. and SKOULIOS, A. E. *Acta Crystallogr.* **16,** 301 (1963).
[56] SPEGT, P. A. and SKOULIOS, A. E. *Acta Crystallogr.* **17,** 198 (1964).
[57] SPEGT, P. A. and SKOULIOS, A. E. *Acta Crystallogr.* **22,** 892 (1966).
[58] SPEGT, P. A., SKOULIOS, A. E. and LUZZATI, V. *Acta Crystallogr.* **14,** 866 (1961).
[59] TARDIEU, A., GULIK-KRZYWICKI, T., REISS-HUSSON, F., LUZZATI, V.

and RAND, R. R. Abstract of Paper read at British Biophysical Society meeting on the Biophysics of Membranes, Birmingham (April 1969).
[60] TARDIEU, A. and LUZZATI, V. *Biochim. biophys. Acta* **219,** 11 (1970).
[61] VANDENHEUVEL, F. A. *Chem. Phys. Lipids* **2,** 372 (1968).
[62] VINCENT, J. M. and SKOULIOS, A. E. *Acta Crystallogr.* **20,** 432, 441, 447 (1966).
[63] VOLD, R. D. *J. Am. chem. Soc.* **63,** 2915 (1941).
[64] VOLD, R. D. and VOLD, M. J. *J. Am. chem. Soc.* **61,** 808 (1939).
[65] WINSOR, P. A. *Chem. Rev.* **68,** 1 (1968).
[66] *International Tables for X-ray Crystallography*, Vol. II, Birmingham (1959).

5

Electrical Properties of Liquid Crystals—Non-Amphiphilic Systems

R. WILLIAMS

Introduction

Nematics and cholesterics combine strong anisotropy with a high degree of deformability, and this gives rise to many striking visible effects on application of an electric field. The comparable effects in normal amorphous isotropic liquids or in anisotropic solids are much smaller and usually detectable only by careful measurements. In the liquid crystals, small steady forces lead to large total deformations, made easily visible by the strong optical anisotropy. We discuss here various electrical effects, but because of the paucity of data on smectics, discussion will be confined to results on nematics and cholesterics.

Effect of Electric Field on Phase Transitions

The very existence of nematics and cholesterics can depend on whether or not an electric field is present. We take up first the effect of the field on the nematic–amorphous isotropic transition and use the terminology and thermodynamic treatment appropriate to a first-order phase transition since the nematic–amorphous isotropic transition is known to be of this kind [1].

It was shown by Helfrich [2] that, at temperatures slightly above the normal nematic–amorphous isotropic transition, it is possible to convert the amorphous isotropic liquid to nematic by application of an electric field (d.c. or a.c.). The results are understandable, as properties of a first-order phase transition. In the absence of a field, at the transition temperature, T_0, nematic and amorphous isotropic phases have the same free energy, G, per mol. At higher temperatures the free energy of the nematic is higher than that of the amorphous isotropic phase and any nematic material present will convert spontaneously into isotropic material with a corresponding decrease in free energy. According to thermodynamic arguments, the excess of free energy, ΔG, of the nematic over the isotropic liquid at the temperature, $T_0 + \Delta T$ is related to the heat of transition, λ, by the equation:

$$\Delta G = \left(\frac{\lambda}{T_0}\right)\Delta T \tag{5.1}$$

In the presence of an electric field the free energies of both phases are lowered. The free energy of a cubic centimetre of material with dielectric constant, ε, is lowered by an electric field $\boldsymbol{E}$ by the amount, $(\varepsilon \boldsymbol{E}^2/8\pi)$. In certain cases* the component, ε_N, of the dielectric constant of the field oriented nematic parallel to the electric field is larger than the dielectric constant of the isotropic material, ε_I. This appears to be the case for *p*-ethoxybenzylidene-*p'*-aminobenzonitrile, investigated by Helfrich. In this case ΔG becomes:

$$\Delta G = \left(\frac{\lambda}{T_0}\right)\Delta T - (\varepsilon_N - \varepsilon_I)\frac{\boldsymbol{E}^2 v}{8\pi} \tag{5.2}$$

where v is the molar volume. The field stabilizes the nematic relative to the isotropic material and the free energy of the two phases becomes equal at a temperature above T_0. Setting ΔG equal to zero and solving equation (5.2) gives the transition temperature for any field, $\boldsymbol{E}$. The effect is very large in comparison with the effect of a field on ordinary phase transitions because the heat of transition is so small for nematic–amorphous isotropic transitions. For *p*-azoxyanisole, λ is 732·7 J mol^{-1}, compared with 21 kJ mol^{-1} for the heat of fusion of organic compounds of comparable complexity. Thus it was observed that the transition temperature was raised by nearly 1° by a field of about 100 kV cm^{-1}, in approximate agreement with the above theory. With a field of 500 kV cm^{-1}, one might anticipate raising the transition temperature by 10°.

(*b*) CHOLESTERIC–NEMATIC TRANSITIONS

A quite different transformation is that from cholesteric to nematic brought about by either a magnetic or electric field. This was predicted independently by de Gennes [3] and Meyer [4] and observed experimentally by Wysocki, Adams and Haas [5], Baessler, Laronge and Labes [6], Heilmeier and Goldmacher [7] and Kahn [8]. An earlier observation of the effect in magnetic fields was reported by Sackmann, Meiboom and Snyder [9].

The basic effect is the action of the field on the helical ordering. The cholesteric structure is viewed [10] as a stack of parallel planes of rod-like molecules. Within each plane the molecules are all parallel. The orientation of the molecules can be specified by a vector, L, lying in the plane and parallel to their long dimension. From one plane to the next the orientation of this vector changes by a small angle, always in the same sense, so that there is an overall helical structure in the liquid in a direction perpendicular to the planes. Fig. 5.1(*a*) shows a side view of the stack of

* In other cases, such as *p*-azoxyanisole, ε_I is slightly larger than the larger component of ε_N.

planes where the arrows indicate the projection of the orientation vector on the plane of the page. The electric field exerts a torque on the liquid because the dielectric constant of a given plane has two values, $\varepsilon_{\parallel}$ and $\varepsilon_{\perp}$, for fields parallel and perpendicular, respectively, to the molecular order vector defined above. For brevity we consider only the case where $\varepsilon_{\parallel} > \varepsilon_{\perp}$. In an electric field, $\boldsymbol{E}$, the free energy of a unit volume of material is lower with L parallel to $\boldsymbol{E}$ than it is with L perpendicular to $\boldsymbol{E}$ by the amount, $(\varepsilon_{\parallel} - \varepsilon_{\perp})\boldsymbol{E}^2/8\pi$. Two distinct effects are seen, depending on whether the field is parallel or perpendicular to the helix axis. Fig. 5.1(*b*) shows qualitatively what happens when it is perpendicular. The field tends to orient all planes so that L is parallel to $\boldsymbol{E}$ but this tendency is resisted by the torsional forces arising from intermolecular interactions between the

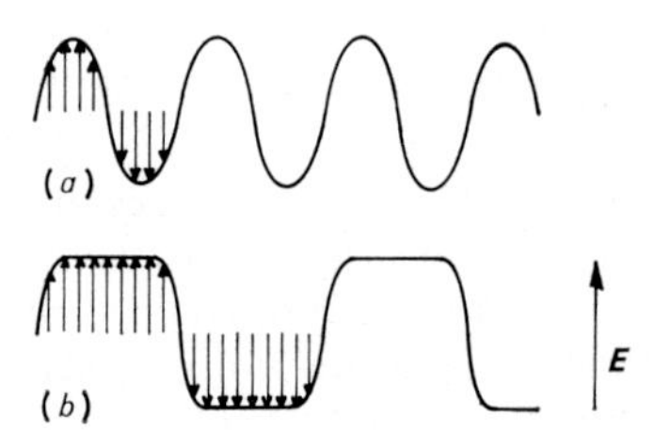

FIG. 5.1. Illustration of the helical ordering of molecules in a cholesteric liquid and how it is changed by an electric field perpendicular to the helix axis. (*a*) No field. Arrows indicate the projection of the molecular long axes in the plane of the page as one moves from left to right along the helix axis. (*b*) An electric field in the plane of the page and perpendicular to the helix axis distorts the helix to favour molecules lying parallel to the field. This leads to an increase in pitch as shown and eventually to complete unwinding of the helix.

asymmetric molecules. The helix is distorted, its pitch increases and ultimately it is unwound completely. At this point the cholesteric has become nematic, and P/P_0 (see Fig. 5.2) is infinite.

The other case of interest is that with $\boldsymbol{E}$ parallel to the helix axis. Here again the field tries to realign the molecules and is resisted by the bending force of the liquid. When the field is large enough, the cholesteric structure breaks up completely and all molecules line up parallel to the field direction, like an umbrella turning inside out in a gale, again giving a nematic. The collapse of the cholesteric structure is more abrupt here than in the previous case.

The earliest results were obtained for the case with $\boldsymbol{E}$ parallel to the helix axis [5]. A mixture of cholesteryl chloride, nonanoate and oleyl carbonate between transparent, tin oxide-coated glass plates had the opalescent appearance characteristic of a cholesteric. On application of an electric field of 100 kV cm^{-1} the liquid becomes clear and the optical properties are now those of a nematic. On removing the field the liquid returns to its original cholesteric state. A particularly favourable system for studying the effect [7] is made by mixing optically active amyl

4-(*p*-cyanobenzylideneamino)cinnamate with structurally similar nematics. The mixtures are cholesteric and have highly anisotropic dielectric constants. With these materials the threshold field for observing the cholesteric–nematic transition is lowered to 10 kV cm^{-1}. The kinetics of the transition indicate a nucleation-dominated process with time constant of the order 20 ms.

Unwinding of the helix by a field perpendicular to the helix axis (Fig. 5.1(*b*)) has been reported [6] for a mixture of cholesteryl chloride and myristate. Measurements were made of the optical rotatory power of a thin layer of material as a function of electric field and the pitch obtained from a theoretical relation due to de Vries [11]. The dependence of the measured pitch on the electric field was in satisfactory agreement with the theory outlined above.

An independent verification of the same theory has been made by

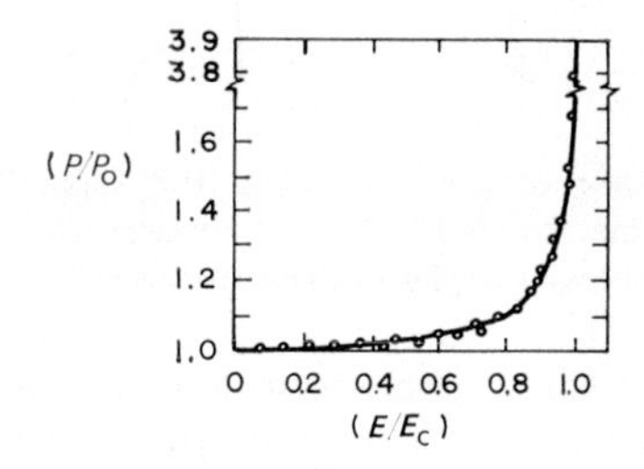

FIG. 5.2. Ratio of the pitch, P, to the pitch, P_0, in the absence of a field, as a function of the ratio of the field, $\boldsymbol{E}$, to the critical field, $\boldsymbol{E}_c$. $\boldsymbol{E}$ is perpendicular to the helix axis. The system is a mixture of cholesteric compounds, taken from ref. [8]. $\boldsymbol{E}_c$ is defined as the field at which the slope of the curve becomes infinite.

measuring the Bragg scattering angle for laser light [8]. The pitch of the cholesteric helix is often between 4000 and 7000 Å. This gives rise to Bragg reflection of visible light. For the first-order reflection of a given wavelength there is a well-defined angle of incidence and reflection. This may be related to the pitch of the helix and used to measure changes of the pitch in an electric field. Kahn [8] worked with a mixture of cholesteryl chloride, nonanoate and oleyl carbonate. With no field applied the helix axis is perpendicular to the glass plates. A field of 20 kV cm^{-1} reorients the entire helix to a new direction approximately parallel to the glass, maintaining the cholesteric structure; it remains so oriented as higher fields are applied. The field is, in all cases, perpendicular to the glass plates. It is in this case that it is possible to observe effects of the field perpendicular to the helix axis. The Bragg angle for reflection of 6328 Å laser light depends strongly on the applied field. Rise and fall times for the change of pitch on application and removal of the field are of the order 1 ms. The data relating the pitch to the applied field are shown in Fig. 5.2. To summarize, an electric field, either parallel or perpendicular to the helix axis, breaks up the cholesteric structure and converts it to a nematic structure. In both cases the magnitude of the field required is about the same. The main difference is that the transition is quite abrupt when the field is parallel to the helix axis and more gradual when the field is perpendicular.

It is not yet clear whether the cholesteric–nematic transition is a phase transition in the usual meaning of the term, though most of the above cited authors refer to it as such. It is a reversible transition and takes place at a well-defined field. Phase diagrams are useful for presenting data relating to transitions, and Fig. 5.3 shows how such a diagram might look.

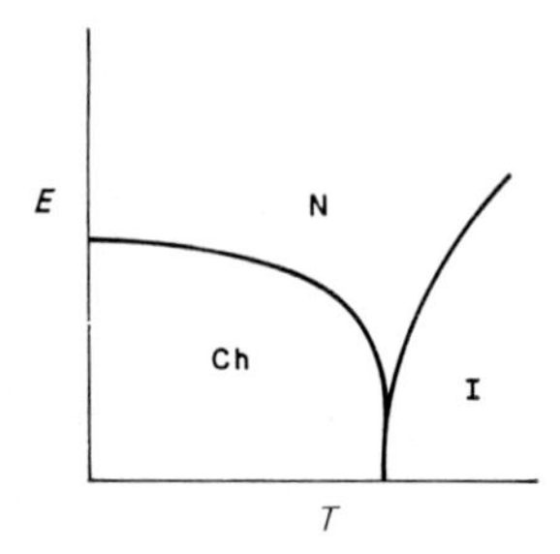

FIG. 5.3. Schematic phase diagram showing the separate regions which can exist as a function of electric field, ***E***, and temperature, *T*. N (nematic), Ch (cholesteric), I (amorphous isotropic).

In zero field there is a direct cholesteric–amorphous isotropic transition as one raises the temperature. There appears to be no evidence that a cholesteric–nematic transition is ever found in the absence of a field. This is reminiscent of the situation in superconductivity where the transition from the superconducting to the normal phase is second order in the absence of but first order in the presence of a magnetic field. Within the temperature range where the cholesteric material exists there is, for any given temperature, a cholesteric–nematic transition at a certain field. This field would be expected to depend in some way on the temperature as shown. Once in the nematic phase, changing the temperature at a given field must ultimately lead to a nematic–isotropic transition as indicated. Since the nematic–isotropic transition temperature itself depends on field, we have a diagram as shown with sharply bounded regions separating physically distinguishable forms. If the line separating cholesteric and nematic regions is not horizontal then it must be possible to go from cholesteric to nematic at a fixed value of the field simply by changing temperature. Experimental data are not yet available to construct a diagram of this kind, but a knowledge of the detailed form of the curves would clarify the nature of the phases involved. A general sketch of possible phase relations for liquid crystals has recently been given by Papoular [12].

Dielectric Constant and Dielectric Loss

The anisotropy of the dielectric constant in oriented nematics underlies many of the effects observed in electric fields. Careful measurements by Maier and Meier [13] on homologues (C_1 to C_8) of *p*-azoxyanisole illustrate the general features. These have been reviewed and analysed by Meier and Saupe [14].

The material was contained in a dielectric measuring cell (thickness, about 3 mm) and placed in a magnetic field of 159 kA m^{-1} (2000 oersted). This lined up all molecules with their long axes parallel to the magnetic field. It was then possible to measure the dielectric constants, $\varepsilon_{\parallel}$, and $\varepsilon_{\perp}$, parallel and perpendicular to the molecular long axes. These were similar for all members of the series with the general form shown in Fig. 5.4. The

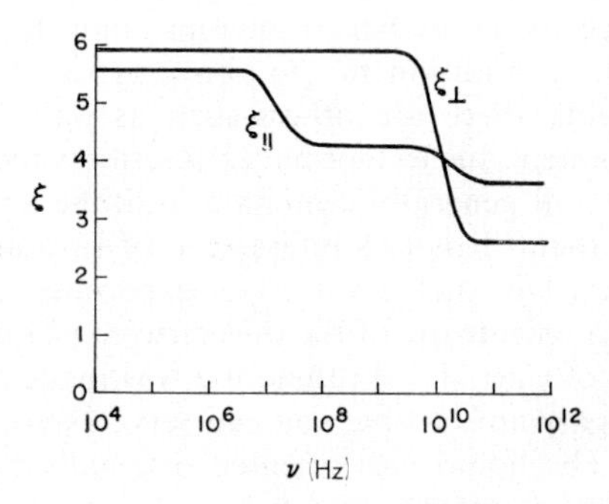

FIG. 5.4. Probable behaviour of the parallel and perpendicular components of the dielectric constant (ε) of *p*-azoxyanisole as a function of frequency (ν). Taken from ref. [13].

features of note are the substantial anisotropy, the dispersion around 10 GHz for both curves and the additional dispersion around 10 MHz in $\varepsilon_{\parallel}$. The dipole moment in these molecules due to the central azoxy group is directed neither along the long axis nor perpendicular to it but somewhere in between. There is an additional component of the dipole moment due to the end groups. The high-frequency dispersion, common to $\varepsilon_{\parallel}$ and $\varepsilon_{\perp}$, is believed due to rotation of the molecules about their long axes. This rotation is nearly as free as the rotations of molecules in normal liquids and a dielectric dispersion is found in most polar liquids at this frequency. The low-frequency dispersion found only in $\varepsilon_{\parallel}$ is interpreted as end-over-end rotation of the molecules. The component of the dipole moment along the molecular axis is the important one in this case. It is believed that the end-over-end motion is severely hindered by the nematic ordering, resulting in the unusual low frequency. This explanation is a reasonable one but a dielectric dispersion in the same frequency range is also found in several liquids, such as octan-1-ol [15], which are not nematic, and where it would seem that it cannot be explained in the same way.

Orientation by Electric Fields

The orientation of nematics by electric fields presents several unsolved puzzles. Early literature has been reviewed by Gray [16] and we confine discussion to some more recent work. Several effects join to make things complicated. The simplest is the tendency for an anisotropic dielectric to align in a field so that the largest component of the dielectric constant coincides with the field direction. This effect is often dominant when the aligning field is a.c., above a certain frequency, usually of the order 100 kHz. In addition to this there is an alignment in d.c. or low-frequency a.c. fields due to any transport of ions within the liquid. This often tends to line up all molecules parallel to the electric field. When the effect due to ion transport gives an alignment different from that dictated by the anisotropy of the dielectric constant it is generally the ion transport that dominates. The mechanism by which the ions align the liquid is not completely understood. In addition to the above types of alignment due to applied electric fields, there are others such as wall alignment, surface alignment and alignment due to thermal convection which are independent of the field. These will generally dominate until the applied field is sufficient to overcome them. It is this interaction of several competing alignment processes that gives such a variety of experimental results.

Alignment due to anisotropy of the dielectric constant is well illustrated in the experiments of Carr [17]. In these use was made of the fact that the dielectric loss is highly anisotropic and can serve as an indicator of molecular orientation. The liquid was oriented originally by a magnetic field and reorientation by a suitably directed electric field was observed. For *p*-azoxyanisole, the dielectric constant has its largest component perpendicular to the nematic axis, i.e. $\varepsilon_{\perp} > \varepsilon_{\parallel}$. In a magnetic field the molecules align parallel to $\boldsymbol{H}$. An a.c. electric field (370 kHz) is applied, also parallel to $\boldsymbol{H}$. For low values of the electric field there is no effect but, at a fairly well-defined threshold field, the long axes of the molecules realign perpendicular to $\boldsymbol{H}$ and $\boldsymbol{E}$, bringing the largest component of the dielectric constant parallel to $\boldsymbol{E}$. The threshold field ($\boldsymbol{E}_{\text{th}}$) depends on the magnitude of $\boldsymbol{H}$. There is rough agreement with the equation [17] relating the electrical and magnetic anisotropy energies.

$$(\varepsilon_{\parallel} - \varepsilon_{\perp})\boldsymbol{E}_{\text{th}}^2 = (\mu_{\parallel} - \mu_{\perp})\boldsymbol{H}^2 \tag{5.3}$$

where $\mu_{\parallel}$, $\mu_{\perp}$ represent the diamagnetic susceptibility components parallel and perpendicular to the nematic axis. The two sides of the equation represent the amount of energy gained or expended by a unit volume of material on realigning the nematic axis through 90° in an electric or magnetic field. The value of $\boldsymbol{E}$ balancing the equation is the threshold field for realignment, $\boldsymbol{E}_{\text{th}}$. When $\boldsymbol{H}$ is approximately 80 kA m^{-1} (1000 oersted) $\boldsymbol{E}_{\text{th}}$ is about 1 kV cm^{-1}. These results are understandable as the simple consequence of dielectric anisotropy.

When a d.c. electric field is used a different result is observed; the alignment, as determined with a high-frequency a.c. field, is one with the nematic axis parallel to the field, contrary to the a.c. case where the high field alignment is with the nematic axis perpendicular to the field. The mechanism is not understood in detail but is probably connected with the anisotropic viscosity of the oriented nematic liquid [18]. More detailed descriptions of similar experiments have been reported by Carr [19].

A more direct observation of the orientation of *p*-azoxyanisole in d.c.

and a.c. electric fields is to be found in the X-ray work of Vainshtein, Chistyakov, Kosterin and Chaikovski [20, 21]. The experiments of Helfrich and others [42] on the reorienting effects of applied electric fields on films of "twisted nematics" of positive dielectric anisotropy are also important.

Domains and Dynamic Scattering

Among the most useful electrical orientation effects are those described by the terms "domains" [22–28] and "dynamic scattering" [29–32]. They are usually observed by applying an electric field to a thin layer of nematic (5 to 100 μ) contained between two glass plates with transparent conductive coatings on their inner faces. With thin cells the wall alignment

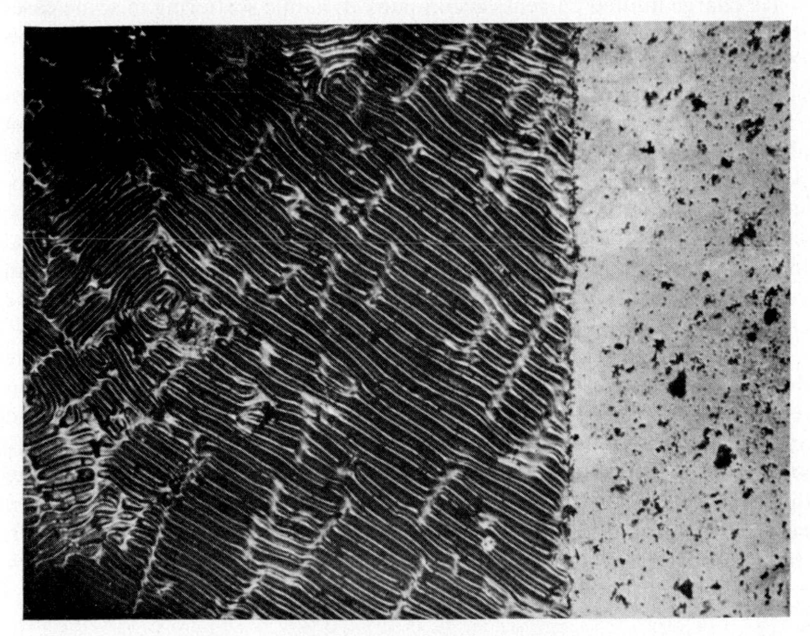

FIG. 5.5. Photograph of the domain structure observed on application of an electric field to a thin layer of *p*-azoxyanisole. There is no field in the region at the right. Details in ref. [22].

of molecules extends into the volume of liquid and, under proper conditions, can give a uniform texture (orientation of the nematic) throughout. Against this uniform alignment a reorientation can give striking optical changes. The two general features observed are a periodic structure, the domains, and stirring of the liquid accompanied by strong light scattering and turbidity, the dynamic scattering. Fig. 5.5 shows domains in *p*-azoxyanisole [22]. These form on application of ten or twenty volts; there is a fairly well-defined threshold voltage for their appearance. As the voltage is increased above this the liquid begins to stir and becomes increasingly turbid. Thus the orientation varies strongly from place to place. This, together with the high optical anisotropy of nematics, gives rise to strong scattering of light. The domain structure nearly always begins to stir and to give way to dynamic scattering as the voltage is increased above the threshold. This suggests that both effects may be different aspects of a single phenomenon, though this is by no means established.

For dynamic scattering, the mechanism was first investigated experimentally by Heilmeier, Zanoni and Barton [30]. Several lines of evidence led to the conclusion that the stirring of the liquid was an electrohydrodynamic effect. This is most easily understood for the case where the conduction in the liquid is due predominantly to ions of a single sign. This happens when there is injection of charge into the liquid from the electrodes, and gives a space charge limited current [33]. There is evidence that space charge limited currents accompany dynamic scattering in some cases though not necessarily in all. Whenever an ion bearing a charge, Q, moves through a liquid under the action of an electric field, $\boldsymbol{E}$, it is acted on by a continuous force of magnitude (QE). Since the ion moves at a constant velocity it experiences no nett acceleration and must therefore transmit all the force to the surrounding liquid. Usually equal charges are carried as positive and negative ions which exert equal forces on the liquid in opposite directions with no nett effect. When there is an excess of ions of one sign there is no such exact cancellation and the ions exert a substantial force on the liquid. This force can be directly demonstrated in certain cases [30]. The ions may be formed by injection of electrons from the cathode, followed by capture by a molecule in the liquid.

Extensive analyses have been made of the problem of formation of the domains in nematic liquids; summaries of the general ideas are given in references [34–36]. Briefly, a separation of ionic charges in the liquid arises from the combined effects of the applied field and the conductivity anisotropy of the liquid. Ions of opposite signs are moved in opposite directions by the field and exert forces on the liquid that give rise to a periodic deformation. Voltage thresholds and effects due to changes in the frequency of the applied field are well accounted for by the theory which is based on continuum hydrodynamics.

Other Effects

A very useful display application of the field alignment effects discussed above derives from the work of Castellano, Heilmeier, Pasierb and McCaffrey [37] on nematics containing dissolved dye molecules. In general, the nematic aligns dissolved molecules. Electronic transitions are generally polarized so that a given absorption band appears only for light polarized along a particular direction with respect to the molecule. Changing the orientation of an array of dissolved, oriented molecules by application of an electric field is then accompanied in certain cases by a substantial colour change in a layer of liquid viewed by transmitted light. By using different dye molecules a wide range of colour changes can be brought about.

Two further effects have been reported by Wysocki, Adams and Haas [38, 39]. In the first of these [38] a cholesteric (a mixture of cholesteryl chloride and nonanoate) has been found to migrate away from a region where there is a high electric field. It has been possible to utilize this effect to produce a distribution of liquid coinciding with an optical image and thus record a picture. The authors attribute the behaviour to an electrophoretic process. A substantial change of viscosity with field is observed in a cholesteric, a field of 36 kV cm^{-1} causing a 12% increase in viscosity at room temperature in a mixture of cholesteryl chloride, nonanoate and oleyl carbonate.

A piezoelectric effect, not yet observed, has been predicted by Meyer [40]. In nematics composed of molecules lacking a centre of symmetry and having a pear shape, for example, mechanical forces could bring about a new alignment with a net electrical polarization. Certain phenomena observed by Meyer in small drops of nematic liquid were cited as a possible example of the effect. A perhaps related phenomenon is the linear electro-optic effect [43] for *p*-azoxyanisole. Such an effect indicates the lack of a centre of symmetry. A d.c. electric field was applied to a sample of nematic *p*-azoxyanisole subjected to a magnetic field. The liquid was viewed between crossed polarizers by transmitted light. For one polarity of the electric field the intensity of transmitted light was increased, and for the opposite polarity it was decreased. The electric field was applied perpendicular to the magnetic field while the polarizer and analyser were oriented with one parallel to the magnetic field and the other perpendicular.

An interesting variation of the dynamic scattering effect was observed by Heilmeier and Goldmacher [41], using a mixture of a nematic (anisylidene-*p*-aminophenyl acetate) and a cholesteric (cholesteryl nonanoate). There is dynamic scattering with a d.c. electric field, but the liquid remains scattering when the field is removed. It reverts gradually to the non-scattering state with a half life of about 9000 s. Application of a 1 kHz a.c. field returns it to the non-scattering state in one second or less. The effect of the d.c. field is apparently to produce small separated regions, differing in crystallographic orientation, which scatter light strongly. These then coalesce gradually to form larger regions and eventually a completely ordered liquid. The non-scattering and scattering states may be considered as the Grandjean and "focal conic" textures of the cholesteric mixture, respectively.

REFERENCES

[1] Barrall, E. M., Porter, R. S. and Johnson, J. F. *J. phys. Chem.* **68,** 2810 (1964).
[2] Helfrich, W. *Phys. Rev. Lett.* **24,** 201 (1970).
[3] de Gennes, P. G. *Solid St. Communs.* **6,** 163 (1968).
[4] Meyer, R. B. *Appl. Phys. Lett.* **12,** 281 (1968).
[5] Wysocki, J. J., Adams, J. and Haas, W. *Phys. Rev. Lett.* **20,** 1024 (1968); *Molec. Crystals Liqu. Crystals* **8,** 471 (1969).

[6] BAESSLER, H. and LABES, M. M. *Phys. Rev. Lett.* **21,** 1791 (1968); BAESSLER, H., LARONGE, T. M. and LABES, M. M. *J. chem. Phys.* **51,** 3213 (1969).
[7] HEILMEIER, G. H. and GOLDMACHER, J. E. *J. chem. Phys.* **51,** 1258 (1969).
[8] KAHN, F. J. *Phys. Rev. Lett.* **24,** 209 (1970).
[9] SACKMANN, E., MEIBOOM, S. and SNYDER, L. C. *J. Am. chem. Soc.* **89,** 5981 (1967).
[10] SAUPE, A. *Angew. Chem. Internat. Edn.* **7,** 97 (1968).
[11] DE VRIES, H. *Acta Crystallogr.* **4,** 219 (1951).
[12] PAPOULAR, M. *Solid St. Communs.* **7,** 1691 (1969).
[13] MAIER, W. and MEIER, G. *Z. Elektrochem.* **65,** 556 (1961); MAIER, W. and MEIER, G. *Z. Naturf.* **16a,** 470 (1961); MAIER, W. and MEIER, G. *Z. Naturf.* **16a,** 262 (1961).
[14] MEIER, G. and SAUPE, A. *Molec. Crystals* **1,** 515 (1966).
[15] Landolt-Börnstein Tables, 6 edn. (Elektrische Eigenschaften I), Springer, Berlin (1959), p. 721.
[16] GRAY, G. W. *Molecular Structure and the Properties of Liquid Crystals*, Academic Press, London and New York (1962).
[17] CARR, E. F. *Adv. Chem. Ser., Ordered Fluids and Liquid Crystals, Am. chem. Soc.* **63** pp. 76–88 (1967).
[18] PORTER, R. S. and JOHNSON, J. F. *J. phys. Chem.* **66,** 1826 (1962).
[19] CARR, E. F. *J. chem. Phys.* **39,** 1979 (1963); **42,** 738 (1965); **43,** 3905 (1965).
[20] VAINSHTEIN, B. K., CHISTYAKOV, I. G., KOSTERIN, E. A. and CHAIKOVSKI, V. M. *Molec. Crystals Liqu. Crystals* **8,** 457 (1969).
[21] KOSTERIN, E. A. and CHISTYAKOV, I. G. *Soviet Phys., Crystallogr.* **14,** 252 (1969), translated from *Kristallografiya* **14,** 321 (1969).
[22] WILLIAMS, R. *J. chem. Phys.* **39,** 384 (1963).
[23] WILLIAMS, R. *Nature, Lond.* **199,** 273 (1963).
[24] ELLIOTT, G. and GIBSON, J. G. *Nature, Lond.* **205,** 995 (1965).
[25] HEILMEIER, G. H. *J. chem. Phys.* **44,** 644 (1966).
[26] WILLIAMS, R. *Adv. Chem. Ser., Ordered Fluids and Liquid Crystals, Am. chem. Soc.* **63,** p. 61 (1967).
[27] HEILMEIER, G. H. *Adv. Chem. Ser., Ordered Fluids and Liquid Crystals, Am. chem. Soc.* **63,** p. 68 (1967).
[28] HELFRICH, W. *J. chem. Phys.* **51,** 2755 (1969).
[29] HEILMEIER, G. H., ZANONI, L. and BARTON, L *Appl. Phys. Lett.* **13,** 46 (1968).
[30] HEILMEIER, G. H., ZANONI, L. and BARTON, L. *Proc. I.E.E.E.* **56,** 1162 (1968).
[31] HEILMEIER, G. H. and HELFRICH, W. *Self Quenching of Dynamic Scattering in Nematic Liquid Crystals*, PTR-2764.
[32] JONES, D., CREAGH, L. and LU, S. *Appl. Phys. Lett.* **16,** 61 (1970).
[33] ROSE, A. *Phys. Rev.* **97,** 1538 (1955).
[34] HELFRICH, W. *J. chem. Phys.* **51,** 4092 (1969).
[35] ORSAY LIQUID CRYSTAL GROUP *Molec. Crystals Liqu. Crystals* **12,** 251 (1971).

[36] Penz, P. A. and Ford, G. W. *Phys. Rev.* **A6**, 414 (1971).
[37] Heilmeier, G. H. and Zanoni, L. *Appl. Phys. Lett.* **13,** 91 (1968); Heilmeier, G. H., Castellano, J. and Zanoni, L. *Molec. Crystals Liqu. Crystals* **8,** 293 (1968).
[38] Haas, W. and Adams, J. *Appl. Optics* **7,** 1203 (1968).
[39] Wysocki, J. J., Adams, J. and Haas, W., *J. appl. Phys.* **40,** 3865 (1969).
[40] Meyer, R. B. *Phys. Rev. Lett.* **22,** 918 (1969).
[41] Heilmeier, G. H. and Goldmacher, J. E. *Proc. I.E.E.E.* **57,** 34 (1969).
[42] Schadt, M. and Helfrich, W. *Appl. Phys. Lett.* **18,** 127 (1971).
and
Boller, A. Scherrer, M., Schadt, M. and Wild, P. *Proc. I.E.E.E.* **60,** 1002 (1972).
[43] Williams, R. *J. chem. Phys.* **50,** 1324 (1969).

6

Electrical Properties of Liquid Crystals: Electrical Conduction by Amphiphilic Mesophases

P. A. WINSOR

Introduction

Although the characteristics of electrical conduction in amorphous solutions of amphiphilic colloidal electrolytes have been the subject of many studies [1], work on conduction in amphiphilic liquid crystalline systems* has been less extensive. The main investigations comprise those of Vold *et al.* on the high-temperature, high-concentration mesophases in soap/water systems [2], Winsor *et al.* on a number of multi-component lamellar liquid crystalline phases [1, 3, 4, 5] as well as on the M_1, V_1 and G phases in the *N,N,N*-trimethylaminododecanoimide/water system [6] and François *et al.* on mesophases in the sodium lauryl sulphate/water and a number of soap/water systems [7–13].

A significant, but little emphasized aspect of the electrical conductivity of anisotropic amphiphilic mesophases, is that electrical conduction in uniformly oriented samples would be expected to be dependent on direction. Thus, in a lamellar phase, the smaller inorganic ions, presumably the principal current carriers, could migrate freely within the $\bar{W}$ lamellae but presumably not within the $\bar{O}$ lamellae (cf. however p. 132). The electrical conductivity in such a phase might therefore be expected to be minimal in the direction perpendicular to the lamellae, along the optic axis. In the M_2 phase, the preferred conducting pathways would be along the polar micellar cores, in the direction along the optic axis. In the M_1 phase the continuous $\bar{W}$ region between the micellar fibres would present pathways for ionic conduction in all directions. However, whereas ions migrating at right angles to the fibres (at right angles to the optic axis) would encounter the hydrocarbon cores as obstacles, ions migrating parallel to the

* For the nomenclature employed, cf. Gray and Winsor, *Liquid Crystals and Plastic Crystals, Vol. 1*, Ellis Horwood Ltd, Chichester (1974).

fibres would not be thus impeded. Conductivity in the M_1 phase would, therefore, as with the M_2 phase, be expected to be maximal in the direction along the optic axis but the variation with direction would be small in comparison with the variations with direction in the G and M_2 phases.

In the optically isotropic cubic mesophases S_{1c}, V_1 and V_2 one might expect the conductivity to be virtually independent of direction.

In the majority of liquid crystals whose electrical conduction has been measured, no uniform orientation has usually been present. The conductivities determined have therefore represented mean values (cf. p. 137 and 142) for each particular preparation.

Conduction in the Amphiphilic Amorphous Solution Phase S_1 under Conditions of Flow

A type of exception to the situation noted in the last paragraph is provided by the investigations by Heckmann and Götz [14, 15, 16] on electrical conduction in solutions of cetyltrimethylammonium bromide and of sodium oleate when subjected to a velocity gradient in a Couette apparatus.

Such solutions, though isotropic when at rest, show streaming birefringence on account of the orientation by the velocity gradient of their fibrous micellar units (M_1 type). The streaming solutions thus, to some extent, simulate an M_1 phase oriented with its optic axis in the direction of flow. The conductivity as measured in the direction of flow (y) was greater than that in the direction perpendicular to this (z) (Fig. 6.1). These results

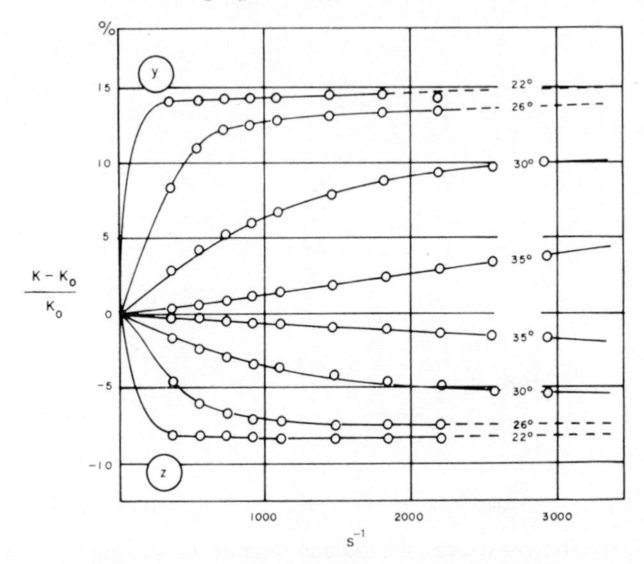

FIG. 6.1. Anisotropy of conductivity in the direction of flow (y) and at right angles to this (z) of a cetyltrimethylammonium bromide solution (19% w) as a function of the velocity gradient at 22, 26, 30 and 35°C. K_0 conductivity of solution at rest; K conductivity in y or z directions.

are in accord with the considerations above, although they were originally interpreted rather differently.

Conduction in some High-Concentration, High-Temperature Soap Mesophases

The high-temperature, high-concentration mesophases produced by the sodium soaps [21], with the exception of the "neat" phase, which is of the fused type, are now believed to be of the semi-crystalline type, the hydrocarbon regions being fused and the polar regions quasi-crystalline.

Studies of the conductivities of these systems were carried out by Vold and collaborators [2] both to establish phase boundaries and to obtain further insight into the structures of the various phases. Measurements were carried out over the temperature range 50°C to 300°C. Absolute reproducibility between runs, particularly at lower temperatures, was not obtained. This was attributed to the possibility of variable electrode contact and submicroscopic cracks in the more solid phases. However, the main emphasis in the investigation lay on the temperatures of changes in the slope of the conductivity/temperature curves. These temperatures were taken to indicate transition points. The temperatures corresponding to these inflexions were reproducible between runs. Further, in cases where the temperature coefficient of conductivity was calculated, the variations in the absolute conductivity values due to irreproducibilities were small in

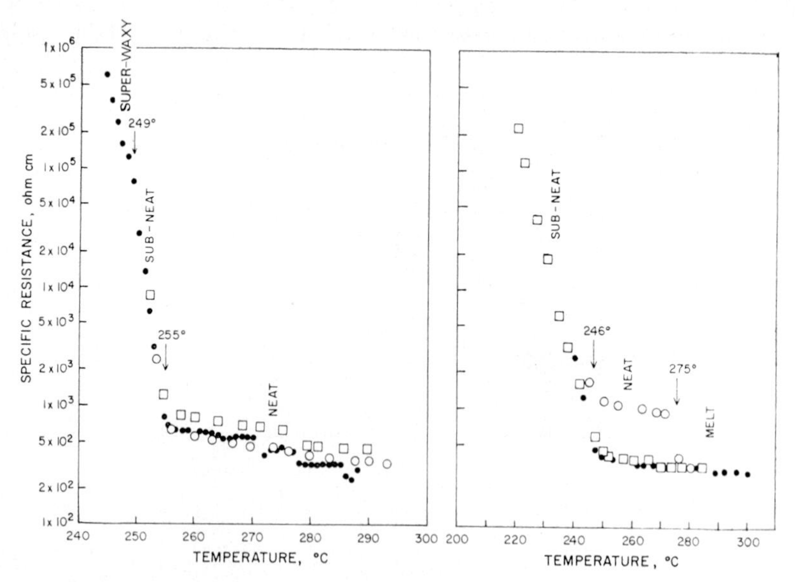

FIG. 6.2. Specific resistances of sodium stearate containing 0·0% w water, determined at 960 cycles. [2] □, first run; ○, second run; ●, third run.

FIG. 6.3. Specific resistances of sodium stearate containing 0·5% w water, determined at 420 cycles. [2] □, first run (cooling); ○, second run (cooling); ●, third run (cooling).

comparison with the large variations in temperature coefficients found at different temperatures. Examples of the curves obtained are in Figs. 6.2, 6.3, 6.4 and 6.5 and a phase diagram for the high-temperature, high-concentration region of the sodium stearate/water system based largely on the conductimetric measurements is in Fig. 6.6. This may be compared with the diagram for the sodium laurate/water system based largely on vapour pressure measurements [17] and [21, Fig. 5.4.] An important feature of the conductivity measurements was that they were considered to demonstrate that many of the anhydrous phases of sodium stearate can incorporate only small amounts of water before undergoing structural rearrangement. Previously, vapour pressure measurements were believed to indicate rather higher water contents [17].

Thus, according to the conductivity measurements, neat and sub-neat soap can incorporate only 4 and 3% of water respectively before conversion to the soap boiler's neat soap phase (G) whereas earlier it had been thought that up to 10% of water could be incorporated. The conductivity of the sub-neat phase, both anhydrous and containing small amounts of water, increases extremely rapidly with temperature while for the neat soap phase the conductivity, though higher, increases much less rapidly. These observations were interpreted as indicating that "within the sub-neat phase, the structure is changing from lattice-like at lower temperatures to a condition of lesser regularity at higher temperatures. . . . This would be

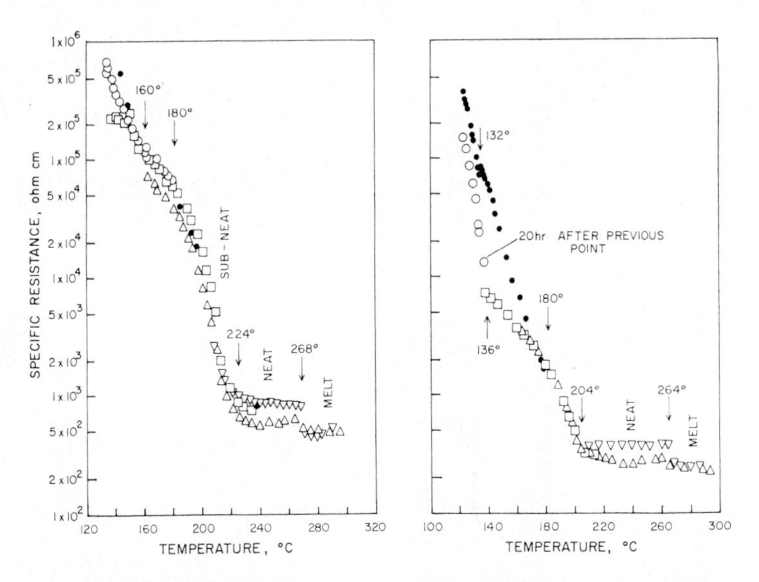

FIG. 6.4. Specific resistances of sodium stearate containing 1·0% w water, determined at 960 cycles. [2] ▽, first run; △, second run; □, third run; ○, fourth run; ●, fifth run.

FIG. 6.5. Specific resistances of sodium stearate containing 2·4% w water, determined at 960 cycles. [2] ▽, first run; △, second run; □, third run (cooling); ○, fourth run; ●, fifth run.

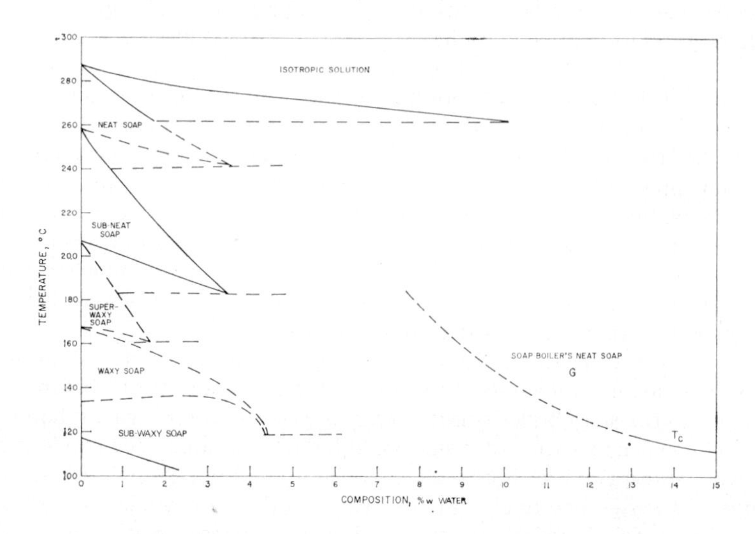

FIG. 6.6. Phase diagram of the high-temperature, high-concentration phases of the sodium stearate/water system based mainly on conductivity measurements [2].

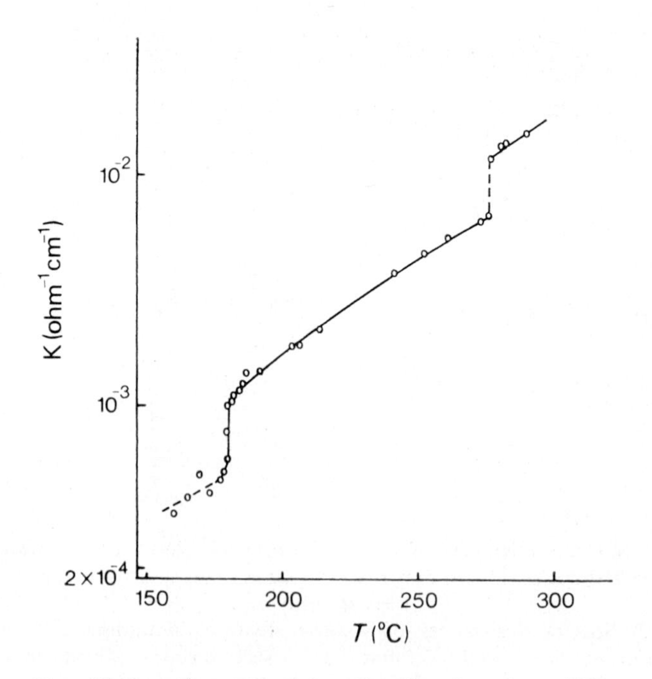

FIG. 6.7. Specific conductivity of sodium isovalerate [19].

the kind of transformation, here occurring over a range of 30–40°C, which ordinarily accompanies the melting of an ionic crystal at a single temperature, except that the residual order in neat soap extends over a larger number of molecules." The conductivity in the neat soap phase itself is essentially similar to that of a fused salt.

In systems of higher water content, the "soap-boiler's neat soap" phase G (28·7% w water) and the middle soap phase M_1 (69·3% w water) showed similar values of the specific resistance so that the equivalent conductivity of the soap in the M_1 phase is considerably greater than in the G phase. This result was interpreted as agreeing with the view [18] that soap-boiler's neat soap had a lattice-like structure, whereas middle soap was micellar. According to present views it agrees with the continuity of the $\bar{W}$ region within the M_1 phase in contrast to its division within the G phase into $\bar{W}$ lamellae separated by $\bar{O}$ lamellae (p. 122).

In a study of liquid crystal formation by the anhydrous sodium salts of the shorter-chain fatty acids Ubbelohde, Michels and Duruz [19] measured the conductance changes with temperature of sodium isovalerate (Fig. 6.7). The jump ($\times$4.5) in specific conductivity on passing (at 187°C) from crystals to liquid crystals (apparently the "neat soap" lamellar phase) is similar to the jump ($\times$4.0) which occurs on passing (at 283°C) from liquid crystal to isotropic melt.

Changes in the Electrical Conductivity on Displacement of the Intermicellar Equilibrium within the Amorphous Solution Phase (S_1, S_2) and on formation of the Liquid Crystalline Solution Phase (G)

As considered elsewhere [21] electrical resistance measurements were used by Bromilow and Winsor [3, 4] to study the displacement of the intermicellar equilibrium with changes in composition and temperature in a large number of amphiphilic solution phases. The electrical resistance changes were measured which occur in passing through the phase sequence

$$S_1 \rightarrow (S_1 + G) \rightarrow G \rightarrow (G + S_2) \rightarrow S_2$$

as produced by gradual changes in composition or temperature.

In the case of the homogeneous amorphous solution phase, S (including both the S_1 and S_2 sections), the electrical resistance is a unique function of temperature and composition and independent of the mechanical history of the sample. Fig. 6.8 records the reversible changes of specific resistance with changes in temperature of a particular amorphous S solution in which the intermicellar equilibrium is highly dependent on temperature. In this system the balance of the micellar equilibrium changes progressively from predominantly S_2 ($R > 1$) at 20°C to predominantly S_1 ($R < 1$) at 30°C. There was apparently no measurable hysteresis in the attainment of equilibrium on change of temperature.

With this particular composition this change with temperature occurs continuously without the intermediate formation of a G phase. With other compositions the continuity of the change with temperature may be

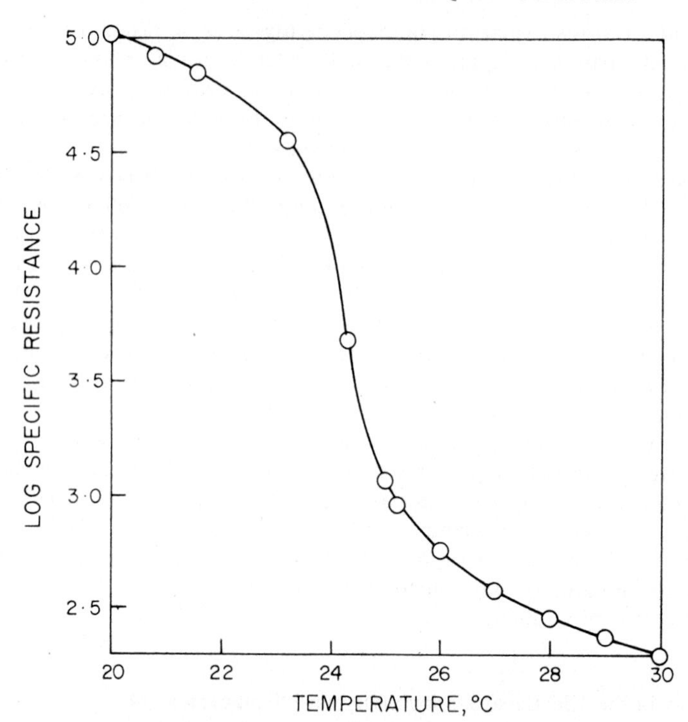

FIG. 6.8. Changes in specific resistance (ohm cm) with change in temperature of a solution containing saturated hydrocarbon* (10 ml), octanol-1 (2·25 ml), undecane-3-sodium sulphate solution (10 ml, 20% w) and sodium sulphate (0·2 g) [3].

* Boiling range 188°–213°C

broken by the intermediate separation of a G phase ($R = 1$) (ref. [21]), Fig. 39; systems containing about 1·7 ml octanol-1). Displacement of the intermicellar equilibrium from S_1 to S_2 may also be effected experimentally by a progressive change in composition at constant temperature. Although in certain systems this change is continuous, in many others it is interrupted by the separation of the liquid crystalline lamellar solution phase, G. In cases where this occurs a break is found in the conductivity/composition curve corresponding to the formation of the G phase. The typical shape of the curve is then that illustrated in Fig. 6.9. The formation of the liquid crystalline G phase is accompanied by an increase in resistance which interrupts what would apparently otherwise be a continuous curve for the amorphous solution phase, S_1–S_2.

Progressive changes in composition, $C_1 \rightarrow C_2$, which give rise to curves of either type may follow the lines of any of the General Methods Ia–Id listed in Table 16 of ref. [21]. Complementarily, the curves may be traversed in the opposite direction through composition changes, C_2–C_1, which follow any of the General Methods IIa–IId of Table 16 in ref. [21].

The absolute levels of the specific resistances vary considerably according

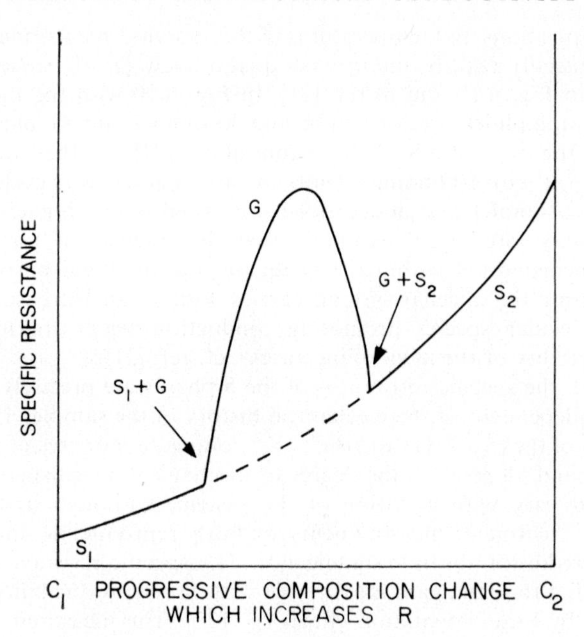

FIG. 6.9. Typical shape of specific resistance vs. composition curves for systems showing the phase sequence $S_1 \rightarrow (S_1 + G) \rightarrow G \rightarrow (G + S_2) \rightarrow S_2$.

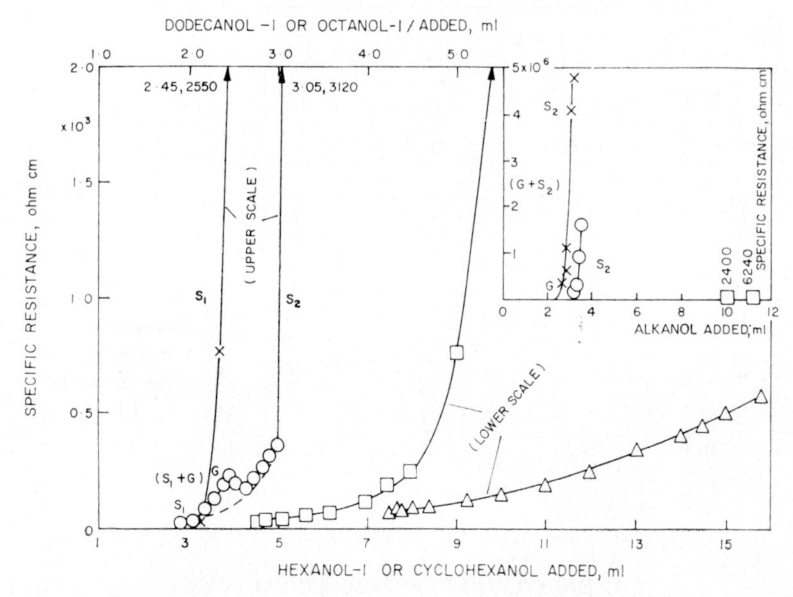

FIG. 6.10. Changes in specific resistance at 20°C on passing through S_1, G and S_2 stages on gradual addition to a mixture of saturated hydrocarbon* (10 ml) and undecane-3-sodium sulphate solution (10 ml, 20% w) of: △ cyclohexanol, or □ hexanol-1 (volume scale at bottom of figure), or ○ octanol-1, or × dodecanol-1 (volume scale at top of figure) [3].

* Boiling range 188°–213° C

to the compositions and temperatures of the systems, in a manner which is, however, usually readily interpreted qualitatively [3, 4]. Some examples are given in Fig. 6.10 and in ref. [21]. In Fig. 6.10 with the more water-miscible amphiphiles, cyclohexanol and hexanol-1, no G phase is produced and the transition $S_1 \rightarrow S_2$ is continuous [21]. Further, the absolute resistances at corresponding points of the curves for cyclo-hexanol, hexanol-1, octanol-1 and dodecanol-1 are progressively higher. This may be reasonably interpreted assuming that the fraction of the respective alcohols incorporated within the $\bar{O}$ region, the principal barrier to conduction within the micellar system, carries with it an increasing fraction of the molecular species promoting conduction (water and ions) as the water-miscibility of the alcohol increases—cf. ref. [21].

As noted, the specific resistances of the S phases are precisely reproducible and independent of the mechanical history of the sample. The specific resistances of the (S_1 + G) and (G + S_2) conjugate mixtures of phases, on the other hand, depend on the degree of mechanical interdispersion of the phases and vary with agitation of the system. Although under similar mechanical treatments measurements are fairly reproducible, these cannot be considered as of absolute significance. The specific resistance of the G phase itself, although again measurable with fair reproducibility, must be influenced by local orientation effects (p. 122). The measured value thus usually represents an average rather than an absolute value.

In the work under discussion only the broad features of the changes in

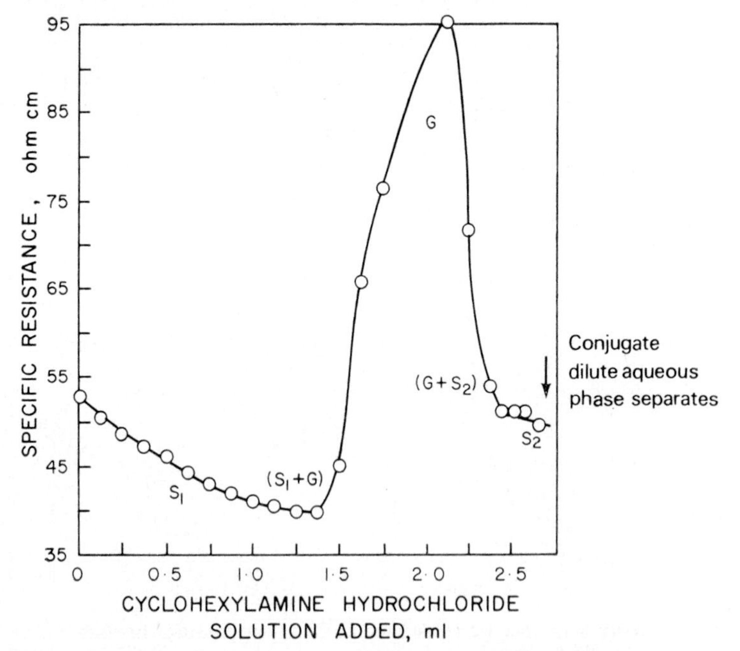

FIG. 6.11. Phase transitions and accompanying specific resistance changes at 20°C on adding aqueous cyclohexylamine hydrochloride (48·6% w/v) to 25 ml of aqueous undecane-3-sodium sulphate (10% w) [4].

conductivity which accompany the displacement of the intermicellar equilibrium were examined. No detailed work was done on the conductivity changes which occur with change of composition within the homogeneous lamellar liquid crystalline phase itself, or on the effect on its conductivity of processes likely to affect its mean orientation, e.g. flow or "recrystallization" following ageing after turbulent agitation. Further, in the systems employed in this series of investigations, the G phase was the only liquid crystalline phase encountered.

To emphasize the very general character of the conductivity changes, the results described for systems containing besides amphiphile (or mixture of amphiphiles) both hydrocarbon and water, may be supplemented by Figs. 6.11 and 6.12 which provide measurements for systems with respectively water only and hydrocarbon only as solvents.

In the former system (Fig. 6.11) the micellar changes are produced by the joint influences of the replacement of a more hydrophilic by a less hydrophilic cation (Table 16, Method Ic in ref. [21]),

$$C_{11}H_{23}{\cdot}SO_4Na + C_6H_{11}{\cdot}NH_3Cl \leftrightharpoons C_{11}H_{23}{\cdot}SO_4{\cdot}NH_3{\cdot}C_6H_{11} + NaCl$$

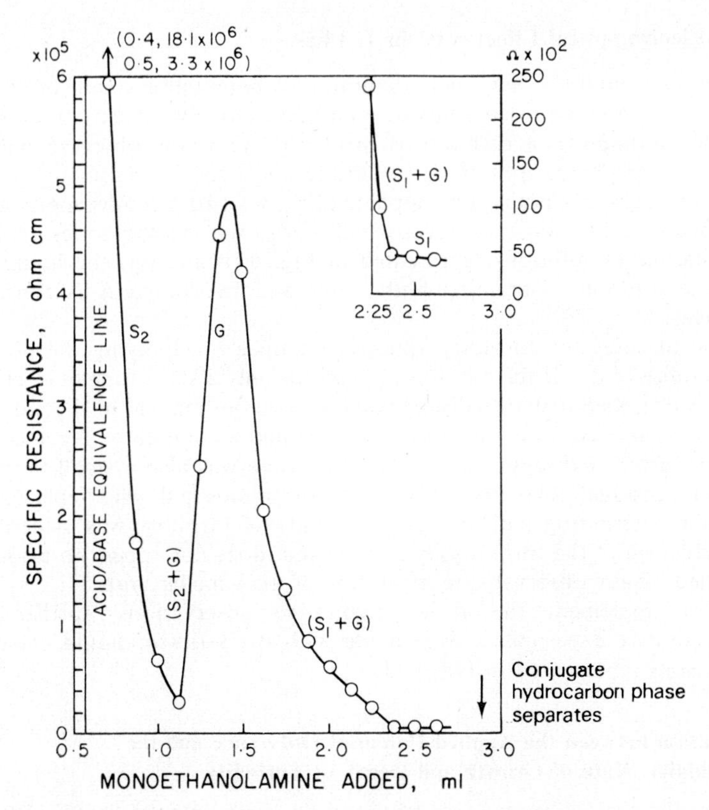

FIG. 6.12. Phase transitions and accompanying specific resistance changes at 20°C on adding monoethanolamine to 25 ml of a 9·90% w/v solution of oleic acid in cyclohexane.

and of the inorganic salt liberated (Table 16, Method Id in ref. [21]). The predominant micellar constitution of the S_2 solution must apparently consist of a dispersion of aqueous regions, $\bar{W}$, enclosed in a foam-like amphiphilic continuum, $\bar{C}$. The relatively high conductivity of this S_2 phase indicates that ionic migration remains facile in spite of the apparent presence of a continuous bimolecular layer of hydrocarbon groups which encloses the ionic groups in association with the aqueous region, $\bar{W}$. This probably reflects the labile character of the intermicellar equilibrium.

In the hydrocarbon system (Fig. 6.12) the micellar changes are produced principally by the conversion of the more lipophilic amphiphile, oleic acid, to the more hydrophilic amphiphile, monoethanolamine oleate (in ref. [21], Table 16, Method IIc). The excess of monoethanolamine also participates. Within the S_1 phase we have the condition complementary to that within the S_2 phase of Fig. 6.11. In the S_1 phase the hydrocarbon regions, $\bar{O}$, must apparently largely be enclosed in a foam-like amphiphilic continuum. The very appreciable conduction of this S_1 phase indicates that ionic migration along this continuum can occur quite readily.

An Electro-optical Effect with the G Phase

It was found [5] that when a continuous potential gradient of 20 to 40 V cm^{-1} is applied to an aqueous liquid crystalline G phase containing an ionic amphiphile, a turbidity is produced. This was observed with the aqueous G phases containing (i) tetradecane-7-sodium sulphate (20% w), (ii) tetradecane-7-potassium sulphate (20% w), (iii) pentadecane-8-sodium sulphate (3·1% w), (iv) Aerosol OT (20% w), (v) the series of liquid crystalline G solutions represented in Fig. 6.10 and (vi) the liquid crystalline solutions containing both water and hydrocarbon as considered above.

In all cases the turbidity appeared at once on applying the potential difference (p.d.). If the p.d. was applied for only a short time (a fraction of a second), then turbidity disappeared at once on removal of the p.d. If the p.d. was applied for a longer time (e.g. for half a minute) on its removal the main turbidity disappeared at once but there was also a small more persistent residual turbidity. When the continuous p.d. was replaced by a similar alternating p.d. (50 cycles/second) no turbidity was evident. The production of the turbidity by a direct p.d. does not appear to be accompanied by any observable reorientation of the lamellar units.

To supplement the above qualitative observations, further more quantitative experiments were made with the series of liquid crystalline solutions represented in Fig. 6.11.

Relation between the Applied Potential Difference and the Turbidity. Rate of Growth and Decay of Turbidity

These relationships were investigated using the liquid crystalline G solution obtained from 100 ml of undecane-3-sodium sulphate solution (10% w) and 7 ml of cyclohexylammonium chloride solution (48·6%

w/total volume). To permit measurements of the turbitity produced a cell (Fig. 6.13) was constructed which could be inserted in an EEL absorptiometer (Evans Electroselenium Ltd). The approximate constant of this cell was 28 and its resistance when charged with the above solution was about 3000 ohms. Some results obtained are shown in Fig. 6.14. The potential differences (from a high-tension battery) were applied for 30 s only, since this was sufficient to reach approximately maximum turbidity. Longer application produced undesirable warming of the solution.

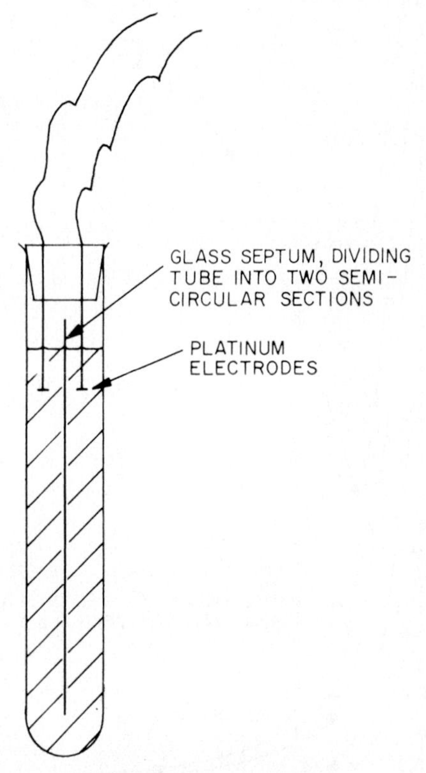

FIG. 6.13. Cell for observation of turbidity; for use in conjunction with EEL absorptiometer [5].

The rate of growth of turbidity was too rapid for measurement. It was however, by no means instantaneous. This was shown by reversing the applied potential difference and varying the frequency of reversal.

With an applied p.d. of ± 48 V, frequencies above 100 cycles/second gave no measurable turbidity on the absorptiometer. With $+48$ V and 100 half-cycles/second a maximum absorptiometer deflection of 15 units was obtained, while with 48 V applied continuously a maximum deflection of 35 units was reached.

The electro-optical effect as observed in the systems described above is specific for the G phase and is not shown by the S_1 or S_2 phases. The

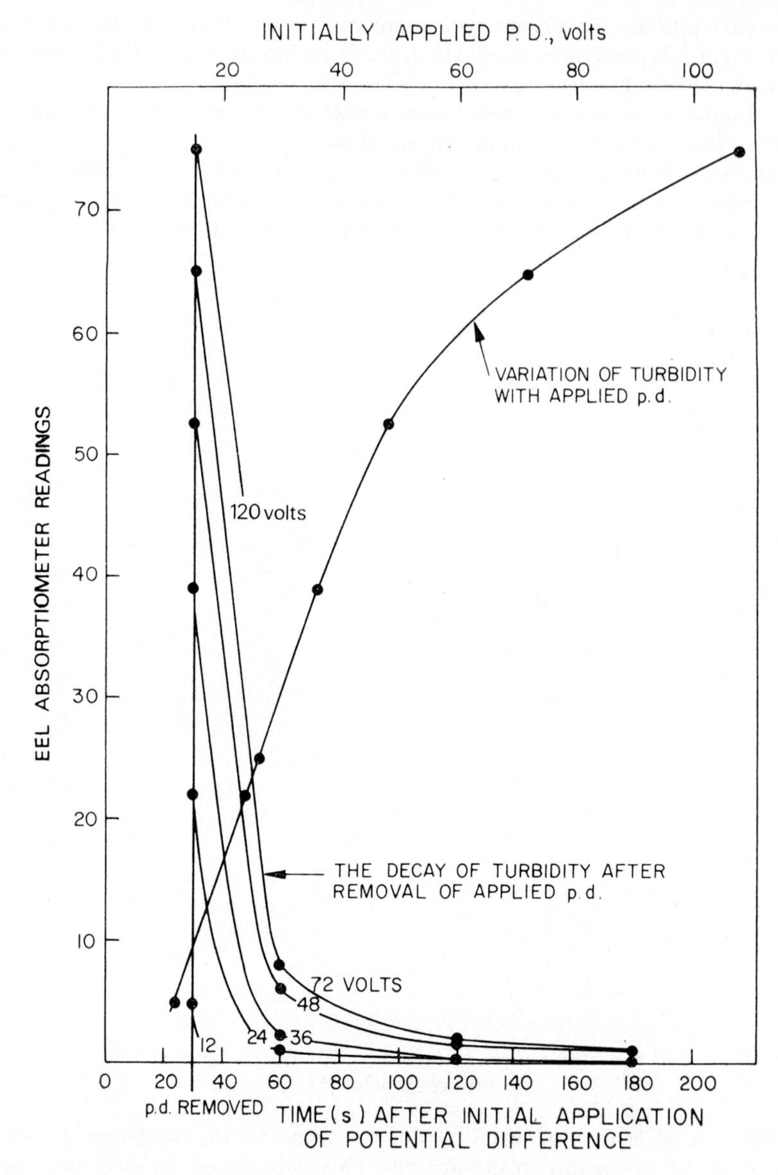

FIG. 6.14. Turbidity produced on application of potential differences (p.d.) to a liquid crystalline solution containing undecane-3-sodium sulphate (10% w, 100 ml), cyclohexylammonium chloride (48·6% w, total volume, 7 ml) [5].

mechanism for its production is uncertain. Suggestions made earlier [5] are open to doubt and are not considered here. An alternative possibility is that in a potential gradient the force fields around individual amphiphile molecules become unbalanced so that the condition for stability of the G phase, namely $R = 1$ in all directions [21], is destroyed. Local limited precipitation, for example, of an M_1 phase, might thus be produced and occasion turbidity. Whether or not comparable turbidity effects may be observable with other amphiphilic liquid crystalline phases, for example, the M_1, V_1, V_2 or M_2 phases has not been extensively investigated. However Tachibana and Tanaka [20] record that in the systems I, sodium-caprylate/caprylic acid (or decanol-1)/water; II, potassium caprylate/decanol-1/water and III, cetyltrimethylammonium bromide/hexanol-1/water (for phase diagrams cf. [21] Chap 5), with the phases B (= G), C, D (= G), E (= M_1) and F (= M_2), the electro-optical effect was recorded only with phase B (systems I and II), phase C (system II) and phase D (system III).

Conductivity Measurements with Systems Forming Additional Mesophases

(1) Soap Water Systems

An extensive series of conductivity measurements with soap/water mesophases was carried out in recent years by François *et al.* [7–13], the results being correlated with the dimensional characteristics of these mesophases as inferred from X-ray diffraction measurements (cf. Chap. 4).

The curves in Figs. 6.15 and 6.16 were obtained by François and Skoulios [7], who developed a technique which ensured both homogeneity of the sample and freedom from air bubbles under the conditions of the measurements. Both these aspects present considerable experimental difficulties with these highly concentrated viscous systems.

It will be noted from Fig. 6.15, that although for the amorphous solution phase, S_1 (aa′), the specific conductivity increases rectilinearly with concentration, in each of the mesomorphous phases, M_1 (bc), intermediate (de) and G (fg) the conductivity is insensitive to concentration. It varies markedly with the phase ratio in the composition ranges where two conjugate mesophases are present. In Curve I the break at a′ along aa′b was attributed by François and Skoulios as probably due to the transition from spherical to cylindrical micelles within the S_1 phase which they considered to be represented over the whole distance ab of the curve. It seems more likely, however, that a′ corresponds to the point of initial separation of the M_1 phase and that a′b corresponds to the conductivity within the two-phase region ($S_1 + M_1$), which otherwise would not be represented. This question is of interest in relation to the continuous character of the conductance/composition curves for the S phases noted in the preceding section.

In Fig. 6.16 the very low conductivities of the "gel" phase, in comparison with the higher conductivity of the M_1 phase (and of the G phase in Fig. 6.15) are striking and, on the basis of the "dissolved" character attributed to the polar groups in this semi-crystalline "gel" mesophase,* are surprising. It appears from the conductivity results that the potassium

* Cf. Gray and Winsor, *Liquid Crystals and Plastic Crystals, Vol. 1*, Chapter 5, Ellis Horwood, Chichester (1974).

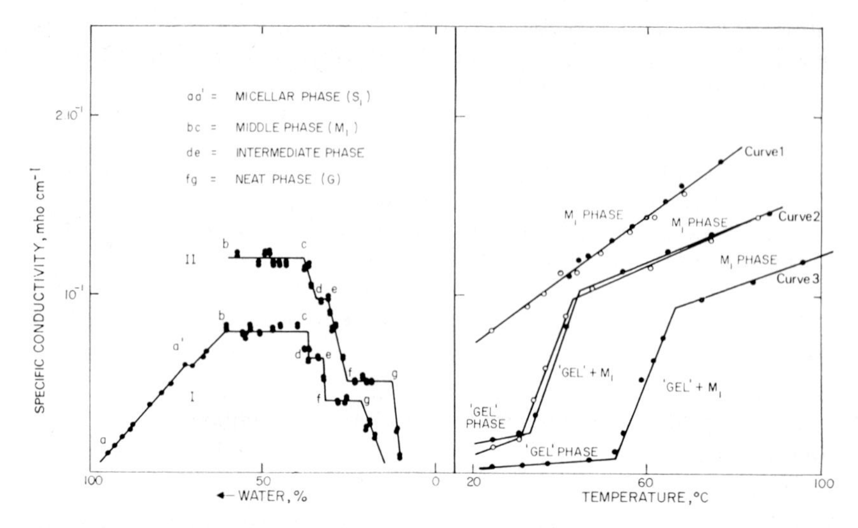

FIG. 6.15. Variation with concentration in the specific conductivity at 50°C of the sodium lauryl sulphate/water (Curve I) and potassium laurate/water (Curve II) systems [7].

FIG. 6.16. Variation with temperature in the specific conductivity of some soap/water systems [7].

	% w water
Curve 1 (potassium laurate)	● 49·2
	○ 47·5
Curve 2 (potassium palmitate)	● 51·3
	○ 47·8
Curve 3 (potassium stearate)	● 58·7

ions of the soap are considerably less mobile in the "gel" than in the M_1 and G phases. In this connection whereas the conductivity of the "gel" mesophase was very sensitive to the presence of small amounts of ionic impurities, this was not the case with the "fused" mesophases, M_1, intermediate and G. These observations apparently imply that the lower conductivity of the pure "gel" phase is due to a lower concentration of mobile potassium ions rather than to any particular obstacle to their passage which might be presented by the "crystalline" hydrocarbon regions of the "gel" phase in comparison with the "liquid" hydrocarbon regions of the fused mesophases (cf. Vol. I, Ch. 5).

These conclusions agree with the results of auto-diffusion measurements by François and Varoqui [8] with the mesophases formed by the caesium soaps using radioactive caesium as tracer. For any particular system the diffusion coefficient for the caesium ion and the electrical conductivity show closely parallel variations with temperature (cf. also ref. [12]).

In studying the conductivity of the G phases in soap/water systems François and Skoulios [9] found that the measured conductivity was dependent on the details of the method of introduction of the phase into the cell and decreased with time, the initial measurement being made shortly after filling the cell. The rate of decrease is greater the higher the temperature (Fig. 6.17). This is probably due to the sample tending gradually to assume a homeotropic arrangement between the electrodes in which the lamellae lie parallel to the electrodes [6]. Locally the particles of the G phase as originally introduced into the cell will tend, like the particles in a crystalline powder or precipitate, to lie oriented in many directions. On standing recrystallization of the particles of the G phase will occur, more rapidly at higher temperatures on account of the higher

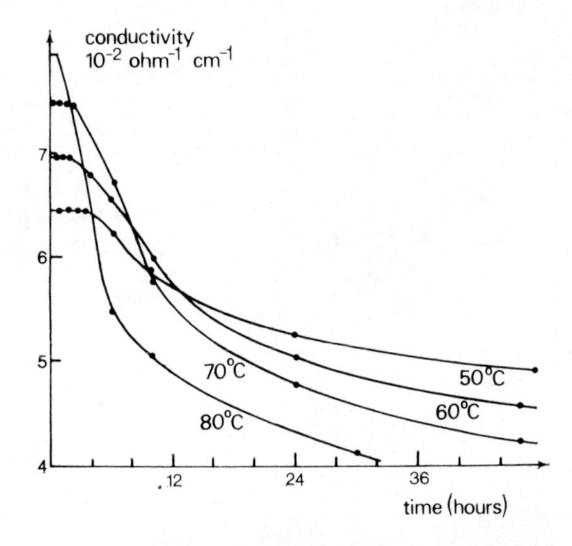

FIG. 6.17. Variation with time in the measured conductivity of the lamellar G phase in the potassium myristate/water system containing 74·8% w amphiphile [9].

mobility of the system, so that the homeotropic condition arising from surface effects close to the electrodes will gradually spread outwards. The sample thus gradually assumes as the preferred orientation between the electrodes the one which offers maximum resistance to the passage of current.

Although the measured electrical conductivity of the G phase is thus found to depend on time and on the method of introducing the sample, this is not found with the M_1 phase (middle phase) in which the conductivity of a uniformly oriented region would be less strongly dependent on orientation as already discussed. With this phase therefore reproducible measurements of the conductivity, probably representing the mean conductivity corresponding to random orientations within the sample, are obtained [9, 10].

Values obtained by François [10] with the M_1 phases of soaps of sodium, potassium, caesium and rubidium are shown in Fig. 6.18 in which the specific conductivity is plotted against N_i, the number of gram ions per litre of aqueous region (metal ions + water), i.e.

$$N_i = \frac{C_a}{(C_i + C_w)M_a} \times 1000 \times d_i$$

where C_a = wt concentration of amphiphile
C_i = wt concentration of metal ions
C_w = wt concentration of water
M_a = molecular wt of amphiphile, and
d_i = the density of the aqueous region calculated from literature values

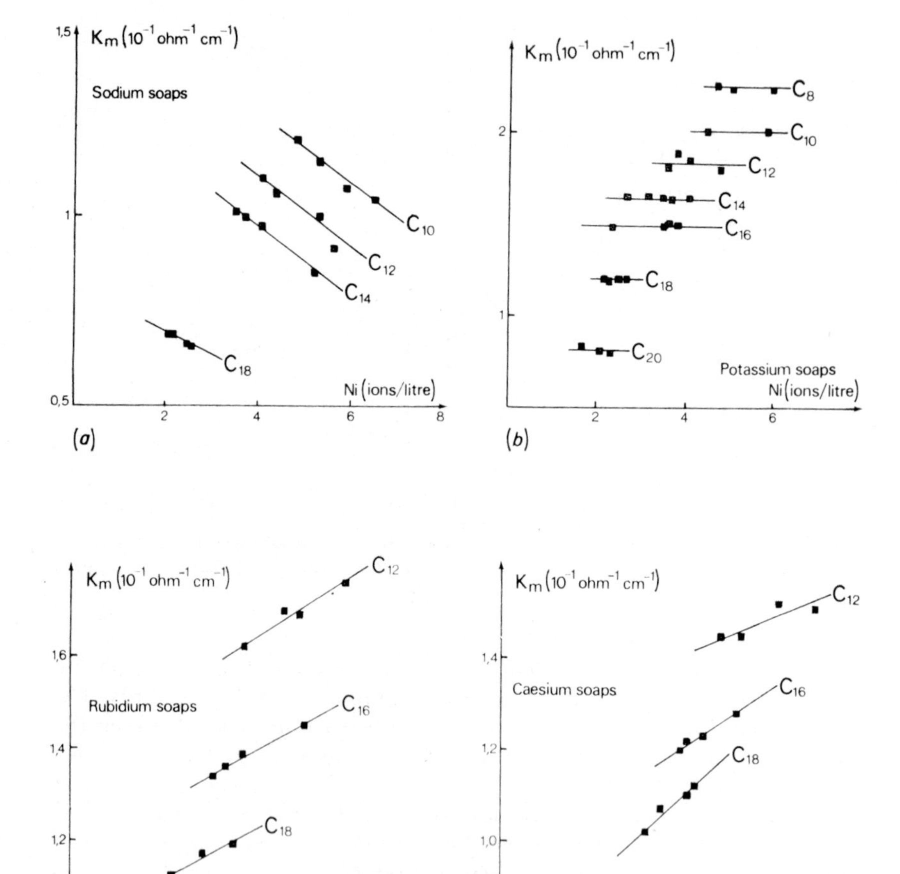

FIG. 6.18. Variation of the specific conductivity, K_m, at 86°C, with the ionic concentration of the aqueous zones in the M_1 phases in some soap/water systems [10]. (Ni should read N_i in (*a*) (*b*) (*c*) (*d*) of this Fig.).

It is clear from Fig. 6.18 that for a given ion at a given ionic concentration the conductivity diminishes with increasing chain length. Further, for any soap, on increase of N_i the specific conductivity falls for the sodium soaps, remains almost constant for the potassium soaps and increases for the soaps of rubidium and caesium.

In attempting to account quantitatively for these results François [13] supposed that the anionic parts of the soap molecules possess virtually no mobility and that only the aqueous region (water + counter ions) participates in electrical conduction. Since this occupies only a very small proportion of the total volume of the phase its real conductivity K_r is not equal to the measured conductivity K_m.

On this view one can introduce a steric factor f, greater than 1, defined by the relation

$$K_r = f K_m$$

The problem is then to determine the factor permitting one to relate the measured to the real conductivity of the aqueous region taking account not only of the structural geometry and dimensions but also of the electrical interactions between the counter ions and the stationary charges of the soap molecules. For details of the solution proposed the reader is referred to the original paper but it is noted that, for the M_1 phase, a reasonably

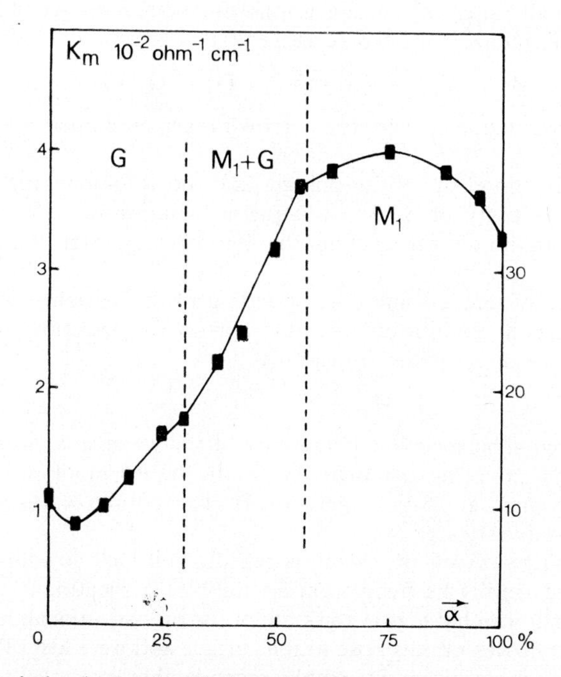

FIG. 6.19. Variation in the electrical conductivity and mesomorphic character of an aqueous system containing the alternate copolymer of maleic acid and hexadecyl vinyl ether on progressive neutralization with NaOH (α = degree of neutralization) [11].

quantitative account is given both of the effects of variations in chain length and in the nature of the counter ions. Interesting tables are provided correlating the dimensional parameters of the numerous M_1 phases studied, as determined by X-ray diffraction measurements, with their observed electrical conductivities. In these phases the thickness of the aqueous regions between the surfaces of two cylinders is of the order of 12 Å. For example in the case of the sodium soaps this distance varies from about 8 Å for the most concentrated M_1 phases from sodium decanoate (60% w) to 20 Å units for the most dilute M_1 phases from sodium stearate (40% w). In comparison the diameter of the hydrated sodium ion is about 6·6 Å. In the lamellar phases the thickness of the water layers is even smaller (usually below 10 Å).

A study of the electrical conductivities of the mesophases which arise in aqueous systems containing the alternate copolymer of maleic acid and hexadecyl vinyl ether at different degrees of neutralization with sodium hydroxide was carried out by François, Varoqui and Schmitt [11]. The specific conductivities, K_m, measured for a concentration of polysoap of 50% w, are shown in Fig. 6.19 as a function of the degree of neutralization α.

(2) THE *N,N,N*-TRIMETHYLAMINODODECANOIMIDE/WATER SYSTEM

Over a small range of concentrations this system shows the following phase sequence with rising temperature [21].

$$M_1 \rightarrow (M_1 + V_1) \rightarrow V_1 \rightarrow (V_1 + G) \rightarrow G \rightarrow (G + S) \rightarrow S.$$

The two-phase regions cover very narrow ranges of temperature and composition.

Figure 6.20 shows the phase changes and the accompanying changes in specific conductivity of a 65% w aqueous solution of *N,N,N*-trimethylaminododecanoimide on cycling the temperature over the range 20–90°C [6].

The ranges of temperature corresponding to the existence of conjugate pairs of phases in equilibrium were too narrow for detection so that what appeared to be direct phase transitions,

$$M_1 \rightarrow V_1 \rightarrow G \rightarrow S,$$

were observed. The specific conductivity of the water used was 5·2 μmho cm^{-1} at 20°C and the conductivity of the liquid imide was very low (0·45 μmho cm^{-1} at 25°C, supercooled). The nature of the conducting ions was not investigated.

The conductivity cell was a miniature dip cell (cell constant 1·5). The sample could readily be freed from air bubbles by heating it initially to a temperature at which it was in the state of the mobile amorphous solution, S, when air bubbles rapidly rose to the surface and were lost. The conductivity measurements were reversibly reproducible over the temperature ranges of the M_1, V_1 and S phases but, on account of supercooling and orientation effects (cf. below), were treatment-dependent in the case of the G phase.

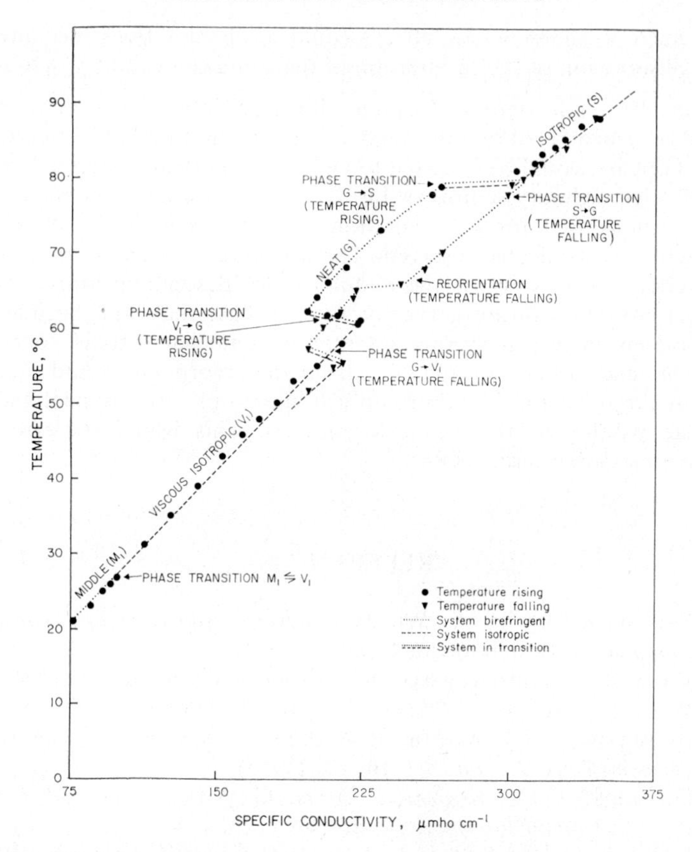

FIG. 6.20. Phase changes and accompanying changes in electrical conductivity shown by an aqueous solution of *N,N,N*-trimethylaminododecanoimide (65% w) with changes in temperature [6].

Noteworthy features of the conductivity measurements are

(i) When the general effect of temperature in increasing ionic conductivity is allowed for, there is no major difference in the order of the conductivities shown by the phases M_1, V_1, G and S.

(ii) In spite of marked differences in the consistencies of the phases, the conductivities of the M_1, V_1 and S phases at different temperatures, within the limits of the experimental accuracy, all lie on a single curve and are reversible with temperature. The curve for the G phase, however, is slightly displaced towards lower conductivities and shows marked hysteresis effects. The supercooling of the G phase over about 5°C, so that the system remains in the G condition when the V_1 phase is the stable form, is remarkable. At the termination of the supercooling the translucent birefringent G phase transformed rapidly into the glass-clear, optically isotropic V_1 phase, growth spreading outwards from some centre of initiation.

Such a supercooling effect would probably itself preclude the observation of the equilibrium of the conjugate G and V_1 phases.

When the conductivities observed with the G phases are considered it should be remembered that the conductivity of a uniformly oriented region of the G phase would be expected to be less across than along the lamellae (p. 122). An ordinary sample of the G phase will usually be made up of many regions in various orientations and the average of directional conductivities would be observed. The accentuated decrease in the electrical conductivity towards the middle of the descending branch of the conductivity/temperature curve for the G phase may well be due to a spontaneous change in average orientation. Lamellar regions, formed in many orientations on precipitation from the amorphous liquid phase S, may, at the inflexion, be taking up a homeotropic arrangement with the lamellae parallel to the electrode surfaces. This would increase their effective measured resistance.

REFERENCES

[1] For review, cf. P. A. WINSOR, *Solvent Properties of Amphiphilic Compounds*, Butterworths, London (1954).
[2] VOLD, R. D. and HELDMAN, M. J. *J. phys. Chem.* **52,** 148 (1948).
[3] WINSOR, P. A. *Trans. Faraday Soc.* **46,** 762 (1950).
[4] BROMILOW, J. and WINSOR, P. A. *J. phys. Chem.* **57,** 889 (1953).
[5] WINSOR, P. A. *J. Coll. Sci.* **10,** 101 (1955).
[6] GILCHRIST, C. A., ROGERS, J., STEEL, G., VAAL, E. G. and WINSOR, P. A. *J. Colloid Interface Sci.* **25,** 409 (1967).
[7] FRANÇOIS, J. and SKOULIOS, A. *Kolloid Z. und Z. Polymere* **219,** 144 (1967).
[8] FRANÇOIS, J. and VAROQUI, R. *C. r. hebd. Séanc. Acad. Sci., Paris,* Série C, **267,** 517 (1968).
[9] FRANÇOIS, J. and SKOULIOS, A. *C. r. hebd. Séanc. Acad. Sci., Paris* **269,** 61 (1969).
[10] FRANÇOIS, J. *J. Phys. Radium* **30,** C4–89 (1969).
[11] FRANÇOIS, J., VAROQUI, R. and SCHMITT, A. *C. r. hebd. Séanc. Acad. Sci., Paris,* Série C, **270,** 789 (1970).
[12] FRANÇOIS, J., *C. r. hebd. Séanc. Acad. Sci., Paris,* Série C, **272,** 876 (1971).
[13] FRANÇOIS, J., *Kolloid Z. und Z. Polymere* **246,** 606 (1971).
[14] HECKMANN, K. and GÖTZ, K. G. *Discuss. Faraday Soc.* **25,** 71 (1958).
[15] GÖTZ, K. G. and HECKMANN, K. *J. Coll. Sci.* **13,** 266 (1958).
[16] GÖTZ, K. G. *J. Coll. Sci.* **20,** 289 (1965).
[17] MCBAIN, J. W. and LEE, W. W. *Oil and Soap,* **20,** 17 (1943).
[18] DOSCHER, T. M. and VOLD, R. D. *J. Phys. Colloid Chem.* **52,** 98 (1948).
[19] UBBELOHDE, A. R., MICHELS, H. J. and DURUZ, J. J. *Nature* **228,** 50 (1970).

[20] Tachibana, T. and Tanaka, M. *Bull. Chem. Soc. Japan* **44,** 1166 (1971).

[21] *Liquid Crystals and Plastic Crystals Vol 1*, Chap. 5 (edited by G. W. Gray and P. A. Winsor), Ellis Horwood, Chichester (1974).

7

Magnetic Resonance Spectroscopy of Liquid Crystals—Non-Amphiphilic Systems

G. R. LUCKHURST

1. Introduction

This chapter is concerned with magnetic resonance spectroscopy and its application to the study of the liquid crystalline phase. Although this branch of spectroscopy has been widely applied in most areas of biology, chemistry and physics, its application to thermotropic liquid crystals (non-amphiphilic mesophases) is still somewhat novel. However, as we shall discover, magnetic resonance is able to probe the nature of the intermolecular interactions in a mesophase as well as to determine the state of its macroscopic alignment.

We begin, in section 2, by briefly reviewing the basic principles of nuclear magnetic resonance and electron resonance spectroscopy. These ideas must however be extended when applied to liquid crystals, and the extensions are considered, in depth, in this section. The remainder of the chapter then deals with the application of these concepts to real problems. Thus the determination of the orientational order of the nematic mesophase and that of a solute dissolved in the mesophase is described in the third section. In many magnetic resonance experiments the macroscopic alignment of the mesophase is controlled by the magnetic field present in the spectrometer. For the nematic mesophase the director is therefore aligned parallel to the magnetic field. This direction can be changed by the application of an electric field, as we see in section 4, or more unusually by spinning the sample; cf. section 5. Because of its high viscosity, special techniques are required when investigating the smectic phase. These techniques are described in section 6 together with some of the results available from such studies. Within the volume of a cholesteric liquid crystal the director cannot be uniformly oriented with respect to any space fixed axis. The complications resulting from this behaviour are discussed in section 7 together with an investigation of the perturbing effect of a magnetic field on the helical structure.

Magnetic resonance spectroscopy can also be used to study dynamic as well as static interactions [1–3]. However, in this chapter we shall be concerned only with the static properties, because magnetic resonance studies have yet to advance our understanding of dynamic processes within a liquid crystal.

2. Magnetic Resonance Spectroscopy

Nuclear magnetic resonance and electron resonance have much in common. However, because nuclear magnetic resonance is the more familiar of the two techniques, we shall start with a description of this. We shall then be able to describe electron resonance by analogy with the ideas developed for nuclear magnetic resonance. For those readers who would like a more detailed account, the introductory text by Carrington and McLachlan [1] is recommended. There are, in addition, more advanced treatments of magnetic resonance spectroscopy by Slichter [2] and Abragam [3].

Nuclear Magnetic Resonance

Certain nuclei possess a characteristic property called spin. This is described by a number I which can be either integral or half integral. For example, protons have spin of one half, whereas deuterons have spin one. Those nuclei which possess a non-zero nuclear spin are collectively known as magnetic nuclei. A magnetic nucleus may exist in one of $2I + 1$ spin states each of which is labelled with a nuclear spin quantum number m taking values from I to $-I$. Because nuclei are also charged, they possess a magnetic moment whose magnitude depends on the value of the nuclear spin quantum number. In many ways the magnetic moment is analogous to an electric dipole moment and consequently the magnetic moment interacts with a magnetic field to change the energy of a spin state. This energy is linear in the magnetic field B

$$E(m) = -\gamma_n \hbar m B, \tag{7.1}$$

where γ_n is the magnetogyric ratio for the particular nucleus and $\hbar$ is Planck's constant divided by 2π. The application of a magnetic field therefore removes the spin degeneracy and this situation is illustrated in Fig. 7.1 for nuclei of spin $\frac{1}{2}$ and 1. Transitions may now be induced between these nuclear spin levels by applying an oscillating magnetic field orthogonal to the static field. The selection rule requires that the nuclear spin quantum number changes only by ± 1 during the transition and so the frequency ν of the oscillating field must be

$$\nu = \frac{\gamma_n B}{2\pi} \tag{7.2}$$

Unlike most branches of spectroscopy, the resonance condition may therefore be satisfied by varying either the frequency ν or the magnetic field B. For wide-line nuclear magnetic resonance and electron resonance

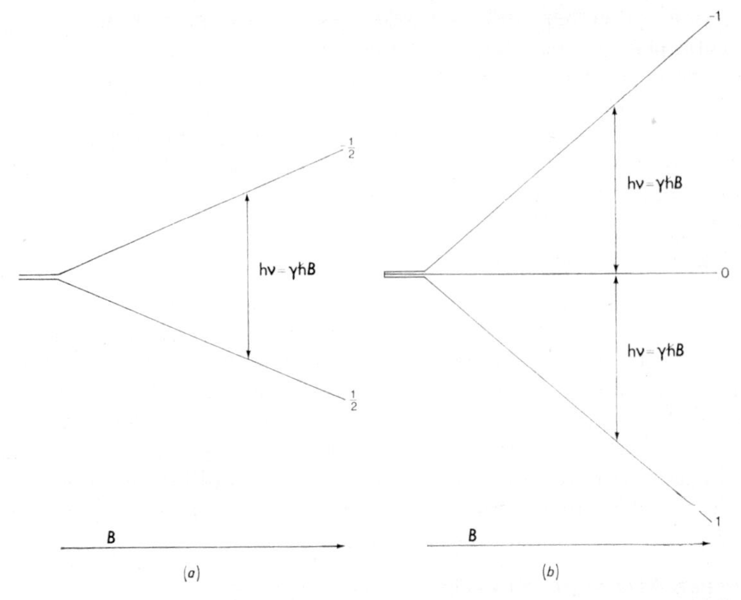

FIG. 7.1. The removal of the spin degeneracy by a magnetic field for nuclei with (a) $I = \frac{1}{2}$ and (b) $I = 1$; the allowed nuclear magnetic resonance transitions are also indicated.

spectroscopy it is more convenient to scan the magnetic field. This form of the spectrum is therefore independent of the nuclear spin, because the $2I$ allowed transitions are degenerate, as we can see from Fig. 7.1, and the spectrum should contain a single line.

In practice, a spectrum rarely contains just a single line, and there are a number of magnetic interactions which may be responsible for the extra lines. For example, if the molecule contains several magnetic nuclei then their magnetic moments can couple in an analogous manner to the interaction between two bar magnets. The effect of this dipolar coupling between the two nuclear spins may be understood in the following way. Consider a molecule possessing just two protons i and j with the magnetic field applied along the inter-proton vector. In the absence of the dipolar coupling, both nuclei would absorb energy at the same frequency, ν_0 say, and the spectrum would contain a single line. However, because of the dipolar coupling, the field experienced by nucleus i is not B but $B + \Delta B$ or $B - \Delta B$ depending on the spin state of nucleus j. The two protons now come into resonance when the frequency of the oscillating magnetic field is $\nu_0 \pm \Delta\nu/2$ where

$$\nu_0 \pm \frac{\Delta\nu}{2} = \frac{\gamma_H}{2\pi}(B \pm \Delta B). \tag{7.3}$$

The dipolar coupling therefore splits the single line into a doublet with a separation $\Delta\nu$ and centred on the original frequency ν_0. The strength of the dipolar coupling depends entirely on the inter-proton separation r, although the spacing between the spectral lines is also dependent on the angle made by the inter-proton vector with the magnetic field. When the field is parallel to the inter-proton vector, say the z-axis, then the splitting may take the value D_{zz}, whereas along the x- or y-axes the splittings will be D_{xx} or D_{yy}. Because of the cylindrical symmetry about the inter-proton vector, the splittings D_{xx} and D_{yy} are identical and in fact equal to $-D_{zz}/2$. The values may be arranged in a square array whose elements correspond to the dipolar splitting for a particular pair of axes. In the xyz co-ordinate system, only the diagonal elements of this array are non-zero:

$$\begin{array}{c|ccc|} & x & y & z \\ x & -D_{zz}/2 & & \\ y & & -D_{zz}/2 & \\ z & & & D_{zz} \end{array}, \tag{7.4}$$

where

$$D_{zz} = -\frac{3\gamma_H^2\hbar}{2\pi r^3}. \tag{7.5}$$

The dipolar splitting is not just a single number but is described by a tensor which may be thought of as a higher-order vector [4]. The co-ordinate system in which the tensor contains only diagonal elements is known as the principal co-ordinate system. The magnitude of an element D'_{ab} in another co-ordinate system is obtained *via* the transformation

$$D'_{ab} = \sum_{\alpha,\beta} l_{a\alpha}\, l_{b\beta}\, D_{\alpha\beta}, \tag{7.6.}$$

where α and β take the values x, y or z. Here $l_{a\alpha}$ is the cosine of the angle between axes a and α; it is usually called the direction cosine. If the magnetic field makes some arbitrary angle with the inter-proton vector, then the dipolar splitting is equal to the component of the tensor D resolved along the field direction. For example, if the field lies along the a-axis then according to equation (7.6) D'_{aa} is related to the principal components of D by

$$D'_{aa} \quad D_{zz}\{l_{az}^2 - \tfrac{1}{2}(l_{ax}^2 + l_{ay}^2)\}. \tag{7.7}$$

However, these particular direction cosines are related [4]:

$$l_{az}^2 + l_{ax}^2 + l_{ay}^2 = 1, \tag{7.8}$$

and so

$$D'_{aa} = D_{zz}\frac{(3l_{az}^2 - 1)}{2}. \tag{7.9}$$

Consequently if this hypothetical molecule was incorporated in a single crystal, the spacing between the spectral lines would vary according to equation (7.9). The maximum line separation is D_{zz} and the minimum is $-D_{zz}/2$, although the sign of the splitting is not available from the spectrum. In addition the dipolar splitting is predicted and observed to vanish when the direction cosine is $1/\sqrt{3}$, that is when the angle between the inter-proton vector and the magnetic field is 54° 46′.

Frequently it is impossible to grow single crystals and so the species must be studied as a powder or polycrystalline sample. In such samples the inter-proton vector will make all possible angles with the magnetic field.

Since each orientation corresponds to a different dipolar splitting the total spectrum will then be a sum of these spectra. In general the form of the spectrum is given by

$$h(B) = \sum_r \int L(B_r, B, T_2^{-1}) f(\omega) \, d\omega, \tag{7.10}$$

where the spectrum is measured by scanning the magnetic field. Here B_r is a resonance field which is a function of the molecular orientation. The shape of the spectral line $L(B_r, B, T_2^{-1})$ is usually taken to be either Lorentzian

$$L(B_r, B, T_2^{-1}) = \frac{T_2}{\pi} \frac{1}{1 + T_2(B - B_r)^2}, \tag{7.11}$$

or Gaussian

$$L(B_r, B, T_2^{-1}) = \frac{T_2}{2\pi} \exp\left\{-\frac{T_2^2}{2}(B - B_r)^2\right\}, \tag{7.12}$$

where T_2^{-1} is a measure of the line width. The weighting function $f(\omega)$ in the integrand gives the probability of finding the orientation between ω and $\omega + d\omega$. When the molecular distribution is uniform, as in a polycrystalline sample formed by freezing conventional fluids,

$$f(\omega) \, d\omega = \sin\theta \, d\theta \, d\phi, \tag{7.13}$$

where θ and ϕ are the spherical polar co-ordinates of the magnetic field vector in a molecular co-ordinate system.

The form of the polycrystalline spectrum corresponding to such a distribution of molecules containing pairs of protons is shown in Fig. 7.2. Since the maximum splitting is D_{zz} the spectrum extends over this range. In addition the magnetic field is more likely to be orthogonal to the inter-proton vector than parallel to it. Consequently the spectrum will be more intense in the central region than in the wings, as the spectrum shows. It is more usual to record the derivative of the absorption spectrum in electron resonance and wide-line nuclear magnetic resonance spectroscopy. The first derivative of the polycrystalline absorption curve is also shown in Fig. 7.2. This demonstrates an advantage of this form of display, for the spectrum apparently contains just two pairs of doublets centred on the same field. The spacing of the less intense pair is D_{zz} and clearly comes from molecules with their inter-proton vector parallel to the magnetic field. The more intense pair of lines has a spacing $D_{zz}/2$ and originates from molecules with their inter-proton vector orthogonal to the magnetic field.

As the solid sample is heated, the rate of molecular motion increases and we must now investigate the effect of this motion on the polycrystalline spectrum. Let us begin however by considering a simpler example in which the resonant frequency for a nucleus may take one of just two values, ω_A and ω_B, with equal probabilities. These two frequencies might be caused by a difference in environment which modifies the magnetogyric ratio. Clearly in the absence of exchange between these two environments, the nuclear magnetic resonance spectrum will contain two lines separated

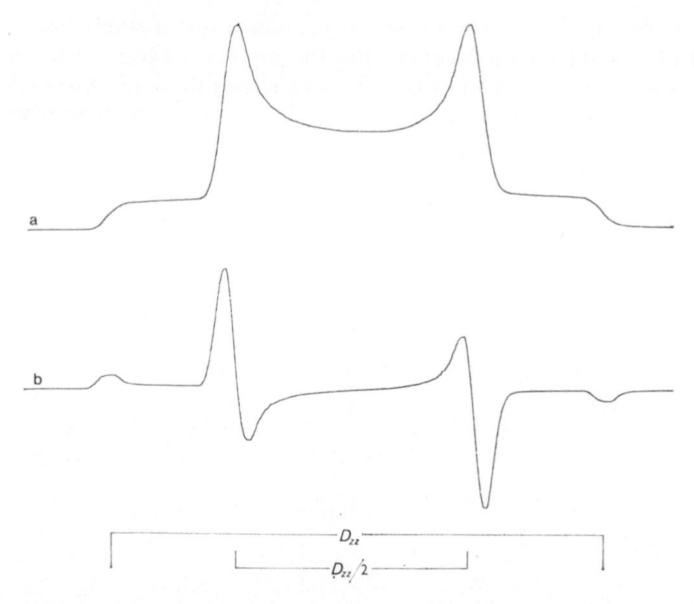

FIG. 7.2. The wide-line proton magnetic resonance spectrum of a polycrystalline sample in which the molecules contain two protons. Spectrum (b) is the first derivative of the absorption curve (a).

by $\omega_A - \omega_B$ or $\Delta\omega$. This situation is known as the slow exchange limit and is observed so long as the life time τ in a given environment satisfies the condition $\Delta\omega^2\tau^2 \gg 1$. At the other extreme, when $\Delta\omega^2\tau^2 \ll 1$, a nucleus will experience both frequencies many times during the time scale of the experiment and so sees their average. The spectrum will therefore contain a single line at the mean frequency in the fast exchange limit. The form of the spectrum for intermediate rates of exchange is complicated and we shall not be concerned with the difficulties encountered in analysis here.

When the different environments correspond to different molecular orientations, the life time τ is equated with the rotational correlation time τ_R. Thus for a dipolar interaction the resonant frequency changes from D_{zz} to $-D_{zz}/2$ when the molecule rotates through $\pi/2$ radians. Consequently the rotational correlation time must satisfy the inequality

$$\left(\frac{3D_{zz}}{2}\frac{2\tau_R}{\pi}\right)^2 \ll 1. \tag{7.14}$$

The factor of $2/\pi$ is introduced because τ_R is strictly the time required to rotate through one radian. Since dipolar splittings are typically 10^5 Hz the rotational correlation time required to average the splitting is less than 10^{-6} s. In the fast exchange limit, the observed dipolar splitting will be the time or ensemble average of the splitting given by equation (7.9). However, the average of l_{az}^2 when all orientations are equally probable is just 1/3 and so in an isotropic fluid $\Delta\omega$ vanishes on average. The nuclear magnetic resonance spectrum therefore contains just a single line and is devoid of any information concerning the dipolar splitting. This loss of information is emphasized in Fig. 7.3 which shows the single line expected from a low viscosity fluid and the structured spectrum obtained when the

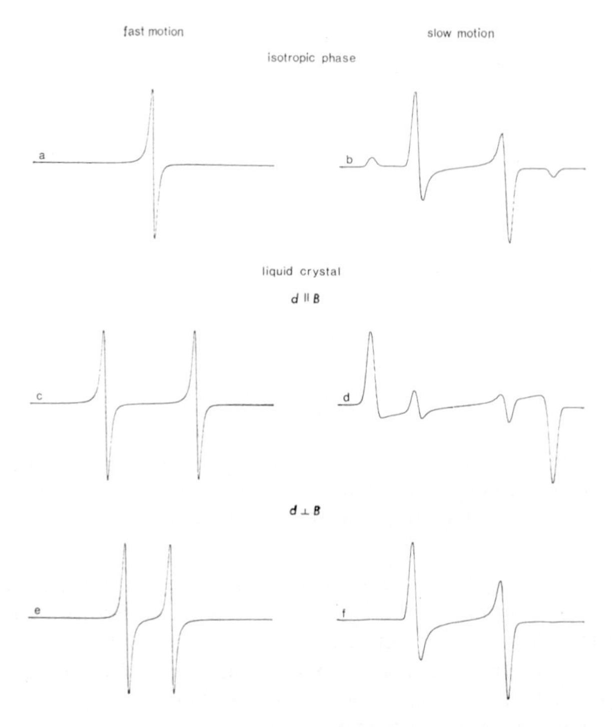

7.3. A selection of idealized proton magnetic resonance spectra for the nematic and isotropic phases of a nematogen.

molecular reorientation is quenched. We shall now see how the sensitivity of the dipolar splitting, both to the rate and nature of the reorientation process, can be exploited to investigate the structure of liquid crystals.

Nematic Liquid Crystals

The liquid crystalline state is characterized by the long-range orientational order which often extends over many thousands of molecules. At any point within a mesophase the long axes of the molecules tend to be parallel to a given direction which is called the director. In a suitably aligned state, the mesophase of a nematogen is found to be uniaxial, and so the system is cylindrically symmetric with respect to the director. The orientation of the long axes fluctuate with respect to the director and so the orientational order within the mesophase is not complete. If the constituent molecules are assumed to be cylindrically symmetric then this order may be defined in terms of the single angle between the molecular long axis and the director. The function which is commonly employed is

$$S = \frac{\overline{(3l_{az}^2 - 1)}}{2}, \tag{7.15}$$

where l_{az} is the direction cosine between the director (z-axis) and the molecular symmetry axis (a-axis) [5]. The bar denotes an average which may be over either time or the ensemble of molecules. Although the microscopic order S is high, the macroscopic order in an unperturbed mesophase is zero since the director is free to adopt all orientations with equal probability. However, a macroscopically oriented sample can be obtained by placing the nematic mesophase in a magnetic field which, because of the anisotropy in the diamagnetic susceptibility, aligns the director parallel to the field. A magnetic field is an inherent part of all nuclear magnetic resonance spectrometers and so we must now consider the effect of the resulting macroscopic alignment on the form of the spectrum of the nematic mesophase.

One of the simplest situations would occur if the molecules contained, amongst others (!) two protons with their inter-nuclear vector parallel to the long axis of the molecule. For many compounds the viscosity of the nematic mesophase is low (~2·5 cP) and so the rate of molecular re-orientation with respect to the director might be expected to be fast. The dipolar splitting in the spectrum is therefore obtained by taking the appropriate time or ensemble average of equation (7.9)

$$\begin{aligned} \overline{\Delta\nu} &= D_{zz}\frac{\overline{(3l_{az}^2 - 1)}}{2}, \\ &= D_{zz}S. \end{aligned} \tag{7.16}$$

Of course, in the isotropic phase the orientational order S is zero and the dipolar splitting therefore vanishes as we discovered earlier. However, in the mesophase, the order is not zero and so the spectrum will contain a pair of lines. The contrast between this spectrum and that from the isotropic phase is shown in Fig. 7.3. Since D_{zz} may be calculated from the molecular geometry the magnitude of the dipolar splitting $\overline{\Delta\nu}$ can be used to measure the orientational order in the mesophase.

In addition to the magnetic field, a variety of constraints may be used to align the director in a nematic mesophase. As we shall now see, the angle which the director makes with the magnetic field is also reflected in the dipolar splitting. The co-ordinate system employed in our calculation is shown in Fig. 7.4. The director lies along the z-axis and makes an angle γ with the magnetic field $\boldsymbol{B}$. The x-axis is then defined to lie in the plane made by the director and $\boldsymbol{B}$; the y-axis is orthogonal to the x- and z-axes. The inter-proton vector a has spherical polar angles θ and ϕ in this co-ordinate system. The partially averaged dipolar splitting is still given by equation (7.16) in which l_{az} is equal to cos α where α is the angle between a and $\boldsymbol{B}$:

$$\overline{\Delta\nu} = D_{zz}\frac{\overline{(3\cos^2\alpha - 1)}}{2}. \tag{7.17}$$

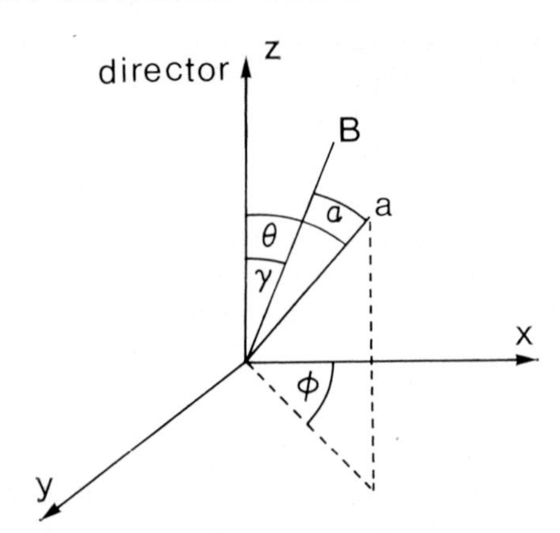

FIG. 7.4. A co-ordinate system for when the director is not parallel to the magnetic field.

The angle α is related to the other angles in the problem by

$$\cos\alpha = \cos\theta\cos\gamma + \sin\theta\sin\gamma\cos\phi, \tag{7.18}$$

and so the observed splitting is determined by the three averages:

$$\overline{\cos^2\alpha} = \overline{\cos^2\theta}\cos^2\gamma + \overline{2\cos\theta\sin\theta\cos\gamma\sin\gamma\cos\phi} + \overline{\sin^2\theta\sin^2\gamma\cos^2\phi}. \tag{7.19}$$

Because the system is cylindrically symmetric with respect to the director, the molecular angular distribution function is independent of ϕ. The averages over θ and ϕ may therefore be evaluated separately using

$$\overline{\cos\phi} = 0,$$

and

$$\overline{\cos^2\phi} = \tfrac{1}{2}. \tag{7.20}$$

These results, when substituted in equation (7.17) give the partially averaged dipolar splitting as

$$\overline{\Delta\nu}(\gamma) = D_{zz}\frac{\overline{(3\cos^2\theta - 1)}}{2}\frac{(3\cos^2\gamma - 1)}{2}. \tag{7.21}$$

The average over θ is just the orientational order S and so the angular dependence of the dipolar splitting is therefore

$$\overline{\Delta\nu}(\gamma) = D_{zz}S\frac{(3\cos^2\gamma - 1)}{2}. \tag{7.22}$$

Rotation of the director so that $\gamma = 90°$ therefore reduces the dipolar splitting to $-D_{zz}S/2$, one half of its original value; the appearance of the resulting spectrum is shown in Fig. 7.3. Clearly if the angle γ is known, then, in this particular example, no new information concerning the structure of the mesophase can be gained. However, as we shall see, there are situations where a knowledge of the angular dependence of other spectral frequencies is of importance. On the other hand, measurement of $\overline{\Delta\nu}$ can be used to determine the angle made by the director with the magnetic field. Observation of the splitting $\overline{\Delta\nu}$ in the absence of any secondary perturbation gives the product $D_{zz}S$. The angle γ adopted by the director when the second constraint is applied can be determined from the value of the new dipolar splitting. Finally comparison of the form of equation (7.22) with equation (7.9) shows that a molecule moving with respect to a cylindrically symmetric ordering potential behaves as a fictitious static species with the largest component of its dipolar tensor parallel to the director. In this chapter such partially averaged quantities will be distinguished by a tilde, for example

$$\tilde{D}_{zz} = D_{zz}S. \tag{7.23}$$

We now consider the form of the liquid crystal spectrum if the molecular motion has been quenched without destroying the ordered structure of the mesophase. As we have seen the spectrum is simply an appropriately weighted sum of spectra from all molecular orientations. If the director is parallel to the magnetic field then the greater number of molecules will have their inter-proton vectors parallel to the field. Consequently the spectral lines separated by D_{zz} will be more intense than in the spectrum observed for the disordered phase. However, this increase in intensity occurs at the expense of the lines separated by $D_{zz}/2$. These changes are illustrated by the spectrum in Fig. 7.3 which was calculated using the distribution function

$$f(\omega)\,\mathrm{d}\omega = \exp(\varepsilon\cos^2\theta)\sin\theta\,\mathrm{d}\theta\,\mathrm{d}\phi. \tag{7.24}$$

This function is consistent with the Maier–Saupe theory of the nematic mesophase [6] which will be discussed in section 3. In the calculations, the parameter ε was set equal to 3·485 since this corresponds to the S value of 0·5 used to simulate the spectrum in the limit of fast exchange.

Usually if the tube containing a nematic mesophase is suddenly rotated, with respect to an applied magnetic field, then the orientation of the director is also changed. However, this new orientation does not persist, for the director is immediately realigned parallel to the field by magnetic forces. When the molecular motion in the mesophase is quenched, this realignment cannot occur and it is possible to observe the spectrum as a function of the angle between the director and the magnetic field. Rotation of the sample does not change the angular distribution function of the molecules with respect to the director. However, the distribution function with respect to the magnetic field does change and it is this function which determines the form of the spectrum. We shall discuss the quantitative aspects of this angular dependence later. Qualitatively it is clear that when the magnetic field is orthogonal to the director there will be a greater probability of finding the field perpendicular rather than parallel to the inter-proton vector. Consequently the lines separated by $D_{zz}/2$ will be more intense than those separated by D_{zz}. This reversal of intensities is shown in Fig. 7.3; the spectra in this figure also demonstrate that when the molecular motion is quenched, the positions of the spectral lines are not angular dependent. However, the intensities are angular dependent and an analysis of this dependence gives the distribution function $f(\omega)$.

The last possibility which we shall consider in this section occurs if the director is not uniformly oriented; the rate of director motion is slow but the rate of molecular motion with respect to the director is fast. Under such conditions, the observed spectrum will be a sum of spectra from all orientations of the director. The spectrum is therefore equivalent, in form at least, to the polycrystalline spectra which have just been described. However, the molecular motion will have partially averaged the dipolar splitting tensor, and so the resonance fields must be calculated using the quantities appropriate for the fictitious static species. Thus

$$h(B) = \sum_r \int L(\tilde{B}_r, B, T_2^{-1}) f(\omega)\, \mathrm{d}\omega, \tag{7.24a}$$

where $f(\omega)\, \mathrm{d}\omega$ is now the probability of finding the director with an orientation between ω and $\omega + \mathrm{d}\omega$. Analysis of such a spectrum could therefore yield the distribution of the director in a mesophase when it is subject to a number of constraints.

Other Magnetic Interactions

In the previous section, we have been concerned with the effect of the dipolar splitting on the nuclear magnetic resonance spectrum of molecules in a nematic mesophase. In this section, we shall review briefly those other magnetic interactions which can influence the appearance of the spectrum.

The most important interaction is the Zeeman coupling between the magnetic field and the nuclear spin which removes the degeneracy of the spin levels. According to equation (7.2) the resonance frequency is determined by γ_n, the magnetogyric ratio, and this is related to the nuclear g factor g_n by

$$\gamma_n = \frac{g_n \beta_n}{h}, \tag{7.25}$$

where β_n is the nuclear Bohr magneton. Although β_n is a constant, the nuclear g factor is determined by the environment of the nucleus in the molecule. The variation in g_n is found to be rather small and so it is convenient to describe the changes in terms of the chemical shift σ which is defined by

$$g_n = g_n^0(1 - \sigma), \tag{7.26}$$

where g_n^0 is the g factor for the free atom. In general the molecular environment will be anisotropic and so we must expect σ to be orientation dependent. This is indeed the case, and σ is correctly represented by a second-rank tensor rather like the dipolar splitting. Since the chemical shift is only partially averaged for a molecule in the nematic mesophase, the resonance frequency will differ from its value in the isotropic phase. However, for pure liquid crystals, this effect is quite overshadowed by the dipolar splitting.

The dipolar interaction between a pair of nuclear spins is averaged to zero when the molecule tumbles in solution. There is, however, a scalar interaction between the spins which does not vanish, and this is called the spin–spin coupling. When the nuclei are magnetically equivalent, as in benzene, this interaction does not affect the nuclear magnetic resonance spectrum which contains a single line. If the equivalence is removed by replacing one hydrogen in benzene by chlorine, then the spin–spin coupling will split the single line into a complex multiplet structure. Fortunately such spin–spin effects are unimportant for pure liquid crystals, although their counterpart in electron resonance spectroscopy cannot be ignored.

The last interaction considered in this section is the quadrupole coupling between the quadrupole moment of a nucleus and the electric field gradient which it experiences. This interaction can only occur for nuclei with spin equal to or greater than one since nuclei with spin $\frac{1}{2}$ do not possess a quadrupole moment. To see the effect of the quadrupole interaction, consider a nitrogen nucleus, which has spin one; there are three possible spin states which are degenerate in the absence of a magnetic field and an electric field gradient. However, if the field gradient is not zero, then the quadrupole coupling removes the degeneracy of the spin states even in the absence of a magnetic field. Consequently the degeneracy of the two allowed transitions in a nuclear magnetic resonance experiment is also removed, and the spectrum consists of two lines with a separation whose magnitude depends on the strength of the quadrupole coupling. However, this coupling is averaged to zero when the molecule tumbles in an isotropic fluid, and so the spectrum contains a single line. In contrast, the spectrum from the nemantic mesophase will contain two lines, since the quadrupole coupling is not completely averaged. The separation between the lines is important since it is directly proportional to the degree of order S and so can be used to investigate the structure of the meophase.

Electron Resonance Spectroscopy

Liquid crystals are diamagnetic and hence cannot be studied directly by electron resonance spectroscopy. However, it has proved possible to employ the technique of doping the mesophase with trace quantities of a paramagnetic solute, commonly called a spin probe. We shall therefore discuss the various factors which influence the electron resonance spectrum of the spin probe when it is dissolved in a nematogen.

The electron, like the proton, has spin one half and so can exist in one of two spin states. The degeneracy of these states is removed by the application of a magnetic field and the energy separation is

$$\Delta E = \gamma_e \hbar B, \tag{7.27}$$

where γ_e is the electron magnetogyric ratio. In electron resonance spectroscopy it is conventional to replace γ_e by the g factor

$$\gamma_e = \frac{g\beta_e}{\hbar}, \tag{7.28}$$

where β_e is the electron Bohr magneton. Transitions can be induced between the two levels by an oscillating magnetic field provided the frequency ν is

$$\nu = \frac{g\beta_e B}{h}. \tag{7.29}$$

The electron resonance spectrum of a sample containing just unpaired electrons will consist of a single line.

This account of the basic electron resonance experiment demonstrates the close similarity to nuclear magnetic resonance spectroscopy and this similarity could be further emphasized by describing the electron resonance spectrum of a species containing two unpaired electrons. However, the number of such triplet states is small and so we shall consider a spin probe containing a single unpaired electron together with a single magnetic nucleus with spin I. Both particles possess magnetic moments which can therefore interact. This coupling is known as the hyperfine interaction and its effect is to split the single line into $(2I + 1)$ equally spaced components. The spacing between the lines is a measure of the strength of the electron–nuclear interaction and, like the coupling between nuclear spins, it is composed of an isotropic and an anisotropic part. Unlike the nuclear spin–spin coupling, the scalar hyperfine interaction can always be observed when the paramagnetic species tumbles in an isotropic solvent. Accordingly the number of lines in the electron resonance spectrum does not change on passing from the amorphous isotropic phase to the nematic mesophase. Instead the spacing between the lines changes and the magnitude of this change will now be described.

In general the total hyperfine interaction may be represented by a second-rank tensor with elements $A_{\alpha\beta}$ where α and β denote Cartesian axes set in the molecule. The hyperfine tensor, like that for the nuclear dipolar interaction, can always be diagonalized, so giving three principal components. When the magnetic field is parallel to a principal axis, then the hyperfine spacing in the electron resonance spectrum is simply the principal component appropriate to that axis. If the magnetic field makes some arbitrary angle with the molecular co-ordinate system, then the hyperfine spacing is essentially equal to the resolved component in the field direction. The magnitude of this component is given by an analogous relationship to equation (7.6):

$$A_{zz} = \sum_{\alpha\,\beta} l_{z\alpha} l_{z\beta} A_{\alpha\beta}. \tag{7.30}$$

The trace of the total tensor is related to the scalar hyperfine interaction a:

$$a = (\tfrac{1}{3}) \sum_{\alpha} A_{\alpha\alpha}. \tag{7.31}$$

Consequently we can write the total hyperfine tensor in terms of the scalar a and an anisotropic tensor $A'_{\alpha\beta}$ whose trace is zero:

$$A_{\alpha\beta} = a\, \delta_{\alpha\beta} + A'_{\alpha\beta}. \tag{7.32}$$

Here $\delta_{\alpha\beta}$ is the Kronecker delta, which is one if α and β are the same, but which vanishes when they differ. The resolved component A_{zz} can now be written as:

$$A_{zz} = a + \sum_{\alpha,\beta} l_{z\alpha} l_{z\beta} A'_{\alpha\beta}, \tag{7.33}$$

and this clearly separates the two contributions to the observed splitting.

When the spin probe is dissolved in some solvent, molecular reorientation will cause the direction cosines to fluctuate in time. If the molecular motion is fast, then the observed hyperfine splitting would be obtained by taking a time or ensemble average of equation (7.33). The anisotropy in a hyperfine interaction is typically 100 MHz and so the rotational correlation time would need to be smaller than 10^{-8} s, to obtain the fast exchange limit. The correlation time in the mesophases of many nematogens is normally less than 10^{-8} s, although this is not the case for most cholesteric liquid crystals. The average $\bar{A}_{zz}$ is normally denoted by $\bar{a}$ and so we have

$$\bar{a} = a + \sum_{\alpha,\beta} \overline{l_{z\alpha} l_{z\beta}} A'_{\alpha\beta}. \tag{7.34}$$

The quantities $\overline{l_{z\alpha} l_{z\beta}}$ could be used to describe the orientational order of the spin probe in the mesophase. However, this formalism would obscure the relationship with the order parameter S which is used to describe the pure mesophase. We therefore define a matrix **S** with diagonal elements equal to

$$S_{\alpha\alpha} = \frac{\overline{(3 l_{z\alpha} l_{z\beta} - 1)}}{2}, \tag{7.35}$$

by analogy with the definition of S, and with off-diagonal elements:

$$S_{\alpha\beta} = \frac{3 \overline{l_{z\alpha} l_{z\beta}}}{2}. \tag{7.36}$$

These two expressions may be combined by using the Kronecker delta to give any element as

$$S_{\alpha\beta} = \frac{\overline{(3 l_{z\alpha} l_{z\beta} - \delta_{\alpha\beta})}}{2}. \tag{7.37}$$

This definition may be incorporated into equation (7.34) for $\bar{a}$ and, by remembering that the anisotropic hyperfine tensor is traceless, we find

$$\bar{a} = a + (\tfrac{2}{3}) \sum_{\alpha,\beta} S_{\alpha\beta} A'_{\alpha\beta}. \tag{7.38}$$

The quantity **S** is known as the ordering matrix and was originally introduced by Saupe to describe the extent of solute alignment in a liquid crystal [7]. We shall now digress slightly to describe some of the properties of the ordering matrix. A three by three matrix contains nine elements but some of these are related in the ordering matrix **S**. For example the averages $\overline{l_{z\alpha} l_{z\beta}}$ and $\overline{l_{z\beta} l_{z\alpha}}$ are clearly identical and so the ordering matrix is symmetric:

$$S_{\alpha\beta} = S_{\beta\alpha}. \tag{7.39}$$

Further the trace of the matrix is zero; this important result follows immediately from the property

$$l_{z\alpha}^2 + l_{z\beta}^2 + l_{z\gamma}^2 = 1, \tag{7.40}$$

of the direction cosines. The largest number of independent elements for **S** is therefore five and this number can often be reduced. Thus the ordering matrix can be diagonalized and this limits the number of independent components to two. Of course this demands a knowledge of the principal co-ordinate system, but this is often determined by the molecular symmetry. If the molecule is cylindrically symmetric, then the principal co-ordinate system clearly contains the symmetry axis and any pair of axes orthogonal to this. The elements of the ordering matrix for these two axes must be identical and, because the trace of **S** vanishes, equal to minus one half the element for the symmetry axis:

$$S_{bb} = S_{cc} = -(\tfrac{1}{2})S_{aa}. \tag{7.41}$$

The ordering matrix is then completely defined by the single element S_{aa} and we see that the order parameter S employed for liquid crystals is just this element. Many of the properties of the **S** matrix are akin to those of the anisotropic hyperfine tensor; they are in fact both second-rank tensors. Accordingly the elements of the matrices change in the same way on transforming from one co-ordinate system to another, and so by analogy with equation (7.6)

$$S'_{ab} = \sum_{\alpha,\beta} l_{a\alpha} l_{b\beta} S_{\alpha\beta}. \tag{7.42}$$

A variety of other functions has been used to describe the orientational order, but these are all related to the ordering matrix and so we shall not discuss them.

Let us now see how the ordering matrix of the spin probe may be determined from its electron resonance spectrum in the nematic mesophase. In the isotropic phase, the ordering matrix, like the order parameter, vanishes and so the hyperfine spacing takes its scalar value a, as expected. On lowering the temperature below the isotropic–nematic transition point the hyperfine spacing will change by an amount

$$\bar{a} - a = (\tfrac{2}{3}) \sum_{\alpha,\beta} S_{\alpha\beta} A'_{\alpha\beta}, \tag{7.43}$$

because **S** is no longer zero. The change therefore gives a sum of products, and if **S** is to be determined, it is necessary to know the magnitude of the anisotropic hyperfine tensor. Unlike the nuclear dipolar interaction, the hyperfine tensor cannot be calculated accurately, but it can be extracted from the solid state electron resonance spectrum of the spin probe. However, there is still insufficient information to determine **S** and in fact the ordering matrix can be obtained from the hyperfine shift only if the spin probe is cylindrically symmetric. With this condition, equation (7.43) reduces to

$$\bar{a} - a = S_{11} A'_{11}, \tag{7.44}$$

where A'_{11} is the component of the anisotropic hyperfine tensor along 1, the molecular symmetry axis. The next most favourable situation occurs when the principal co-ordinate system is known from the molecular symmetry, for then equation (7.43) becomes

$$\bar{a} - a = S_{11} A'_{11} + (\tfrac{1}{3})(S_{22} - S_{33})(A'_{22} - A'_{22}). \tag{7.45}$$

However, another experimental quantity is clearly needed if the two independent components of **S** are to be determined. The additional quantity is the change in the g factor on passing from the isotropic to the nematic phase. The g-shift is related to the principal components of the ordering matrix by an expression analogous to equation (7.45):

$$\bar{g} - g = S_{11} g'_{11} + (\tfrac{1}{3})(S_{22} - S_{33})(g'_{22} - g'_{33}), \tag{7.46}$$

where $g'_{\alpha\beta}$ is the anisotropic g tensor. Provided the g and hyperfine tensors are not cylindrically symmetric about a common axis, equations (7.45) and (7.46) may be solved to give the principal components of **S**.

The magnitude of the solute ordering matrix is determined, as we shall see in the next section, by the strength of the solute–solvent interaction and only reflects the orientational order of the pure mesophase. However, an electron resonance investigation can yield properties characteristic of the pure mesophase because the spectrum of the spin probe is also influenced by the orientation of the director. Strictly these properties are those of the solution of the spin probe in the mesophase but the solute concentration is so small that they might be identified with those of the pure solvent. In the discussion of the hyperfine shift, the director was taken to be parallel to the magnetic field; however, when the director is perpendicular to the field, the shift is reduced to minus one half this value:

$$\bar{a}(90°) - a = -(\tfrac{1}{3}) \sum_{\alpha,\beta} S_{\alpha\beta} A'_{\alpha\beta}. \tag{7.47}$$

There is a comparable change in the g-shift when the director is rotated through 90°:

$$\bar{g}(90°) - g = -(\tfrac{1}{3}) \sum_{\alpha,\beta} S_{\alpha\beta} g'_{\alpha\beta}. \tag{7.48}$$

It is helpful to think of these results in the following way. When the director is parallel to the magnetic field, the hyperfine spacing and the g factor, determined from the spectrum, are really the components of the appropriate, partially averaged tensor. In accord with the notation that was introduced earlier in this section, they should therefore be denoted by $\tilde{A}_{\parallel}$ and $\tilde{g}_{\parallel}$. Similarly when the director is perpendicular to the magnetic field, the observed hyperfine spacing and g factor are really the components of the relevant partially averaged tensors perpendicular to the director, i.e. $\tilde{A}_{\perp}$ and $\tilde{g}_{\perp}$. These components are related to the ordering matrix by

$$\tilde{A}_{\parallel} = a + (\tfrac{2}{3}) \sum_{\alpha,\beta} S_{\alpha\beta} A'_{\alpha\beta}, \tag{7.49}$$

$$\tilde{A}_{\perp} = a - (\tfrac{1}{3}) \sum_{\alpha,\beta} S_{\alpha\beta} A'_{\alpha\beta}, \tag{7.50}$$

$$\tilde{g}_{\parallel} = g + (\tfrac{2}{3}) \sum_{\alpha,\beta} S_{\alpha\beta} g'_{\alpha\beta}, \tag{7.51}$$

and

$$\tilde{g}_{\perp} = g - (\tfrac{1}{3}) \sum_{\alpha,\beta} S_{\alpha\beta} g'_{\alpha\beta}. \tag{7.52}$$

Thus far these results are analogous to those developed for the dipolar splitting. However, the analogy ends here, for the angular dependence of both the g factor and the hyperfine spacing is more complicated than that for the dipolar splitting. Thus the g factor is

$$\bar{g}(\gamma) = \{\tilde{g}_{\perp}^{2} + (\tilde{g}_{\parallel}^{2} - \tilde{g}_{\perp}^{2}) \cos^{2}\gamma\}^{1/2}, \tag{7.53}$$

when the director makes an angle γ with the magnetic field and the hyperfine spacing is

$$\bar{a}(\gamma) = \frac{\{\tilde{A}_{\perp}^{2}\tilde{g}_{\perp}^{2} + (\tilde{A}_{\parallel}^{2}\tilde{g}_{\parallel}^{2} - \tilde{A}_{\perp}^{2}\tilde{g}_{\perp}^{2}) \cos^{2}\gamma\}^{1/2}}{\bar{g}(\gamma)}. \tag{7.54}$$

Frequently the anisotropy in the partially averaged g tensor is small, and so equation (7.54) may be simplified to

$$\tilde{a}(\gamma) = \{\tilde{A}_{\perp}^2 + (\tilde{A}_{\parallel}^2 - \tilde{A}_{\perp}^2)\cos^2\gamma\}^{1/2}. \tag{7.55}$$

Clearly if both components of $\tilde{A}$ are known, then measurement of the hyperfine spacing in the presence of several constraints would give the angle γ which the director is forced to make with the field. The component $\tilde{A}_{\parallel}$ is simply the hyperfine spacing measured in the absence of the constraints, for then the director is aligned parallel to the magnetic field. The scalar coupling a can be determined from measurements in the isotropic phase and can be obtained with $\tilde{A}_{\parallel}$ to give the other component $\tilde{A}_{\perp}$ since

$$a = (\tfrac{1}{3})(\tilde{A}_{\parallel} + 2\tilde{A}_{\perp}). \tag{7.56}$$

In describing the principles of electron resonance studies of nematogens we have dealt exclusively with the fast exchange limit. Of course, when the molecular motion is quenched, the observed spectrum will simply be the weighted sum of spectra from all orientations of the spin probe. Since the concepts involved in understanding such spectra are the same as those described for nuclear magnetic resonance spectra we shall not discuss them further. Instead we shall see how the ideas developed in this section have been applied in actual studies of thermotropic (non-amphiphilic) liquid crystals.

3. Orientational Order in the Nematic Mesophase

We shall begin this section by analysing the proton magnetic resonance spectrum observed for the nematogen 4,4′-dimethoxyazoxybenzene, since the problems encountered are common to studies of other nematogens. In addition, 4,4′-dimethoxyazoxybenzene was the first liquid crystal to be studied by magnetic resonance spectroscopy [8] and has since been the subject of numerous investigations. A molecule of 4,4′-dimethoxyazoxybenzene, whose structure is shown in Fig. 7.5, contains a large number of

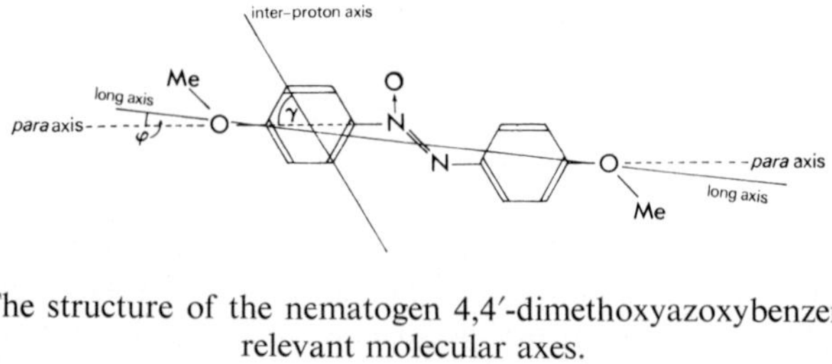

FIG. 7.5. The structure of the nematogen 4,4′-dimethoxyazoxybenzene and the relevant molecular axes.

non-equivalent nuclei and so the nuclear magnetic resonance spectrum from even the isotropic phase might be expected to be complex. This is indeed the case [7], although under the low resolution inherent in a wide-line experiment only one line is observed. However, the spectrum of the nematic mesophase contains a strong central peak flanked by a pair of lines with a lower intensity [8]. The central peak is thought to come from protons in the two methoxy groups and this suspicion has been confirmed by replacing the protons with deuterons. Since the magnetogyric ratio for deuterons is quite different to that for protons, the effect of this substitution is to remove the lines caused by the methoxy groups from the proton spectrum. As expected, the central peak is absent from the spectrum of the partially deuteriated nematogen [9]. However, the spectrum, shown in Fig. 7.6, is now found to consist of a large doublet each component of

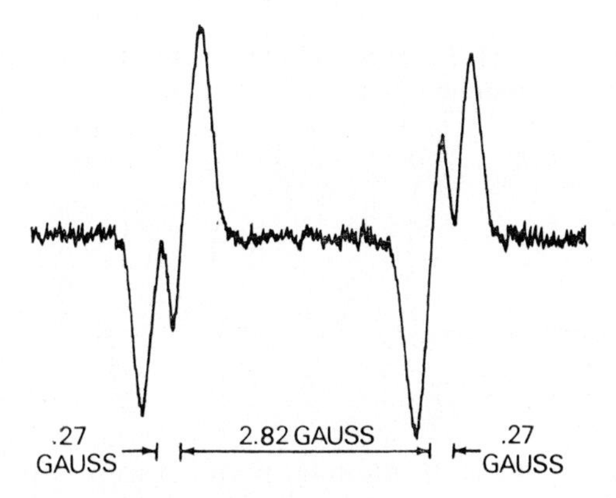

FIG. 7.6. The proton magnetic resonance spectrum of the nematic mesophase of 4,4′-di-trideuteriomethoxyazoxybenzene. [Reproduced with permission from *J. chem. Phys.* **43,** 3442 (1965).]

which is split into a smaller doublet. According to equation (7.5), the components of the dipolar tensor are proportional to r^{-3}, and so the tensor for the protons *para* to one another will be considerably smaller than that for the *ortho* protons. Consequently, the larger spacing comes from the dipolar coupling between the *ortho* protons and the smaller from the *para* protons.

Measurement of either dipolar splitting then gives the element of the ordering matrix corresponding to the appropriate inter-proton vector. Since neither of these vectors is parallel to the molecular long axis, it is necessary to see how the experimental quantity is related to the order parameter S. A glance at the structure in Fig. 7.5 shows that 4,4′-dimethoxyazoxybenzene, like all nematogens, is not strictly cylindrically symmetric. However, it is conventional to assume that the ordering matrix does possess cylindrical symmetry about an axis which is usually determined by inspection of the molecular structure. The element S_{zz}, where z is the inter-proton axis, can then be related to the order parameter simply by transforming from the co-ordinate system containing z to one involving the long molecular axis with the aid of equation (7.42). It is informative to perform this transform in two steps rather than one. The first transformation is of S_{zz} to a co-ordinate system containing the *para* axis shown in Fig. 7.5:

$$S_{zz} = \sum_{\alpha,\beta} l_{z\alpha} l_{z\beta} S'_{\alpha\beta}. \tag{7.57}$$

The benzene rings in the molecule are thought to execute rapid rotation about the *para* axis and so it is necessary to average the direction cosines in equation (7.57). The situation is somewhat similar to that encountered in our discussion of the ordering matrix, and so equation (7.57) can be written as

$$S_{zz} = \left(\frac{2}{3}\right) \sum_{\alpha,\beta} \frac{\overline{(3l_{z\alpha}l_{z\beta} - \delta_{\alpha\beta})}\, S'_{\alpha\beta}}{2}. \tag{7.58}$$

Although the rotation of the benzene ring is subject to a barrier which has a low symmetry it is assumed that

$$\frac{\overline{(3l_{z\alpha}l_{z\beta} - \delta_{\alpha\beta})}}{2} \equiv \begin{array}{c|ccc|} & 1 & 2 & 3 \\ 1 & X & & \\ 2 & & -X/2 & \\ 3 & & & -X/2 \end{array}. \tag{7.59}$$

Here 1 denotes the *para* axis,

$$X = \frac{3\cos^2\gamma - 1}{2}, \tag{7.60}$$

and γ is the angle made by the inter-proton vector with the *para* axis and is therefore independent of the rotation of the benzene ring. The element S_{zz} can now be written as

$$\begin{aligned} S_{zz} &= \left(\frac{2}{3}\right) \frac{(3\cos^2\gamma - 1)}{2} \frac{\{S'_{11} - S'_{22} + S'_{33}\}}{2}, \\ &= \frac{(3\cos^2\gamma - 1)S'_{11}}{2}, \end{aligned} \tag{7.61}$$

because the ordering matrix is traceless. Finally we transform from the co-ordinate system containing the *para* axis to one including the long axis. In this co-ordinate system, the ordering matrix is cylindrically symmetric, and so the single element S_{11} is obtained from equation (7.57) as

$$S'_{11} = \frac{S_{aa}\{l^2_{1a} - (l^2_{1b} + l^2_{1c})\}}{2}, \tag{7.62}$$

where a is parallel to the symmetry axis. However, because of the property of the direction cosines expressed in equation (7.8) we can write S_{zz} as

$$S_{zz} = \frac{(3\cos^2\gamma - 1)}{2} \frac{(3\cos^2\phi - 1)S}{2}, \tag{7.63}$$

where S_{aa} has been replaced by the order parameter and ϕ is the angle between the long axis and the *para* axis.

This is the final result, and we now proceed with our analysis of the magnetic resonance spectrum of 4,4′-dimethoxyazoxybenzene. The angle ϕ is estimated to be 10° from molecular models [10], and because this angle is close to zero, the error in S incurred by any uncertainty in ϕ is likely to be negligible. This is not so for the angle γ required to extract the order parameter from the dipolar splitting for the *para* protons. At first sight it would seem reasonable to assume that γ is 60°, and so the factor $(3\cos^2\gamma - 1)/2$ is $-0{\cdot}1250$. However, if γ is taken to be just 2° smaller, then this factor is reduced to $-0{\cdot}0778$, and so an error as small as 2° would introduce an error of 50% in S. Since the molecular geometry of most nematogens is not known to this accuracy, it is necessary to exercise considerable care in the choice of the interaction used to determine the order parameter. In addition, values of γ close to 54° 46′ greatly diminish the dipolar spacing in the magnetic resonance spectrum. For example, the three protons in a methoxy group are in close proximity and so have a strong dipolar interaction. This is not observed in the spectrum,

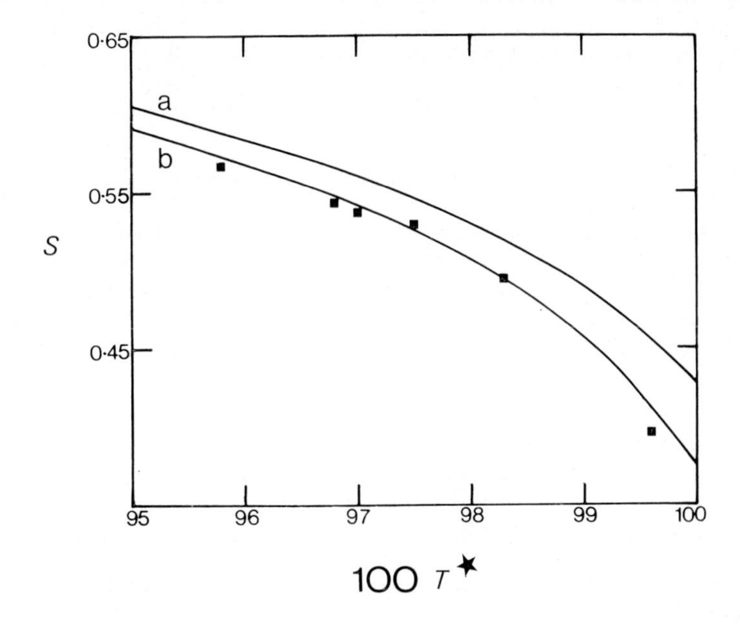

FIG. 7.7. The temperature dependence of the order parameter S for the nematic mesophase of partially deuteriated 4,4′-dimethoxyazoxybenzene. Curve (a) is calculated from the Maier–Saupe theory and curve (b) from the Humphries–James–Luckhurst model.

because internal rotation of the methoxy group averages the dipolar splitting to zero. Analysis of the dipolar splitting from the *ortho* protons does not suffer from the same difficulties as that from *para* protons. The angle γ is expected to be zero, and so any deviations from this value will have a negligible effect on the calculated order parameter. The temperature dependence of S calculated in this way from the *ortho* dipolar spacing in the spectrum of partially deuteriated 4,4′dimethoxyazoxybenzene is shown in Fig. 7.7.

A variety of theories has been devised to account for the temperature dependence of the order parameter, and of these the Maier–Saupe theory [6] is possibly the most realistic. We shall therefore outline this theory together with the extension necessary to achieve complete agreement with experiment [11]. Maier and Saupe begin by assuming that the molecules are cylindrically symmetric and so the pairwise intermolecular potential may be written formally as [12]

$$U_{12}(r;\theta_1,\phi_1;\theta_2,\phi_2) = 4\pi \sum_{L_1,L_2;\,n} u_{L_1L_2:n}(r)\, Y_{L_1,n}(\theta_1,\phi_1)\, Y_{L_2,-n}(\theta_2,\phi_2). \tag{7.64}$$

Here $Y_{L,n}(\theta,\phi)$ is the nth component of the Lth spherical harmonic [13]. The aim of the theory is to obtain the orientational energy of a single molecule resulting from its interactions with all other molecules by averaging over their co-ordinates. Then, given this energy or pseudo-potential, it is possible to calculate any orientational property of the mesophase by taking the appropriate Boltzmann average; for example

$$S = \frac{\int \frac{(3\cos^2\theta - 1)}{2} \exp\{-U(\cos\theta)/kT\}\, \mathrm{d}\cos\theta}{\int \exp\{-U(\cos\theta)/kT\}\, \mathrm{d}\cos\theta}. \tag{7.65}$$

The Maier–Saupe theory implicitly ignores all but those terms with $L_1 = L_2 = 2$ in the expansion of the intermolecular potential. The intermolecular vector is then assumed to be uniformly distributed within the mesophase and the molecular field approximation [14] is invoked to obtain the average over all molecular orientations. The resulting pseudo-potential is

$$U(\cos\theta) = \bar{u}_2 \bar{P}_2 P_2(\cos\theta), \tag{7.66}$$

where θ is the angle between the director and the molecular symmetry axis, $P_2(\cos\theta)$ is the second Legendre polynomial [13]

$$P_2(\cos\theta) = \frac{(3\cos^2\theta - 1)}{2}, \tag{7.67}$$

and its average $\bar{P}_2$ is just the order parameter S. The coefficient $\bar{u}_2$ is defined as

$$\bar{u}_2 = \left(\frac{1}{\rho}\right) \sum_n \int u_{22:n}(r) n^{(2)}(r)\, \mathrm{d}r, \tag{7.68}$$

where ρ is the number density and $n^{(2)}(r)$ is the pair distribution function. Since $n^{(2)}(r)$ is volume dependent, the coefficient $\bar{u}_2$ will also be a function of V which, according to Maier and Saupe, takes the form

$$\bar{u}_2 = \bar{u}_2^0 V^{-2}. \tag{7.69}$$

The theory appears to contain the single parameter $\bar{u}_2^0$ but this is related to the nematic–isotropic transition temperature T_K by

$$\bar{u}_2^0 = -4{\cdot}542 kT_K V_K^2. \tag{7.70}$$

Consequently the order parameter S is then a universal function of the reduced variable $TV^2/T_K V_K^2$ and hence the reduced temperature T/T_K or T^*. This universal curve is plotted as line (a) in Fig. 7.7 as a function of T^* and is in reasonable but not complete agreement with experiment. This discrepancy led Humphries *et al.* [11] to extend the Maier–Saupe theory to include all terms in the expansion of U_{12}. The resulting pseudo-potential now depends on all of the Legendre polynomials:

$$U(\cos\theta) = \sum_{L\,(\text{even})} \bar{u}_L \bar{P}_L P_L(\cos\theta), \tag{7.71}$$

although agreement with experiment is achieved by retaining just the first two terms in the expansion. The agreement was further improved by assuming a different volume dependence of the coefficients $\bar{u}_L$:

$$\bar{u}_L = \bar{u}_L^0 V^{-4}. \tag{7.72}$$

The second line, (b), in Fig. 7.7 was computed from this extension of the Maier–Saupe theory and is clearly in complete accord with experiment.

SOLUTE ALIGNMENT

Humphries *et al.* have also extended the Maier–Saupe theory to include multi-component mixtures of rod-like molecules [15]. This theory shows that the orientational order of the components is easier to interpret when the concentration of all but one is infinitely small. Thus for a binary mixture, the pseudo-potential for the solvent is identical to that, given in equation (7.71), for the pure mesophase. The pseudo-potential for the solute is

$$U^{(2)}(\cos\theta) = \sum_{L} \bar{u}_L^{(12)} \bar{P}_L^{(1)} P_L^{(2)}(\cos\theta), \tag{7.73}$$

where $\bar{P}_L^{(1)}$ is the orientational order of the solvent and $\bar{u}_L^{(12)}$ is a solute–solvent interaction parameter analogous to $\bar{u}_L$ for the pure mesophase. At present the solute concentration required to obtain a nuclear magnetic resonance spectrum is rather high and so does not provide the best way of studying solute–solvent interactions in liquid crystals. In contrast, electron resonance spectroscopy is a particularly sensitive technique and the solute concentration can be taken to be vanishingly small. Electron resonance determinations of the ordering matrix for the spin probe therefore provide a straightforward method for investigating solute–solvent interactions, and we shall consider one such investigation [16].

The spin probe used to investigate these interactions within the nematic mesophases formed by the 4,4′-di-n-alkoxyazoxybenzenes was (3-spiro-[2′-*N*-oxyl-3′,3′-dimethyloxazolidine])-5α-cholestane whose structure is given in Fig. 7.8. The unpaired electron in the probe interacts exclusively

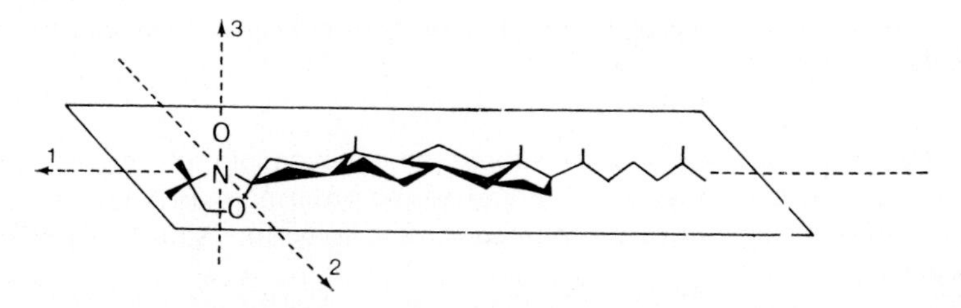

FIG. 7.8. The molecular structure of the spin probe (3-spiro-[2′-*N*-oxyl-3,3′-dimethyloxazolidine])-5α-cholestane and the principal axes for the nitrogen hyperfine tensor.

with the spin of the nitrogen and so the electron resonance spectrum contains just three lines for both the isotropic phase and the nematic mesophase. There is however a marked difference in the hyperfine spacing for the two phases, as the spectra in Fig. 7.9 demonstrate for the nematogen

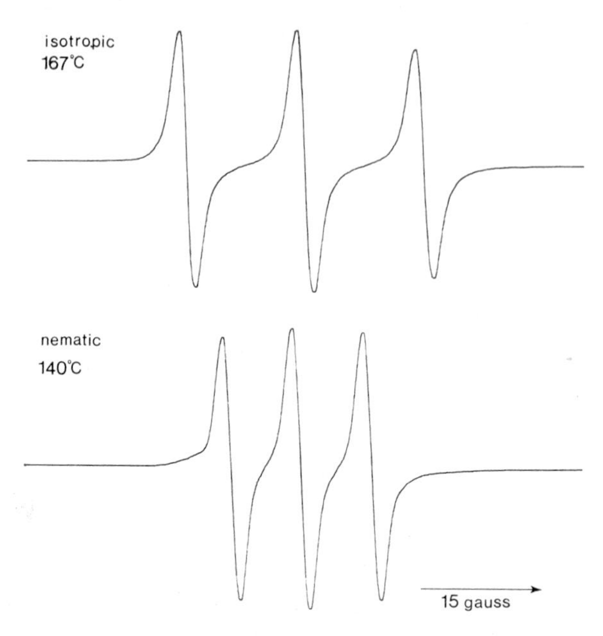

FIG. 7.9. The electron resonance spectra of the nitroxide spin probe in the isotropic and nematic phases of 4,4′-diethoxyazoxybenzene.

4,4′-diethoxyazoxybenzene. The nitrogen hyperfine tensor for the spin probe is cylindrically symmetric about the axis, 3, orthogonal to the oxazolidine ring, and so the hyperfine shift calculated from equation (7.43) is formally the same as in equation (7.44):

$$\bar{a} - a = S_{33}A'_{33}.$$

The g tensor does not possess cylindrical symmetry but we might expect the ordering matrix to be cylindrically symmetric about the long axis 1 which is parallel to the plane of the oxazolidine ring. Consequently, the g shift is simply

$$\bar{g} - g = S_{11}g'_{11}, \tag{7.74}$$

and the ratio $(\bar{a} - a)/(\bar{g} - g)$ should be independent of temperature. This ratio is found to be constant and so the ordering matrix for the spin probe must be cylindrically symmetric. The hyperfine shift is therefore given by:

$$\bar{a} - a = -\frac{\bar{P}_2 A'_{33}}{2}, \tag{7.75}$$

where $\bar{P}_2$ denotes the ordering matrix element for the long axis. Since the anisotropic hyperfine tensor is available from the solid state spectrum, equation (7.75) can be used to calculate the degree of solute alignment $\bar{P}_2$ and the temperature dependence of this quantity is shown in Fig. 7.10.

By analogy with the analysis of corresponding results for a pure mesophase, we might hope to interpret the temperature dependence of the solute alignment by retaining just the first two terms in the pseudo-potential for both solute and solvent. This calculation would therefore involve four adjustable parameters and it would be possible to fit almost any result. It is clearly necessary to have an alternative method for determining some of these four parameters. For example, the solvent parameters $\bar{u}_2^0$ and $\bar{u}_4^0$ can be obtained if the temperature dependence of the

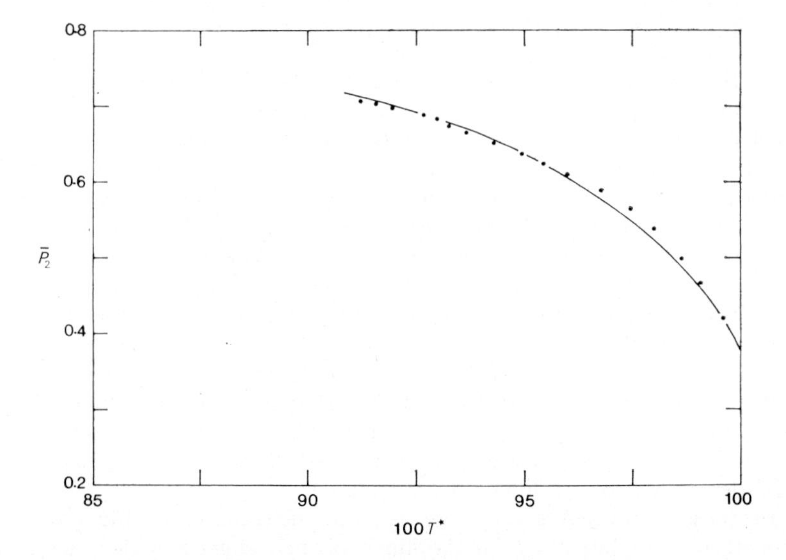

FIG. 7.10. The temperature dependence of the order parameter $\bar{P}_2$ for the nitroxide spin probe dissolved in the nematic mesophase of 4,4′-diethoxyazoxybenzene. The curve is calculated from the Humphries–James–Luckhurst theory.

order in the pure mesophase is known. These parameters may then be used to calculate $\bar{P}_4^{(1)}$ and, since $\bar{P}_2^{(1)}$ is known experimentally, the only unknowns in the solute pseudo-potential are $\bar{u}_2^{(12)}$ and $\bar{u}_4^{(12)}$. Values for these solute–solvent parameters can now be obtained by fitting the temperature dependence of the solute order $\bar{P}_2$. It is again found necessary, in these calculations, to allow for the volume dependence of $\bar{u}_L^{(12)}$ and the line in Fig. 7.10 was obtained using the same dependence as for the pure solvent:

$$\bar{u}_L^{(12)} = \bar{u}_L^{0(12)} V^{-4}. \tag{7.76}$$

The theory is clearly in complete agreement with experiment. However, meaningful values of the solute–solvent parameters may only be determined from such experiments if the solvent order is known [16].

The Frozen Mesophase

Measurement of the temperature dependence of the order for both solvent and solute provides one important method of testing statistical theories of liquid crystals and their mixtures. However, although the agreement just described is impressive, it does not permit us to see how many terms are really needed in the expansion of the pseudo-potential. The angular distribution function $f(\omega)$ or $f(\cos\theta)$ can, as we saw in section 2, be determined from the spectrum, provided the molecular motion is quenched and this function is directly related to the pseudo-potential:

$$f(\cos\theta) = \exp\left\{\frac{-U(\cos\theta)}{kT}\right\}. \tag{7.77}$$

Since the absolute spectral intensity is extremely difficult to determine, the distribution function has not been normalized. Measurement of the angular dependence of the spectrum from the frozen mesophase gives $f(\cos\theta)$ and hence the shape of the pseudo-potential. Although there are several nematogens in which the molecular motion may be quenched without destroying the macroscopic structure of the mesophase, they have yet to be investigated by nuclear magnetic resonance. In contrast, several electron resonance investigations of frozen nematic mesophases have been reported [17–19]. All of these studies have employed vanadyl acetylacetonate as a spin probe, even though the cylindrical symmetry of the g and hyperfine tensors about a common axis makes it impossible to ascertain the symmetry of the ordering matrix. Consequently, the analysis of the spectra may be somewhat involved [19]. We shall therefore describe and discuss the results for the same nitroxide spin probe used in the previous study since its ordering matrix is cylindrically symmetric.

In these experiments [20], the nematic mesophase of the solvent, 4′-methoxybenzylidene-4-n-butylaniline, was first aligned by the magnetic field of the spectrometer. The temperature was then lowered until the molecular motion was quenched and the frozen mesophase was obtained with the director parallel to the magnetic field. The electron resonance spectrum was then measured as a function of the angle γ made by the director with the magnetic field; the two extreme spectra with γ equal to 0° and 90° are shown in Fig. 7.11. These spectra may be understood in the following way. Although the spectrum observed from a frozen sample comes from all orientations of the spin probe, it is dominated by those spectra associated with the principal components of the g and hyperfine tensors. The spectra corresponding to the three principal axes for the nitroxide spin probe are given in Fig. 7.12. The line separations in spectra (a) and (b) are identical because the nitrogen hyperfine tensor is cylindrically symmetric, but the centres differ in all spectra because the g tensor has a lower symmetry. When the director is parallel to the magnetic field, there is a high probability of finding the long axis, 1, also parallel to the field. The observed spectrum is therefore expected to be dominated by spectrum (a) associated with the 1-axis and so should consist of just three

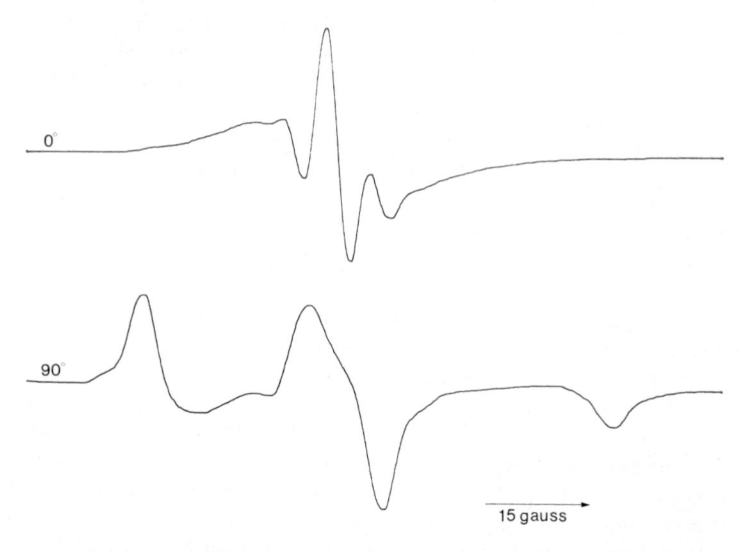

FIG. 7.11. The electron resonance spectrum of the nitroxide spin probe in the frozen nematic mesophase of 4′-methoxybenzylidene-4-n-butylaniline when the director is either parallel or perpendicular to the magnetic field.

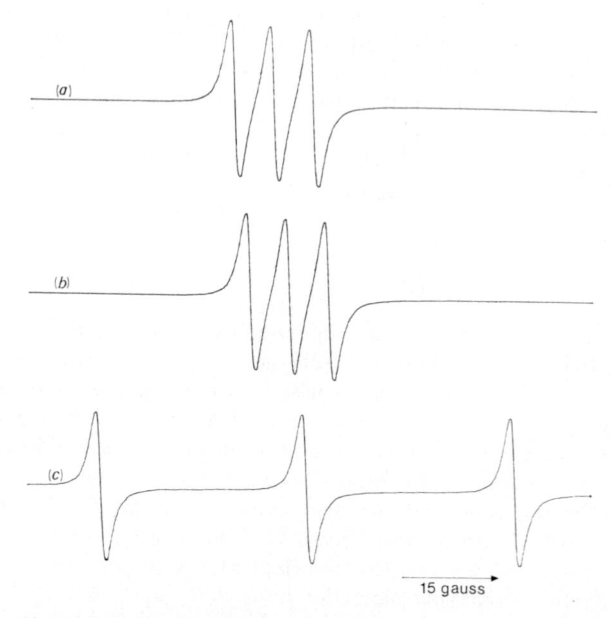

FIG. 7.12. The electron resonance spectra of the spin probe when the magnetic field is applied along the three principal axes of the g and hyperfine tensors.

lines separated by the perpendicular component of the hyperfine tensor which is six gauss. Inspection of Fig. 7.12 shows that this is indeed the case. When the director is perpendicular to the magnetic field, there is an equally high probability of finding axes 2 and 3 parallel to the field, because the ordering matrix is cylindrically symmetric. Consequently the observed spectrum will appear to be the sum of spectra (b) and (c) which are associated with these principal axes. Since the components of the g tensor for these two axes are not the same, the middle lines of the two spectra will not coincide. The central region of the spectrum will therefore have a complicated appearance and will be flanked by two lines separated by 64 gauss, twice the parallel component A_{33} of the hyperfine tensor. This prediction is again in accord with the spectra shown in Fig. 7.11.

We shall now see how the distribution function ($f(\cos\theta)$ may be extracted from the angular dependence of the spectra. We are only concerned with the lines at the spectral extremities, because the resolution of the central region is impaired by line overlap. The intensity of an outside line is approximately proportional to the number of spin probes with the symmetry axis of their hyperfine tensor parallel to the magnetic field. If the director makes an angle γ with the field, then this intensity is also proportional to the probability distribution function $F(\gamma)$ for finding axis 3 at an angle γ with respect to the director. The experimental intensities normalized to unity when γ is 90° are plotted as a function of γ in Fig. 7.13. However, we really require the angular distribution for the long axis, since it is this function, $f(\cos\gamma)$, which is simply related to the pseudo-potential. The calculation relating the two distribution functions is fairly lengthy, although straightforward, and so we shall just quote the result. If the pseudo-potential is restricted to the first term, as in the Maier–Saupe theory, the distribution function for the symmetry axis of the nitrogen hyperfine tensor is

$$F(\gamma) = \exp\left\{-\frac{3\bar{u}_2\bar{P}_2}{4kT}\sin^2\gamma\right\}I_0\left\{-\frac{3\bar{u}_2\bar{P}_2}{4kT}\sin^2\gamma\right\} \tag{7.78}$$

The modified Bessel function I_0 is defined by

$$I_0(p) = \left(\frac{1}{\pi}\right)\int_0^{\pi} e^{p\cos\phi}\,d\phi. \tag{7.79}$$

The curve shown in Fig. 7.13 was calculated using the distribution function given by equation (7.78) and is in very good agreement with experiment. On reflection, this agreement is surprising, for in the previous section we saw that to account for the temperature dependence of $\bar{P}_2$ for this spin probe, it is necessary to include the quartic terms in the pseudo-potential. The apparent success of the Maier–Saupe theory can be understood by estimating the magnitudes of the coefficients $\bar{u}_2\bar{P}_2$ and $\bar{u}_4\bar{P}_4$ for the angular dependent terms $P_2(\cos\theta)$ and $P_4(\cos\theta)$. Using the parameters obtained from the analysis of the temperature dependence of the order parameter for 4,4′-dimethoxyazoxybenzene, the ratio $\bar{u}_4\bar{P}_4/\bar{u}_2\bar{P}_2$ is only 0·07 even when the order is as high as 0·6. Since the solute–solvent parameters should be close to those for the mesophase of 4,4′-dimethoxyazoxy-benzene, we see that the higher-order terms in the pseudo-potential should

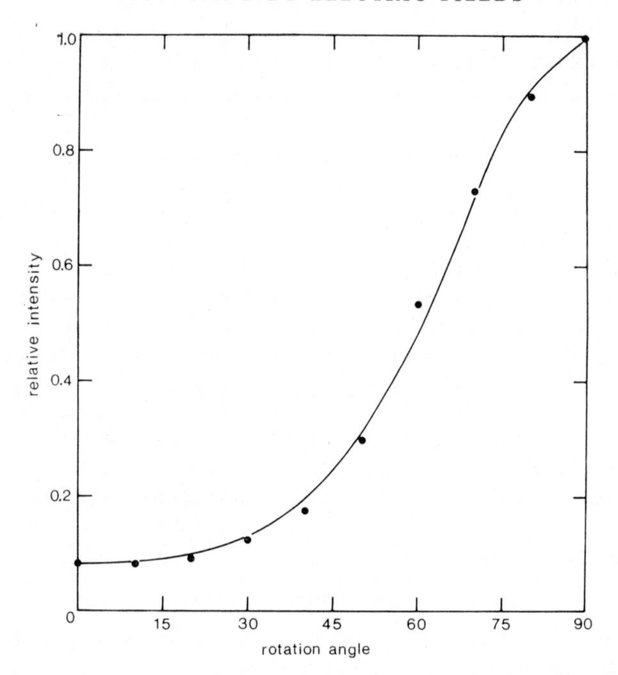

FIG. 7.13. The angular dependence of the intensity of the low field line in the electron resonance spectrum of the spin probe dissolved in a frozen nematic mesophase. The solid line represents a least-squares fit to equation (7.78).

make a negligible contribution to the angular distribution function. Extremely accurate measurements of the line intensities as well as a rigorous treatment of their relationship to the distribution function [18] are therefore required to detect the quartic terms.

4. Alignment by Electric Fields

There has been considerable controversy concerning the nature of the interaction between a liquid crystal and an applied electric field. The apparent contradiction implied by some of the published results stems in part from the presence of two quite different mechanisms by which the director is aligned. For high-frequency electric fields the dominant mechanism is the interaction with the anisotropy in the dielectric tensor and this is analogous to the alignment of the director by magnetic fields. At low frequencies a mechanism associated with the motion of ionic impurities in the mesophase may dominate and produce a shear induced alignment of the director. In this section we shall describe a nuclear magnetic resonance study of alignment at high frequencies and an electron resonance investigation of orientation by a static electric field.

High-Frequency Electric Fields

When a magnetic field is applied to a nematic mesophase the magnetic force,

$$F_{\text{magnetic}} = -\Delta\chi B^2 \sin\gamma \cos\gamma, \tag{7.80}$$

is normal to the director. Here γ is the angle made by the director with the magnetic field and $\Delta\chi$ is the anisotropy $\chi_{\parallel} - \chi_{\perp}$ in the diamagnetic susceptibility attributed to the director. Since $\Delta\chi$ is positive for most nematogens, the force tends to align the director parallel to the field. In much the same way, if the magnetic field is replaced by a high-frequency (~20 kHz) electric field, then the director experiences an electric force

$$F_{\text{electric}} = -\Delta\varepsilon E^2 \sin\gamma \cos\gamma. \tag{7.81}$$

If $\Delta\varepsilon$ the anisotropy $\varepsilon_{\parallel} - \varepsilon_{\perp}$ in the dielectric constant is positive, the director will be aligned parallel to the electric field. Let us now consider what happens if both fields are applied simultaneously and make an angle α with one another. If γ is the angle between the director and the magnetic field, then the electric force becomes

$$F_{\text{electric}} = -\Delta\varepsilon E^2 \sin(\alpha - \gamma)\cos(\alpha - \gamma). \tag{7.82}$$

At equilibrium, the magnetic and electric forces balance one another, and so the angle γ is

$$\tan 2\gamma = \frac{\sin 2\alpha}{\cos 2\alpha + \Delta\chi B^2/\Delta\varepsilon E^2}, \tag{7.83}$$

provided $\Delta\varepsilon$ is positive. This result does not apply when the two fields are orthogonal, for then the director is aligned parallel to whichever field gives the greatest contribution to the orientational energy.

The equation for the dependence of γ on the angle α and the strength of the electric field has been tested by measuring the proton magnetic resonance spectrum of the nematic mesophase of anisylidene-4-aminoazobenzene in the presence of a high-frequency electric field [21]. The observed spectrum always contained a doublet splitting which originates from the dipolar interaction between the *ortho* protons. In these experiments, the electric field strength and α were varied and their magnitudes used to predict the angle made by the director with the magnetic field. Such predictions are possible because the ratio $\Delta\chi/\Delta\varepsilon$ is known from other measurements on the same mesophase [22]. The observed dipolar splitting $\overline{\Delta B}$ is plotted against the predicted angle in Fig. 7.14. According to the analysis in section 2, the angular dependence of the dipolar splitting is given by

$$\overline{\Delta B}(\gamma) = \frac{\tilde{D}(3\cos^2\gamma - 1)}{2}. \tag{7.84}$$

This dependence is shown as the curve in Fig. 7.14 and is clearly in good agreement with experiment thus confirming equation (7.8). The interaction of a high-frequency electric field with the director appears to be well understood, but this is not the case for static fields.

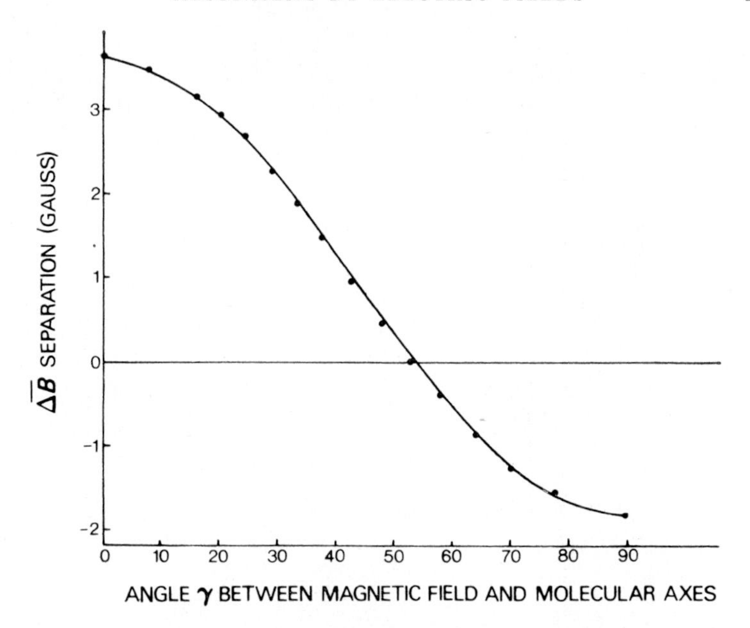

FIG. 7.14. The angular dependence of the dipolar splitting in the proton magnetic resonance spectrum of the nematic mesophase of anisylidene-4-amino-azobenzene. The angle γ was calculated from equation (7.83) and the curve from equation (7.84). [Reproduced with permission from *Molec. Crystals Liqu. Crystals* **18**, 369 (1972).]

STATIC ELECTRIC FIELDS

One of the more complete studies of alignment by static electric fields was performed using electron resonance spectroscopy [23]. The nematogen investigated was 4,4′-dimethoxyazoxybenzene with vanadyl acetylacetonate as the spin probe. This transition metal complex contains just one unpaired electron interacting with the spin of the vanadium nucleus, which is 7/2. The spectrum of the probe in both the isotropic melt and the nematic mesophase consists therefore of eight hyperfine lines. A static electric field of about 5 kV cm^{-1} was employed to align the director, since this overcomes the effect of the magnetic field. The vanadium hyperfine spacing $\bar{a}(\gamma)$ was measured for the nematic mesophase as a function of the angle γ between the two fields with the results shown in Fig. 7.15. The director is thought to be oriented parallel to the electric field and so the angular dependence of $\bar{a}(\gamma)$ should be given by equation (7.54) which becomes

$$\bar{a}(\gamma) = \frac{\{\tilde{A}_{\perp}^2 \tilde{g}_{\perp}^4 + (\tilde{A}_{\parallel}^2 \tilde{g}_{\parallel}^4 - \tilde{A}_{\perp}^2 \tilde{g}_{\perp}^4) \cos^2 \gamma\}^{1/2}}{\{\tilde{g}_{\perp}^2 + (\tilde{g}_{\parallel}^2 - \tilde{g}_{\perp}^2) \cos^2 \gamma\}}, \tag{7.85}$$

when the hyperfine couplings are written in field rather than frequency units. The conversion factor depends on the g factor and is

$$1 \text{ gauss} = 1{\cdot}3996g \text{ MHz}. \tag{7.86}$$

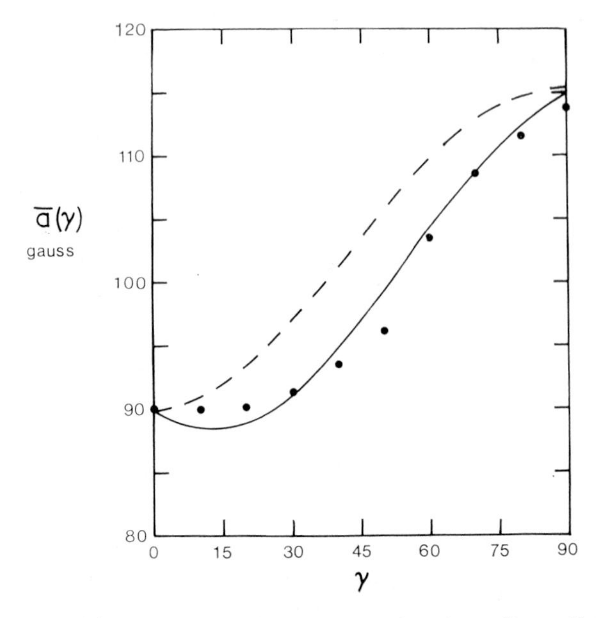

FIG. 7.15. The angular dependence of the vanadium hyperfine splitting for the spin probe vanadyl acetylacetonate dissolved in the nematic mesophase of 4,4′-dimethoxyazoxybenzene.

Since the isotropic vanadium coupling constant is known and $\tilde{A}_{\parallel}$ is determined from the spacing when γ is zero, the other component $\tilde{A}_{\perp}$ can be calculated. The partially averaged g tensor can be determined experimentally, but as this may be difficult, it is more convenient to calculate $\tilde{g}$. The components of the cylindrically symmetric anisotropic hyperfine tensor are known and so the hyperfine shift can be used to calculate the element of the ordering matrix along the symmetry axis. In addition, the total g tensor has been measured, and so the components $\tilde{g}_{\parallel}$ and $\tilde{g}_{\perp}$ can be calculated from equations (7.51) and (7.52). The theoretical angular dependence of $\bar{a}(\gamma)$ may then be calculated with these partially averaged tensors using equation (7.85). The results, shown as the dashed curve in Fig. 7.15, are found to be in rather poor agreement with experiment.

One possible explanation of this discrepancy is suggested by the increase in the vanadium hyperfine spacing even when the electric field is applied parallel to the magnetic field. There are two effects which could cause this increase. The first is a decrease in the solute alignment and the other is a change in the orientation of the director. The action of an electric field is unlikely to alter the microscopic order in the mesophase and so the electric field must align the director at an angle to the magnetic field. The magnitude of this angle θ can be estimated from the increase in $\bar{a}$ when the two fields are parallel by using equation (7.85); this gives θ as about 13° [24]. We shall now see if this idea is in accord with the unusual angular dependence of the vanadium coupling constant.

As long as the electric field is parallel to the magnetic field the director will make a common angle θ with the magnetic field despite the distribution of the director over the surface of a cone. The axis of this cone is of course the electric field, and the cone angle is just θ. Consequently when the magnetic field is inclined at an angle α to the electric field, the director may adopt a range of angles with the magnetic field which go from $\alpha - \theta$ to $\alpha + \theta$. Not all these orientations occur with equal probability because the interaction of the director with the magnetic field favours the smaller angle. Indeed the field of 3300 gauss, typically employed in an electron resonance spectrometer, should cause the director to adopt just the minimum angle which is $\alpha - \theta$ so long as $0 \leqslant \alpha < \pi/2$. When the fields are orthogonal, two angles, $90° - \theta$ and $90° + \theta$, minimize the orientational magnetic energy and will occur with equal probability. Fortunately, this will not complicate the observed spectrum since the spectra from both orientations are identical. According to this model the angular dependence of the hyperfine spacing should be given by equation (7.85), but with the angle γ replaced by $\alpha - \theta$. The theoretical angular dependence is calculated from this equation in much the same manner as before, but with $\tilde{A}_{\parallel}$ equal to the hyperfine spacing in the absence of the electric field and with θ equal to 13°. The results of the calculation are shown as the solid line in Fig. 7.15 and are clearly in very reasonable agreement with experiment. The origin of the angle adopted by the director with respect to the static electric field is not certain. It may however be significant that when the director is aligned by flowing the nematic mesophase of 4,4′-dimethoxyazoxybenzene, the angle made by the director with the direction of flow is 10° [25].

5. Magnetohydrodynamics

The orientation of the director with respect to an applied magnetic field can also be varied by spinning the nematic mesophase about an axis orthogonal to the field [26]. Under such conditions, the director experiences two opposing forces; one is the magnetic force encountered previously and the other is a viscous force caused by the motion of the mesophase relative to the director. Below a critical spinning speed, the two forces can be balanced provided the director adopts an angle γ with the magnetic field. Continuum theory predicts [27] this angle to be

$$\sin 2\gamma = \frac{\Omega}{\Omega_c}, \tag{7.87}$$

where the critical speed is

$$\Omega_c = -\frac{\Delta\chi B^2}{2\lambda_1}. \tag{7.88}$$

The parameter λ_1 is one of the Leslie viscosity coefficients which is negative and vanishes in the isotropic melt [28]. The maximum value of γ is 45° for, as equation (7.80) shows, the magnetic force is a maximum at this point. We shall see shortly what happens when the critical speed has been exceeded.

It is clear, from the discussion in section 4, that both nuclear magnetic resonance and electron resonance spectroscopy could be employed to test the validity of equations (7.87) and (7.88). Indeed several such investigations have been reported [27, 29–31], although we shall only be concerned with one of these [31]. The nematogen employed in this study was Merck Phase IV [32] which exists as the mesophase at room temperature. Although the nuclear magnetic resonance spectrum of the mesophase depends on γ, the accuracy with which this may be determined is limited primarily by the large line widths. In contrast, the widths of the spectral lines for dissolved solutes may be as small as a few Hz, and so the dipolar splittings may be determined with considerable accuracy. The choice of a diamagnetic spin probe is dictated by a variety of factors. For example, its spectrum in the liquid crystal should be simple, it must be well aligned in the mesophase and its molecular geometry must be known. Acetonitrile satisfies all of these conditions and, in addition, is cylindrically symmetric, so that only one element S of the ordering matrix is required to describe its alignment. The three methyl protons are equivalent and so the spectrum in the isotropic phase contains a single line. In the uniformly aligned mesophase, this peak is split into three equally spaced lines with intensities 1 : 2 : 1 by the partially averaged dipolar interaction. When the director makes an angle γ with the field, the line separation $\overline{\Delta\nu}(\gamma)$ is still given by equation (7.22):

$$\overline{\Delta\nu}(\gamma) = \frac{\tilde{D}(3\cos^2\gamma - 1)}{2}.$$

However, now the quantity $\tilde{D}$ is related to the inter-proton separation r and the order parameter S by

$$\tilde{D} = -\frac{3\hbar\gamma^2}{4\pi r^3}S. \tag{7.89}$$

The predicted dependence of the dipolar splitting on the spinning speed may be obtained from equations (7.22) and (7.87) as

$$\frac{\overline{\Delta\nu}(\Omega)}{\tilde{D}} = \left(\frac{1}{4}\right)\left[1 + 3\left\{1 - \left(\frac{\Omega}{\Omega_c}\right)^2\right\}^{1/2}\right]. \tag{7.90}$$

The splitting should therefore decrease from its maximum value $\tilde{D}$ with increasing Ω until, at the critical speed, it takes the limiting value $\tilde{D}/4$.

The experimental results for two concentrations of acetonitrile in Merck Phase IV do exhibit this predicted behaviour, as we can see from Fig. 7.16. According to equation (7.90), the reduced dipolar splitting $\overline{\Delta\nu}/\tilde{D}$ should be a universal function of the reduced spinning speed Ω/Ω_c. The results are therefore plotted using these reduced variables in Fig. 7.16 and do confirm this prediction. However, although $\tilde{D}$ is readily measured from the spectrum of the static sample, the critical angular velocity is not so easy to obtain. In fact Ω_c is best determined by fitting the experimental line separations to equation (7.90). The theoretical dependence of the reduced splitting on Ω/Ω_c, shown as the solid line in Fig. 7.16, is found to be in good agreement with experiment, thus confirming equation (7.87).

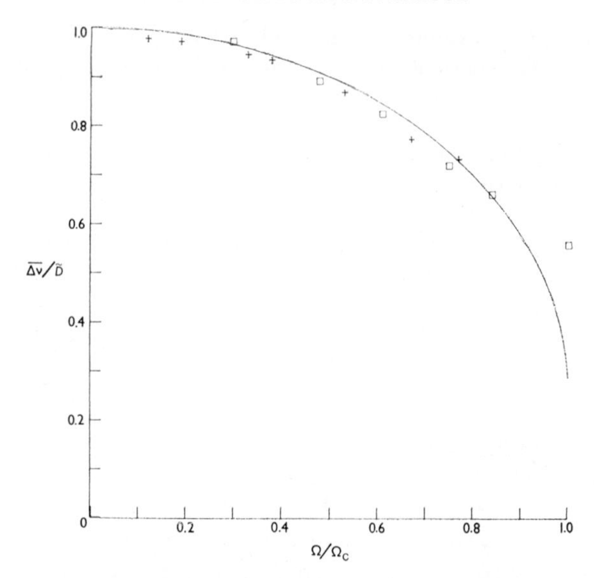

FIG. 7.16. The dependence of the dipolar splitting of acetonitrile dissolved in the nematic mesophase of Merck Phase IV on the spinning speed. The curve is calculated from equation (7.90).

Measurement of the critical speed is clearly of some importance since it provides a method for determining one of the viscosity coefficients. Of course the values of Ω_c for the two solutions of acetonitrile are not of immediate use because the results are characteristic of the solution and not the pure mesophase. In contrast electron resonance determinations of Ω_c are not open to such objections [27].

Above the critical rotation speed the director is predicted to spin with a constant angular velocity

$$\omega = (\Omega^2 - \Omega_c^2)^{1/2}, \tag{7.91}$$

which is slightly less than that of the mesophase, because of the frictional retardation by the magnetic field [27]. This calculation ignores the effects of director inertia since their inclusion makes it impossible to obtain an analytic solution for the equations of motion for the director. The complete equation has been solved numerically for the similar situation in which the mesophase is subject to a rotating magnetic field. The director is still predicted to rotate, but now the angular velocity is time dependent [33]. Neither version of the theory includes the interaction of the director with the surface of the container, although this interaction can lead to the creation of inversion walls in the bulk mesophase [34].

The behaviour of the mesophase above the critical speed has been investigated using electron resonance spectroscopy [27]. If equation (7.91) is correct, then the angle γ made by the director with the magnetic field has the following time dependence

$$\gamma = \omega t, \tag{7.92}$$

and so the hyperfine spacing should also be time dependent. Consequently, the intensity of the spectrum at an appropriate field strength will fluctuate in time as the director rotates and causes the line separations to oscillate. The output from the spectrometer, monitored at a field close to a hyperfine line, will therefore oscillate with a frequency equal to 2ω. The frequency is not simply ω, because the line separations and indeed the resonance fields are even functions of γ. A typical time spectrum obtained in this manner for a nitroxide spin probe dissolved in the mesophase of Merck Phase IV is shown in Fig. 7.17. The oscillations clearly demonstrate that the director

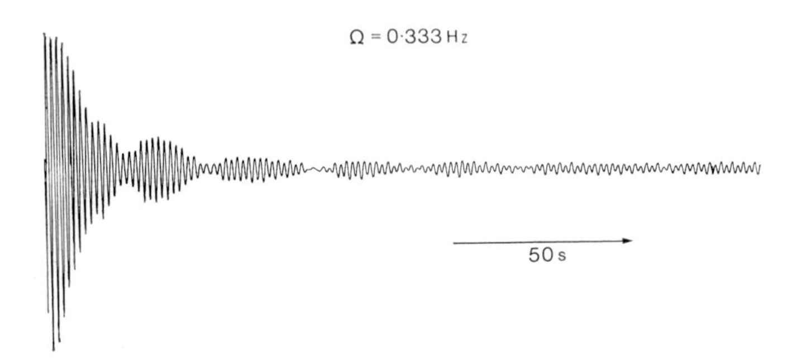

FIG. 7.17. A time spectrum for a nitroxide spin probe dissolved in the nematic mesophase of Merck Phase IV.

does rotate and with an angular velocity close to Ω. More accurate measurements from the initial region of the time spectrum tend to confirm the validity of equation (7.91). However, the decay of these oscillations shows that the stable state of the system is not dynamic, but one in which the director adopts a static configuration. The decay to the static situation is clearly a complex process, and the resulting configuration of the director is not known with any certainty. We shall not therefore consider this aspect of magnetohydrodynamics in the mesophase further.

Another intriguing feature of the time spectrum is the occurrence of a beat pattern which is normally associated with the presence of several periodic processes. It is preferable to investigate the possibility of the director rotating with several angular velocities by taking the Fourier transform of the time spectrum:

$$F(\omega) = \int_{-\infty}^{\infty} (t) \exp(\mathrm{i}\omega t)\,\mathrm{d}t. \tag{7.93}$$

For example, the Fourier transform of the cosine function $\cos \omega t$ is simply a delta function centred at ω in the frequency spectrum. However, not all periodic functions have such simple Fourier transforms; thus the transform of a square wave with periodicity $1/\omega$ is a set of delta functions separated by ω and with a complicated intensity distribution [35]. The Fourier transform of the time spectrum in Fig. 7.17 is shown in Fig. 7.18. There appears to be a dominant peak at a frequency near to 2ω, but close inspection reveals the presence of several peaks corresponding to rotation of the director at slightly different angular velocities. It is clearly this group of velocities which is responsible for the beat pattern in the time spectrum. The origin of these different angular velocities is not readily explained, although it is associated with the rapid acceleration of the sample from rest to some velocity above Ω_c. Thus when the sample is accelerated slowly to a velocity above the critical speed, few oscillations are observed in the time spectrum and the static distribution of the director is obtained

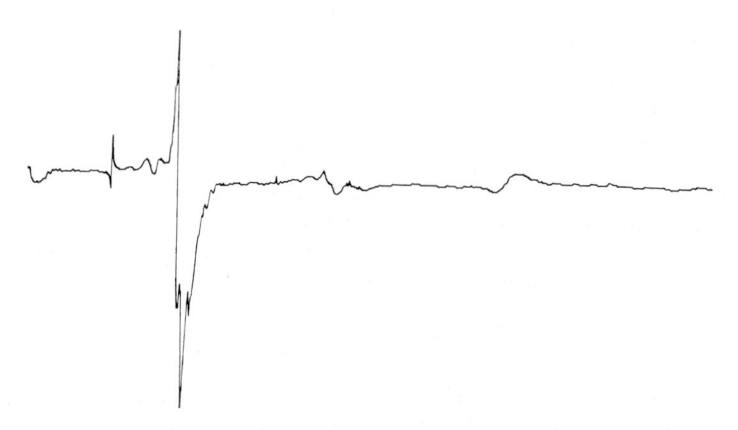

FIG. 7.18. The Fourier transform of the time spectrum shown in Fig. 7.17.

almost immediately [36]. The frequency spectrum also contains small peaks at frequencies of approximately 4Ω and 6Ω. Although these peaks might indicate rotation of the director at multiples of Ω, they are more likely to be associated with the shape of the periodic function and are not therefore of physical significance.

6. The Smectic Mesophase

The success of magnetic resonance studies of the nematic mesophase stems, in part, from the ability of a magnetic field to produce a monodomain sample of the mesophase. The same procedure cannot be employed with the smectic phase, for although there is a strong interaction with a magnetic field, the high macroscopic viscosity makes it impossible for the director to change its orientation. Consequently, if the director is not already uniformly aligned, a magnetic field is unable to create a monodomain sample. This stability of the smectic phase can also be an advantage for the orientation of a monodomain sample may be varied with respect to a magnetic field without subsequent realignment. Possibly the most convenient method of obtaining a uniform sample is to choose a smectogen which also produces a nematic phase. A magnetic field can then be employed to align the nematic mesophase and, by lowering the temperature, the alignment should be preserved on passing to the smectic phase. The technique requires that any disrupting effects caused by nucleation at the surface of the sample container be minimized. This can usually be achieved by lowering the temperature slowly (at about 1° per minute), by employing the maximum magnetic field and if necessary by appropriate treatment of the container surface. The difficulties encountered when preparing uniform samples of the smectic phase have possibly inhibited magnetic resonance studies of this phase. We shall therefore describe just two investigations, both employing electron resonance spectroscopy, one of a smectic A [37] and the other of a smectic C mesophase [38].

The Smectic A Phase

The mesogen 4-n-butyloxybenzylidine-4′-acetoaniline exists in the nematic state from 110° to 98°C and then forms a smectic A phase which is readily supercooled below the freezing point of 84°C. The spectrum of the nitroxide spin probe (3-spiro-[2′-*N*-oxyl-3′,3′-dimethyloxazolidine])-5α-cholestane in both the isotropic and nematic phases contains just three lines, as expected. On passing into the smectic mesophase, there is a decrease in the line separation. In addition, the line shape is asymmetric and two low-intensity lines appear at the extremities of the spectrum. In other words, the spectrum resembles the polycrystalline spectrum shown in Fig. 7.11 and this demonstrates that the uniform alignment of the nematic phase has not been preserved in the smectic phase. The disturbing effects of the glass surface were therefore removed by coating the tube with plastic. The spectrum obtained from the smectic phase produced in this tube consisted of just three symmetric lines expected for a monodomain sample. This form of the spectrum is important, because it also shows that, although the bulk viscosity is high, the spin probe and presumably the molecules of the host are reorienting rapidly within the smectic layers.

Because the spin probe experiences a cylindrically symmetric ordering potential in the smectic as well as the nematic mesophase, the orientational order can be calculated from the splittings in both phases using equation (7.75). The temperature dependence of the order parameter $\bar{P}_2$ for the solute in the two phases is shown in Fig. 7.19. In the nematic phase $\bar{P}_2$ is strongly temperature dependent, but there is a discontinuity at the first-order nematic–smectic transition and then the orientational order in the smectic phase changes only slightly with temperature. Although these measurements are for a spin probe, and not the pure mesogen, they do provide strong support for a theory of the smectic state based on an extension of the Maier–Saupe theory to include spatial order [39]. The results may also be understood in a simple phenomenological manner. Suppose that the pseudo-potential in both mesophases conforms to the Maier–Saupe theory, but because the pair distribution function differs in the two phases, the parameters $\bar{u}_2$ will also be different. The magnitude of the parameter $\bar{u}_2^{(n)}$ for the nematic phase can, of course, be calculated from the nematic–isotropic transition temperature using equation (7.70) which becomes

$$\bar{u}_2^{(n)} = -4{\cdot}542kT_k, \qquad (7.94)$$

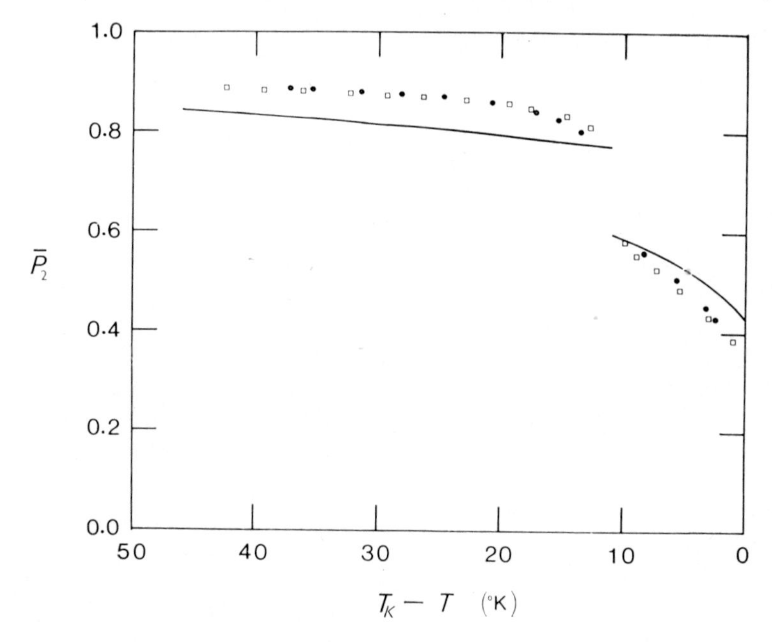

FIG. 7.19. The temperature dependence of the order parameter $\bar{P}_2$ for the nitroxide spin probe dissolved in the nematic and smectic A phases of 4-n-butyl-oxybenzylidene-4′-acetoaniline.

if the volume dependence is ignored. The other parameter $\bar{u}_2^{(s)}$ may be determined by fitting the change in $\bar{P}_2$ at the nematic–smectic transition point. These calculations give a value of 435°K for $\bar{u}_2^{(s)}/4{\cdot}542k$ in contrast with the nematic–isotropic transition of 383°K where the higher temperature corresponds to the hypothetical smectic–isotropic transition. This increase in the strength of the interaction parameter $\bar{u}_2$ for the smectic phase is presumably caused by a decrease in the average molecular separation in this phase. The temperature dependence of $\bar{P}_2$ calculated with these two parameters is shown in Fig. 7.19; since no attempt was made to allow for quartic terms or for solute–solvent interactions the agreement with experiment is encouraging.

Rotation of the smectic sample also rotates the director with respect to the magnetic field and so has a marked effect on the spacing between the hyperfine lines. The observed angular dependence of $\bar{a}$ is shown in Fig. 7.20 where γ is the angle through which the sample has been rotated after its production from the nematic phase. Consequently if the director was originally parallel to the magnetic field then γ is the angle made by the director with the field. Since the partially averaged g tensor is virtually isotropic for this spin probe, the angular dependence of the mean line separation is

$$\bar{a}(\gamma) = K + \frac{\tilde{Q}^2}{K^5}\tilde{A}_{\parallel}^2\tilde{A}_{\perp}^2 \sin^2\gamma \cos^2\gamma + \frac{\tilde{Q}^2}{K^5}\tilde{A}_{\perp}^4 \sin^4\gamma, \qquad (7.95)$$

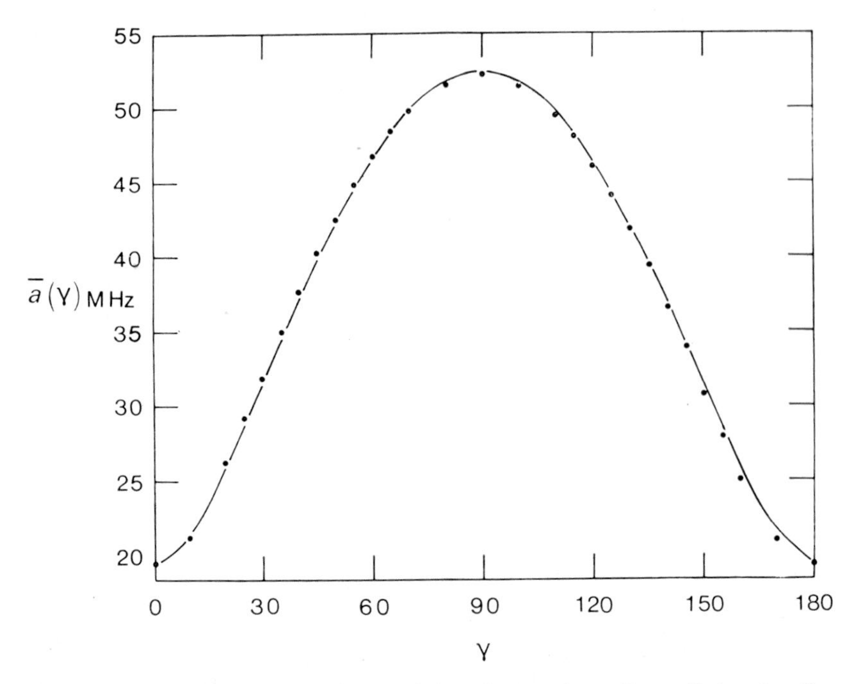

FIG. 7.20. The angular dependence of the nitrogen hyperfine splitting for the spin probe dissolved in the smectic A phase of 4-n-butyloxybenzylidene-4′-acetoaniline. The curve was calculated from equation (7.95).

where K is equal to $\bar{a}(\gamma)$ in equation (7.55). The additional, small terms in equation (7.95) result from the partially averaged nitrogen quadrupole coupling which does not vanish in the mesophase. The quantity $\tilde{Q}$ is related to the ordering matrix and the total quadrupole tensor by

$$\tilde{Q} = (\tfrac{2}{3}) \sum_{\alpha\,\beta} S_{\alpha\beta} Q_{\alpha\beta}. \tag{7.96}$$

The solid line in Fig. 7.20 shows a least squares fit to equation (7.95) using the parameters

$$\tilde{A}_{\parallel} = 19{\cdot}31 \text{ MHz},$$
$$\tilde{A}_{\perp} = 52{\cdot}21 \text{ MHz}$$

and

$$\tilde{Q} = 3{\cdot}8 \text{ MHz}.$$

The theoretical curve is clearly in excellent agreement with experiment and so confirms the notion that the alignment of the director parallel to the magnetic field is maintained at the transition to the smectic phase. The experimental value of the partially averaged hyperfine tensor $\tilde{\mathbf{A}}$ may be used, with equation (7.56), to determine the scalar hyperfine coupling a in the mesophase. This calculation gives a as 41·24 MHz which is close to the value of 41·54 MHz obtained from measurements on the isotropic phase of the mesogen. This agreement therefore justifies the general procedure of determining the scalar interaction from spectra measured for the isotropic phase.

The Smectic C Phase

Knowledge of the orientation of the layers in the smectic A mesophase automatically defines the orientation of the director. This is not the case for the smectic C phase, as we can see from the co-ordinate system shown in Fig. 7.21. Here the xy plane defines a smectic layer and z is the normal to this layer; the spherical polar co-ordinates of the director in this system are θ and ϕ. The angle θ is simply the tilt angle of the smectic phase and is apparently insensitive to external fields [40], although not necessarily to temperature [41]. At a given temperature the tilt angle may be considered as

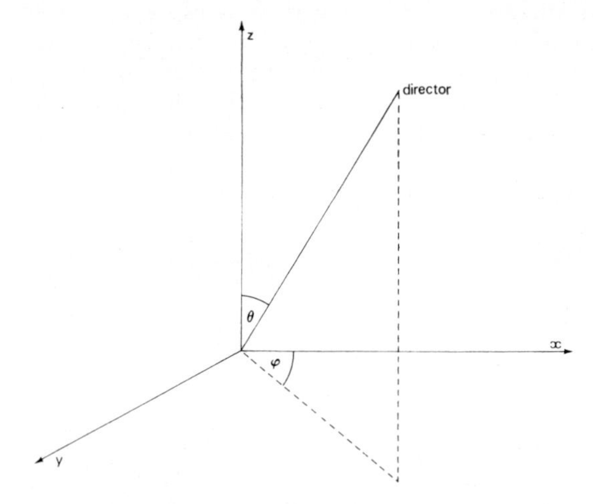

Fig. 7.21. The orientation of the director in a smectic C mesophase.

fixed, although the azimuthal angle ϕ is free to adopt any value providing external fields are absent. In preparing a monodomain sample of a smectic C mesophase it is therefore necessary to obtain a uniform orientation of the smectic layers and of the angle ϕ.

Magnetic resonance studies of the smectic C phase formed by 4,4′-di-n-heptyloxyazoxybenzene have shown that it is not possible to obtain a truly monodomain specimen if the sample is held in a glass tube [38, 42, 43]. However, a uniform sample of this smectic C phase can be prepared if both magnetic and surface forces are employed [38]. Surface forces are used to align the smectic planes by placing a thin sample between two glass plates which have been suitably prepared. This usually involves washing them in a mixture of concentrated nitric and sulphuric acids, followed by water and finally ethanol which is removed by flaming. Such plates align the smectic phase with the layers parallel to the glass surface, but with no restriction on the value of ϕ which the director may assume. A particular value or range of values for ϕ may be selected by applying a magnetic field to the sample when it passes from the nematic to the smectic mesophase. As we have seen, the effect of the magnetic field is to force the director to adopt a configuration which minimizes the angle between the director and the field. However, in the case of a smectic C phase this minimization is subject to the constraint that the tilt angle remains constant.

Let us now consider two examples of a monodomain sample prepared in this manner. In the first, the magnetic field is applied in the xz plane at an angle to z which is similar, but not necessarily equal to the tilt angle. The orientational magnetic energy can only be minimized if ϕ takes the value zero, and consequently, a monodomain sample should be formed when the smectic phase is obtained from the nematic mesophase. The electron resonance spectrum of a nitroxide spin probe dissolved in the smectic phase shows that this is indeed the case. Thus the spectrum always contains just three symmetric hyperfine lines, although the nitrogen coupling constant exhibits a pronounced angular dependence. When the magnetic field makes an angle α with the z-axis, then the director must make an angle $|\alpha - \theta|$ with the magnetic field. The angular dependence of the hyperfine spacing is therefore given by equation (7.55) with γ replaced by $|\alpha - \theta|$. The situation is therefore analogous to that encountered when the director was aligned by a static electric field, and so the angular dependence is given by

$$\tilde{a}(\alpha) = \{\tilde{A}_{\perp}^2 + (\tilde{A}_{\parallel}^2 - \tilde{A}_{\perp}^2)\cos^2|\alpha - \theta|\}^{1/2}. \qquad (7.97)$$

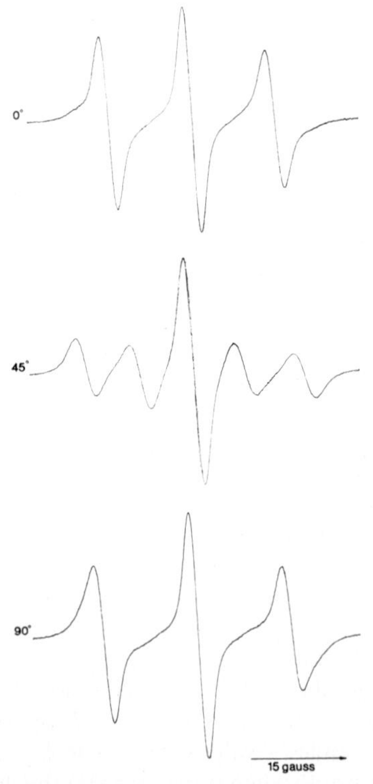

FIG. 7.22. Electron resonance spectra of a nitroxide spin probe dissolved in the smectic C phase of 4,4′-di-n-heptyloxyazoxybenzene. The angle is that between the magnetic field and the normal to the smectic layers.

Since the components of the partially averaged hyperfine tensor are known, measurements of $\bar{a}(\alpha)$ can lead to a determination of the tilt angle θ. In the case of 4,4′-di-n-heptyloxyazoxybenzene the angular dependence of $\bar{a}$ is in complete agreement with equation (7.97) and gives a value for the tilt angle of 38·5° $\pm$ 1°, in accord with less precise estimates [44].

In the second experiment, the magnetic field was applied along the x-axis, and so two values of ϕ can now minimize the magnetic energy. One is 0° and the other is 180°, but because the magnetic energy is identical for these angles, both values of ϕ will occur with equal probability. The sample will therefore behave rather like a single crystal which contains two magnetically non-equivalent sites. In other words, for a given orientation of the sample, the spectra from these two sites are not necessarily identical. For example when the magnetic field makes an angle of 45° with the z-axis, the director for the site with ϕ equal to zero makes an angle of $45° - \theta$ with the field, but for the other site, the angle is $45° + \theta$. The observed spectrum would be a sum of two spectra with different line separations but the same intensities. In contrast, when the magnetic field is either parallel or perpendicular to the z-axis, calculation of the angle made by the director with the magnetic field shows that the two sites are now magnetically equivalent. The experimental spectra for these two orientations of the mesophase do contain the expected three hyperfine lines as we can see from Fig. 7.22. In addition, the spectrum when α is 45° does consist of two spectra with equal intensity, as predicted. These experiments clearly demonstrate the ability of surface and magnetic forces to produce uniformly aligned samples of a smectic C mesophase.

7. The Cholesteric Mesophase

The helical structure of the cholesteric phase clearly makes it impossible to obtain a monodomain sample in that the director may not be uniformly aligned with respect to a particular direction. The spectrum observed for a magnetic resonance study of the cholesteric phase would therefore be a weighted sum of spectra from all orientations of the director. Because the polycrystalline nuclear magnetic resonance spectra resulting from such samples can be quite complicated, this technique has not been applied with any success to the cholesteric phase. However, electron resonance spectroscopy has been used to study the molecular organization within the cholesteric phase [45]. The spectral analysis requires an understanding of the effect of a magnetic field on the helical structure, and so we begin by reviewing this aspect of the problem.

We shall consider a system where the anisotropy $(\chi_{\parallel} - \chi_{\perp})$ in the diamagnetic susceptibility is positive and so would cause the director to align parallel to a magnetic field. This is the case for the majority of nematogens. The component of the susceptibility along the helix axis would then be just $\chi_{\perp}$ since the director is always orthogonal to this axis. The component perpendicular to the helix axis is the mean $(\chi_{\parallel} + \chi_{\perp})/2$, because the director adopts all orientations with respect to such an axis with equal probability. Consequently the component $\chi_{\perp}$ along the helix axis is less than the component $(\chi_{\parallel} + \chi_{\perp})/2$ perpendicular to this axis, and so the helix will be aligned orthogonal to the magnetic field. However, this configuration is potentially unstable for the director is forced to be perpendicular to the field, which increases the magnetic energy. The helix is able to minimize the magnetic energy by distorting in such a way as to preserve the periodic structure while increasing the probability of finding the director parallel to the field [46]. In fact the probability $f(\phi)\,\mathrm{d}\phi$ of finding the director between the angles ϕ and $\phi + \mathrm{d}\phi$ with respect to the magnetic field is

$$f(\phi)\,\mathrm{d}\phi = \frac{(1 - k^2 \cos^2 \phi)^{-1/2}\,\mathrm{d}\phi}{\int_0^{2\pi} (1 - k^2 \cos^2 \phi)^{-1/2}\,\mathrm{d}\phi} \tag{7.98}$$

The parameter k in this distribution function is

$$\frac{k}{E(k)} = \frac{Z_0 B}{\pi^2}\left(\frac{\Delta\chi}{K_{22}}\right)^{1/2}, \tag{7.99}$$

where $E(k)$ is an elliptic integral of the second kind:

$$E(k) = \int_0^{\pi/2} (1 - k^2 \sin^2 \phi)^{1/2}\,\mathrm{d}\phi. \tag{7.100}$$

The pitch of the unperturbed helix is Z_0 and K_{22} is the twist elastic constant. As the field strength is increased so the pitch increases until, at some critical field B_c, it becomes infinite and a nematic phase results. The parameter k is related to this critical field by

$$\frac{k}{E(k)} = \frac{B}{B_c}. \tag{7.101}$$

When the variable k is zero, the director is free to adopt all values of ϕ with equal probability. Deviations from this distribution only become appreciable when the magnitude of k is close to unity. Further, k is only close to one when the magnetic field is approximately equal to the critical value B_c. Consequently, to study the magnetic perturbation of the helix, it is essential to choose a cholesteric mesophase where B_c is close to 3300 gauss, the field employed in most electron resonance spectrometers. The critical field for the cholesteric phase obtained from derivatives of cholesterol is in excess of 20 k gauss and so they are unsuitable for study by electron resonance spectroscopy. However, the cholesteric phase formed by adding an optically active solute to a nematic mesophase has a critical field which may only be several thousand gauss [47, 48]. These systems have another advantage, for the pitch of the helix is inversely proportional to this solute concentration, and so, from equations (7.99) and (7.101), the critical field will be directly proportional to this concentration.

The system used in the electron resonance investigation was a mixture of cholesteryl chloride and 4,4′-dimethoxyazoxybenzene with (3-spiro-[2′-*N*-oxyl-3′,3′-dimethyloxazolidine])-5α-cholestane as the spin probe. Of course this particular probe is itself optically active and would convert the nematic mesophase to a cholesteric phase. However, the probe concentration required in the experiments is so small that the critical field is far below 3300 gauss, and consequently, the optical activity of the spin probe will not affect the analysis. The electron resonance spectrum of the probe in the nematic phase of 4,4′-dimethoxyazoxybenzene is shown in Fig. 7.23 and contains the three lines expected for a nitroxide radical. The spectrum undergoes a pronounced change when cholesteryl chloride is added to the nematic phase for the number of lines increases to five. The appearance of

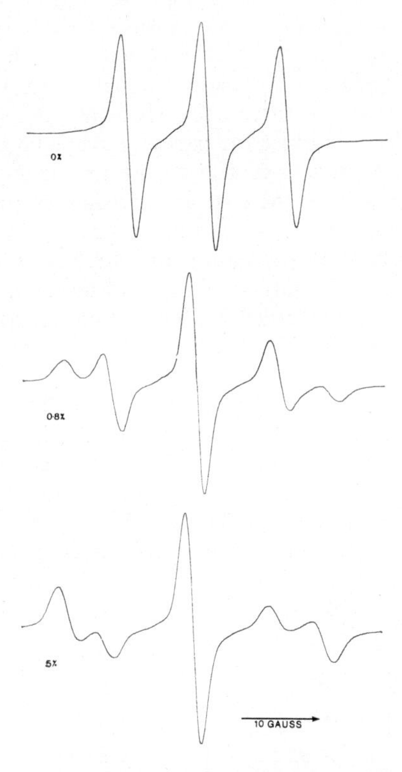

Fig. 7.23. The electron resonance spectrum of the nitroxide spin probe dissolved in the nematic phase of 4,4′-dimethoxyazoxybenzene and the cholesteric mesophase formed on addition of cholesteryl chloride (0·8% and 5%).

these lines is also shown in Fig. 7.23. Further, the intensity of these additional lines increases, at the expense of the inner pair, as the concentration of cholesteryl chloride is increased.

We shall now attempt to interpret these spectral changes. The observation of three lines in the spectrum from the nematic phase with a spacing less than the isotropic value is important for it shows that the probe is reorienting rapidly with respect to the director. Consequently, the spin probe may be thought of in terms of a fictitious static species with partially averaged magnetic tensors given by equations (7.49–7.52). In addition the symmetry of each line demonstrates the uniform alignment of the director which is, as we have seen, parallel to the magnetic field. The addition of cholesteryl chloride produces a cholesteric mesophase for which $\Delta\chi$ is positive as for the parent nematic phase. As a result the helix axis will be aligned orthogonal to the magnetic field and the director will adopt a range of orientations to the field. Now if the motion of the director with respect to the magnetic field is fast, then three spectral lines should be seen, but since five are found, the motion of the director must be slow. The observed spectrum is then a sum of spectra from all orientations of the director, and the theory of such polycrystalline spectra was referred to briefly in section 2. The spectrum is dominated by spectra associated with the director either parallel or perpendicular to the magnetic field and so, in general, will seem to contain two sets of three lines. However, the spectra in Fig. 7.23 do not contain six lines because the partially averaged g tensor is virtually isotropic and so the middle lines of both spectra are superimposed to give a strong central peak. The increase in intensity of the outer lines with increasing solute concentration can now be seen to demonstrate an increase in the probability of finding the director perpendicular to the field. Such an increase is only to be expected, for as B_c increases, the helical structure is less perturbed by the magnetic field. In the limit, the probabilities of finding the director parallel or perpendicular to the field are equal.

A quantitative analysis of the spectra demands a simulation of the polycrystalline spectra using the angular distribution function in equation (7.98). By matching such simulations with the experimental spectra it is possible to determine the parameter k and hence the ratio $(K_{22}/\Delta\chi)^{1/2}$ since both Z_0 and B are known. The starting point of the calculation is equation (7.24a) for the form of a polycrystalline spectrum. This requires an expression for the resonant field which, for a line with nuclear quantum number m, is

$$\tilde{B}_r(\phi) = B_0 - \frac{hKm}{\bar{g}\beta} - \frac{h^2\tilde{A}_\perp^2}{4\bar{g}^2\beta^2B_0}\left\{\frac{\tilde{A}_\parallel^2 + K^2}{K^2}\right\}\{I(I+1) - m^2\} - \frac{h^2m^2}{2\bar{g}^2\beta^2B_0}\left\{\frac{\tilde{A}_\parallel^2 - \tilde{A}_\perp^2}{K}\right\}^2\left\{\frac{\tilde{g}_\parallel^2\tilde{g}_\perp^2}{\bar{g}^4}\right\}\cos^2\phi\sin^2\phi. \quad (7.102)$$

In this expression

$$B_0 = \frac{h\nu}{\bar{g}\beta}, \quad (7.103)$$

$\bar{g}$ is given by equation (7.53) and K is equal to $\bar{a}$ in equation (7.54). The partially averaged g and hyperfine tensors may be determined from the line positions in the polycrystalline spectrum. For example, the spacing between the outer pair of lines is twice the component $\tilde{A}_\perp$ and that between the inner pair is $2\tilde{A}_\parallel$. The line shape $L(\tilde{B}_r, B, T_2^{-1})$ for this particular spin probe is properly taken to be a mixture of Gaussian and Lorentzian, but may be approximated by a Gaussian function, cf. equation (7.12). The width of a hyperfine line often varies with the nuclear quantum number [49] and this variation can be discerned in the nematic spectrum shown in Fig. 7.23 In addition, the line widths may also depend on the angle made by the director with respect to the magnetic field [50]. For this problem, the angular dependence of the line widths was approximated by

$$T_2^{-1}(m = \pm 1) = T_\perp^{-1} + (T_\parallel^{-1} - T_\perp^{-1}) \cos^2 \phi. \qquad (7.104)$$

and the width of the central peak, with m equal to zero, was taken to be orientation independent. The magnitude of the linewidth parameters has to be determined by fitting the shape of a particular hyperfine line to the observed shape. Of course the linewidth parameters contain considerable information concerning the dynamic as well as the static properties of the mesophase [50], but we shall not discuss this aspect of the problem.

Spectra can now be simulated for a range of values for the single variable k in the distribution function and its magnitude determined by comparing the theoretical spectra with their experimental counterparts. The agreement between experiment and theory was generally found to be very good as we can see from the simulated and observed spectra shown in Fig. 7.24. This procedure was employed to determine the temperature dependence of k and hence $(K_{22}/\Delta\chi)^{1/2}$ for a mixture of cholesteryl chloride (1%) in 4,4′-dimethoxyazoxybenzene. For example, at 129°C $(K_{22}/\Delta\chi)^{1/2}$ is found to be 1·63 dyn$^{1/2}$ which is in good agreement with a value of 1·69 dyn$^{1/2}$ determined for the same system by measuring the critical field using optical techniques [48]. Since the concentration of cholesteryl chloride is so low, the properties of the mixture might well be

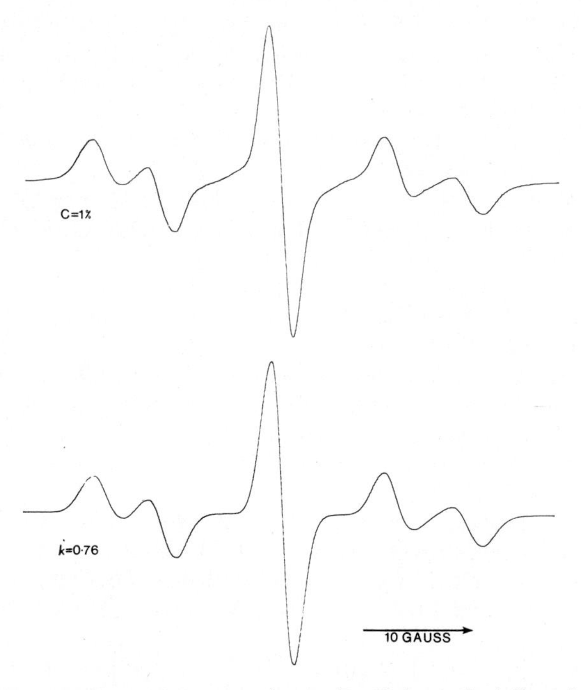

FIG. 7.24. A comparison of the experimental and theoretical (k = 0·76) spectra of the nitroxide spin probe dissolved in the cholesteric mesophase formed from 4,4′-dimethoxyazoxybenzene and 1% of cholesteryl chloride.

identical to those of the pure nematic mesophase. The ratio $(K_{22}/\Delta\chi)^{1/2}$ has been determined for the nematic mesophase of 4,4′-dimethoxy-azoxybenzene by studying the deformations induced in a thin slab of material by a magnetic field and is found to be 1·50 dyn$^{1/2}$ at 129°C [51]. Although the origin of the difference has yet to be diagnosed, it is clear that electron resonance spectroscopy provides another valuable technique for determining elastic constants.

REFERENCES

[1] CARRINGTON, A. and MCLACHLAN, A. D. *Introduction to Magnetic Resonance*, Harper & Row, New York (1967).

[2] SLICHTER, C. P. *Principles of Magnetic Resonance*, Harper & Row, New York (1963).

[3] ABRAGAM, A. *The Principles of Nuclear Magnetism*, Clarendon Press, Oxford (1961).

[4] JEFFREYS, H. *Cartesian Tensors*, University Press, Cambridge (1965).

[5] ZWETKOV, V. *Acta Physicochim.* (*U.S.S.R.*) **16,** 132 (1942).

[6] MAIER, W. and SAUPE, A. *Z. Naturf.* **13a,** 564 (1958); **14a,** 882 (1959); **15a,** 287 (1960).

[7] SAUPE, A., *Z. Naturf.* **19a**, 161 (1964).

[8] SPENCE, R. D., MOSES, H. A. and JAIN, P. L. *J. chem. Phys.* **21,** 380 (1953); SPENCE, R. D., GUTOWSKY, H. S. and HOLM, C. H. *J. chem. Phys.* **21,** 1891 (1953); JAIN, P. L., LEE, J. C. and SPENCE, R. D. *J. chem. Phys.* **23,** 878 (1955).

[9] ROWELL, J. C., PHILLIPS, W. D., MELBY, L. R. and PANAR, M. *J. chem. Phys.* **43,** 3442 (1965).

[10] MAIER, W. and SAUPE, A. *Z. phys. Chem.* **6,** 327 (1956).

[11] HUMPHRIES, R. L., JAMES, P. G. and LUCKHURST, G. R. *J. chem. Soc. Faraday Trans. II* **68,** 1031 (1972).

[12] POPLE, J. A. *Proc. R. Soc.* **A221,** 498 (1954).

[13] ROSE, M. E. *Elementary Theory of Angular Momentum*, John Wiley & Sons, New York (1957).

[14] KRIEGER, T. J. and JAMES, H. M. *J. chem. Phys.* **22,** 796 (1954).

[15] HUMPHRIES, R. L., JAMES, P. G. and LUCKHURST, G. R. *Symposium of the Faraday Soc.*, **5,** 107 (1971).

[16] LUCKHURST, G. R. and SETAKA, M. *Molec. Crystals Liqu. Crystals* **19,** 279 (1973).

[17] SCHWERDTFEGER, C. F. and DIEHL, P. *Molec. Phys.* **17,** 417 (1969); DIEHL, P. and SCHWERDTFEGER, C. F. *Molec. Phys.* **17,** 423 (1969).

[18] JAMES, P. G. and LUCKHURST, G. R. *Molec. Phys.* **20,** 761 (1971).

[19] HEPPKE, G. and LIPPERT, E. *Ber. Bunsenges. physik. Chem.* **75,** 61 (1971).

[20] BROOKS, S. A. and LUCKHURST, G. R. (unpublished results).

[21] CARR, E. F. and MURTY, C. R. K. *Molec. Crystals Liqu. Crystals* **18,** 369 (1972).

[22] CARR, E. F. *J. chem. Phys.* **43,** 3905 (1965).

[23] SCHARA, M. and SENTJURC, M. *Solid St. Commun.* **8,** 593 (1970).
[24] LUCKHURST, G. R. *Chem. Phys. Lett.* **9,** 289 (1971).
[25] Orsay Liquid Crystal Group. *Molec. Crystals Liqu. Crystals* **13,** 187 (1971); HELFRICH, W. *J. chem. Phys.* **50,** 100 (1969).
[26] ZWETKOV, V. *Acta Physicochim.* (*U.S.S.R.*) **10,** 557 (1939).
[27] LESLIE, F. M., LUCKHURST, G. R. and SMITH, H. J. *Chem. Phys. Lett.* **13,** 368 (1972).
[28] LESLIE, F. M. *Quart. J. Mech. Appl. Math.* **19,** 358 (1966).
[29] LIPPMANN, H. *Annln Phys.* **1,** 157 (1958).
[30] DIEHL, P. and KHETRAPAL, C. L. *Molec. Phys.* **14,** 283 (1967).
[31] EMSLEY, J. W., LINDON, J. C., LUCKHURST, G. R. and SHAW, D. *Chem. Phys. Lett.* **19,** 345 (1973).
[32] Merck Phase IV is a mixture of the two isomers 4-methoxy-4′-n-butylazoxybenzene and 4′-methoxy-4-n-butylazoxybenzene.
[33] LEE, J. D. and ERINGEN, A. C. *J. chem. Phys.* **55,** 4504 (1971).
[34] DE GENNES, P. G. *J. Phys.* (*Paris*) **32,** 1297 (1971).
[35] BETTS, J. A. *Signal Processing, Modulation and Noise*, English Universities Press, London (1970).
[36] LUCKHURST, G. R. and SMITH, H. J. (unpublished results).
[37] LUCKHURST, G. R. and SETAKA, M. *Molec. Crystals Liqu. Crystals* **19,** 179 (1972).
[38] LUCKHURST, G. R., PTAK, M. and SANSON, A. *J. chem. Soc. Faraday II* (in the press).
[39] MCMILLAN, W. L. *Phys. Rev.* **4A,** 1238 (1971); **6A,** 936 (1972).
[40] SAUPE, A. *Molec. Crystals Liqu. Crystals* **7,** 59 (1969).
[41] TAYLOR, T. R., ARORA, S. L. and FERGASON, J. L. *Phys. Rev. Lett.* **25,** 722 (1970).
[42] GELERINTER, E. and FRYBURG, G. C. *Appl. Phys. Lett.* **18,** 84 (1971).
[43] MEIBOOM, S. and LUZ, Z. *Molec. Crystals Liqu. Crystals* (in the press).
[44] DE VRIES, A. *Acta Crystallogr.* **A25,** s1 135 (1969); *Molec. Crystals Liqu. Crystals* **10,** 31 (1970); TAYLOR, T. R., FERGASON, J. L. and ARORA, S. L. *Phys. Rev. Lett.* **24,** 359 (1970).
[45] LUCKHURST, G. R. and SMITH, H. J. *Molec. Crystals Liqu. Crystals* (in the press).
[46] DE GENNES, P. G. *Solid St. Commun.* **6,** 163 (1968); MEYER, R. B. *Appl. Phys. Lett.* **12,** 281 (1968).
[47] MEYER, R. B. *Appl. Phys. Lett.* **14,** 208 (1969).
[48] DURAND, G., LEGER, L., RONDELEZ, F. and VEYSSIÉ, M. *Phys. Rev. Lett.* **22,** 227 (1969).
[49] HUDSON, A. and LUCKHURST, G. R. *Chem. Rev.* **69,** 191 (1969).
[50] LUCKHURST, G. R. and SANSON, A. *Molec. Phys.* **24,** 1297 (1972).
[51] FRÉEDERICKSZ, V. and ZWETKOV, V. *Physik. Z. Sowjetunion* **6,** 490 (1934).

8

Nuclear Magnetic Resonance Spectroscopy of Liquid Crystals—Amphiphilic Systems

Å. JOHANSSON and B. LINDMAN

Introduction

This article presents an account of the application of nuclear magnetic resonance (NMR) to studies of liquid crystals formed by amphiphilic compounds. The liquid crystals may consist of a pure amphiphile, a mixture of amphiphiles or of an amphiphile or mixture of amphiphiles with one or more solvating agents (e.g. water). In the preceding chapter Luckhurst reviewed NMR studies on non-amphiphilic systems and also gave an outline of the theoretical aspects of the method, which is equally relevant in the case of amphiphilic systems. In the present treatment the significance of the NMR parameters considered will therefore only be briefly stated; for a more detailed description the reader is referred to standard textbooks on NMR [1–4]. A previous review covering the subject of this article has been given by Flautt and Lawson [5].

The majority of NMR investigations on amphiphilic systems involve the determination of line widths and/or second moments of proton magnetic resonance (PMR) signals. The information obtained by these studies is discussed in section 1.

The PMR spectrum in most cases does not exhibit fine structure, due for example to different chemical surroundings for different protons (chemical shifts), but in those cases where a well-resolved spectrum is obtained, additional information may be gained from the positions of the peaks in the spectrum. This is discussed in section 2.

In section 3 the use of proton magnetic relaxation methods is reviewed.

By nuclear magnetic resonance experiments it is possible to determine self-diffusion coefficients and this is treated in section 4.

Studies with ^{13}C and ^{19}F NMR are discussed in section 5.

In section 6 the relaxation of nuclei with an electric quadrupole moment (e.g. ^{2}H, ^{23}Na, ^{81}Br) is reviewed.

In section 7 the fine structure due to quadrupole interactions is considered.

Macroscopic alignment* of liquid crystals may conveniently be studied by NMR, and all such investigations, irrespective of which of the NMR methods has been used, are grouped together in the final section. In this section the use of amphiphilic liquid crystals as solvents for structure determination of organic molecules by NMR is also considered.

A large number of investigations has been carried out with biological systems containing amphiphilic liquid crystals, e.g. biological membranes. In this article examples of the use of NMR for studies of this type are mentioned, but an extensive discussion is not given. More complete reviews of this subject have been published [6–11]. For biological systems a further difficulty often is to establish to what extent the liquid crystalline systems are responsible for the observed effects.

1. Width and Shape of Proton Magnetic Resonance Signals

In an amphiphilic liquid crystal there are normally several types of protons (e.g. water protons, methylene protons). In some cases the signals may be so broad that, as with most solids, only one PMR signal is observed. In other cases, each different type of proton gives a separate peak, as with most isotropic solutions. Intermediate cases also exist, in which one or more sharp peaks (e.g. due to H_2O) are observed superimposed on the background of a broad absorption (normally due to the amphiphile). In this section the relevance of the line width and related parameters are discussed irrespective of the appearance of the spectrum.

The line width may be defined either as the width at half-height of the absorption signal, $\Delta B_{1/2}$ or $\Delta\nu_{1/2}$, or as the distance between maximum and minimum slope of the absorption curve, ΔB_{pp} or $\Delta\nu_{pp}$, where $\Delta B_{1/2}$ and ΔB_{pp} refer to magnetic field units and $\Delta\nu_{1/2}$ and $\Delta\nu_{pp}$ to frequency units. Another important property of a resonance line is its second moment, $\langle\Delta B^2\rangle$, defined (in magnetic field units) by

$$\langle\Delta B^2\rangle = \frac{\int_0^\infty (B - B_0)^2 f(B)\,\mathrm{d}B}{\int_0^\infty f(B)\,\mathrm{d}B} \tag{8.1}$$

where $f(B)$ is the absorption line shape function.

The line width or second moment may be studied both to detect phase transitions and to obtain information on mesophase structure and molecular motion.

* To avoid confusion we will throughout this article use the term "alignment" to denote *macroscopic* mesophase orientation as produced by external magnetic or electric fields or solid boundaries whereas the term "orientation" will be reserved for effects on the molecular level.

PHASE TRANSITIONS

Many experiments have shown that phase transitions are usually accompanied by changes in the line width parameters. For a pure amphiphile there is in most cases a regular decrease in these parameters when the temperature is increased, with discontinuous changes at the phase transitions. In amphiphile–water systems there is often a more irregular temperature variation of line width and second moment, but the transition points are again clearly depicted.

The first investigation of this type for amphiphilic systems was that of Grant, Hedgecock and Dunell [12] on anhydrous $C_{17}COONa$.* These authors detected the following phase transitions by studying the temperature variation of ΔB_{pp}: super-curd → sub-waxy, sub-waxy → waxy and waxy → super-waxy. The transition temperatures reported are in good accordance with those determined by other methods. Dunell *et al.* also used the line width and second moment to determine phase transitions for the following anhydrous soaps: $C_{17}COOLi$ [13, 14], $C_{17}COONa$ [15, 16], $C_{17}COOK$ [17], $C_{17}COORb$ [18], $C_{17}COOCs$ [18], $C_{15}COOK$ [19], $C_{13}COOK$ [19], $C_{11}COOK$ [19], C_9COOK [19], C_7COOK [20] and C_5COOK [20]. Recently Ripmeester and Dunell [21] detected phase transitions in anhydrous alkali oleates by studying the temperature variation of the second moment and line width. The phase transitions in $(C_{17}COOCH_2)_2CHOOCC_{17}$ have been studied by Barrall and Guffy [22].

Phase transitions in $C_{17}COOLi$ were observed by Janzen and Dunell [23] using saturation narrowed spectra, a technique which gives a great improvement in signal-to-noise ratio. In their studies [13, 14, 23] on $C_{17}COOLi$ Dunell *et al.* observed that the thermal history of the sample has a great influence on the NMR spectrum. The explanation for this effect has not yet been found.

An extensive study of $\Delta B_{1/2}$, ΔB_{pp} and $\langle \Delta B^2 \rangle$ at various temperatures (90–473° K) for some anhydrous saturated sodium soaps was carried out by Lawson and Flautt [24]. In Fig. 8.1 [25] the line widths (on a logarithmic scale) are given as a function of temperature. Lawson and Flautt detected all known phase transitions and one not previously observed for $C_{13}COONa$. The transition temperatures usually agree well (except for $C_{17}COONa$) with those obtained by other methods [24, 25]. Since the wings of the absorption signal are more sensitive to phase transitions than the centre of the signal, $\Delta B_{1/2}$ is normally a more useful quantity than ΔB_{pp} in this type of study [16, 24]. The experimental NMR curves were broader in the wings than expected for a Lorentzian type curve and the shape was termed "super-Lorentzian" by Lawson and Flautt [24]. A discussion of the shape in terms of a distribution of motion was given [5].

In amphiphile–water systems a large variety of phases occurs at different temperatures and compositions. Reviews of these phenomena are given by Winsor [26] and Ekwall *et al.* [27]. We here use the notation proposed by Winsor [26] cf., *Liquid Crystals and Plastic Crystals*, Vol. 1 (Editors

* Here and elsewhere throughout this chapter C_n is used to indicate the saturated n-alkyl group C_nH_{2n+1}.

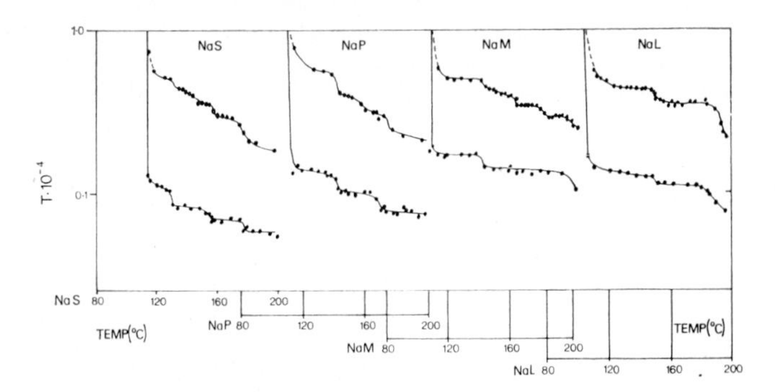

FIG. 8.1. Proton magnetic resonance line widths for anhydrous sodium soaps as a function of temperature. (T stands for Tesla, 1 T = 10^4 gauss; ▲ denotes $\Delta B_{1/2}$ and ● ΔB_{pp}; cf. text.)

The following soaps are represented: $C_{17}COONa$(NaS), $C_{15}COONa$(NaP), $C_{13}COONa$(NaM) and $C_{11}COONa$(NaL) [25].

Gray, G. W. and Winsor, P. A.), Ellis Horwood Limited, Chichester (1974).

The first study of such systems by NMR was by McDonald [28], who recorded the PMR spectra for some mesophases in the systems C_8NH_2–H_2O, $C_{12}SO_4Na$–C_8OH–H_2O and $C_{12}SO_4Na$–C_7COOH–H_2O. The alkyl protons did not give a high-resolution spectrum whereas the water protons gave narrow peaks.

Lawson and Flautt [25, 29] made a thorough study of the PMR line widths in mixtures of D_2O (used to eliminate the water PMR signal) with $C_{17}COONa$, sodium oleate, sodium elaidate, $C_{15}COONa$, $C_{13}COONa$, $C_{11}COONa$ or $C_{12}NO(CH_3)_2$. The line width in the middle phase (M_1) was smaller than that in the neat phase (G). A remarkable observation first made in this work is that a viscous optically isotropic cubic mesophase, like an amorphous isotropic solution,* gives a high resolution spectrum, i.e. the line widths are an order of magnitude less than those of the M_1 and G phases. An example is given in Fig. 8.2. "Super-Lorentzian" lines were observed with the middle and neat phases. Flautt and Lawson [5] introduced the parameter $R(8/2)$, the ratio of the line width at one eighth-height to the line width at half-height, as a sensitive measure of the line shape. Of the three parameters, ΔB, $\langle \Delta B^2 \rangle$ and $R(8/2)$, the last was the most useful in distinguishing between the neat and middle phases. From their NMR data Lawson and Flautt constructed a phase diagram for the system $C_{12}NO(CH_3)_2$–D_2O, which is in good agreement with that obtained by other techniques [30, 31].

Gilchrist *et al.* [32] investigated the two-component systems

$$C_{11}CON^-N^+(CH_3)_3\text{–}H_2O$$

* In order to distinguish between truly isotropic fluids and optically isotropic mesophases we will in the former case use the notation "amorphous isotropic" throughout this article.

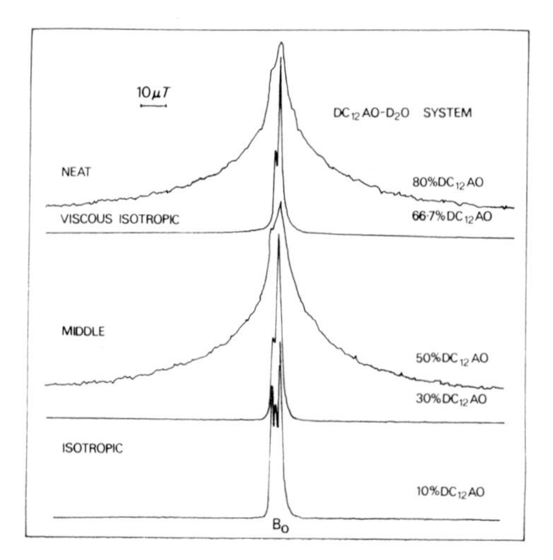

FIG. 8.2. Proton magnetic resonance spectra in different phases in the $C_{12}NO(CH_3)_2$ [$DC_{12}AO$]–D_2O system at 30°C. B_0 denotes the magnetic field strength and μT stands for microtesla [25].

and Aerosol OT–H_2O at various compositions and temperatures. They observed narrow hydrocarbon group signals in the optically isotropic mesophases (V_1 and V_2) but broad signals in the lamellar (G) and hexagonal (M_1 and M_2) phases. The phase diagram constructed for $C_{11}CON^-N^+(CH_3)_3$–H_2O from NMR data is in Fig. 8.3 together with a phase diagram constructed from results obtained by other methods.

More recently the cubic phases S_{1c} and V_1 in the $C_{12}N^+(CH_3)_3Cl^-$–H_2O system were found to give high-resolution spectra [33] and this also applies to a cubic phase in the $C_{10}SO_4Na$–$C_{10}OH$–Na_2SO_4–D_2O system [34].

Phase transitions in the systems $C_{11}COOK$–H_2O [35],

$$C_{11}COOCH_2CH(OH)CH_2OH\text{–}H_2O \text{ [36]}$$

and

$$C_7COOCH_2CH(OH)CH_2OH\text{–}H_2O \text{ [37]}$$

have been detected by abrupt changes in the PMR spectra.

Phase transitions have also been studied with phospholipids both under anhydrous conditions and in the presence of water using the methods described above [38–40]. These studies have led to the identification of crystal → crystal, crystal → mesophase and mesophase → mesophase transitions.

PHASE STRUCTURE

Although the value of PMR line widths for the construction of phase diagrams has been well recognized for some years, only recently, with the application of relaxation methods (see section 3), has a better understanding of the line-broadening mechanisms in these systems been achieved. The difficulties encountered are primarily due to several different effects, which are not easily separated and may contribute to the line

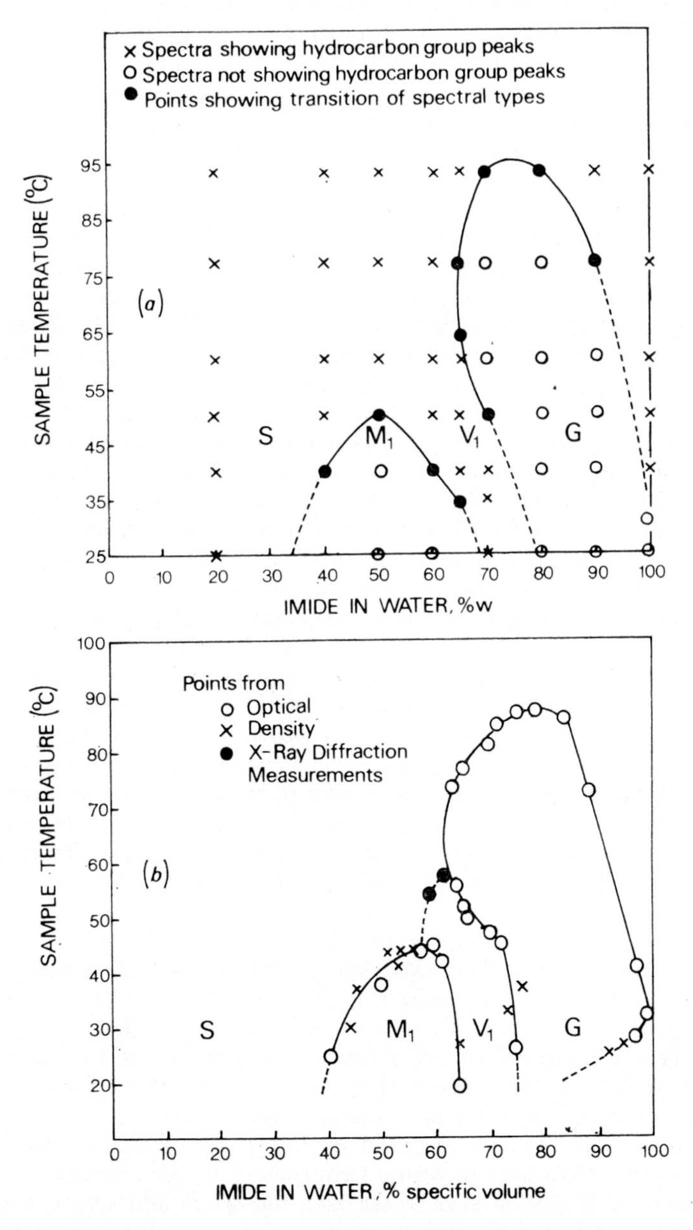

FIG. 8.3 Phase diagrams for $C_{11}CON^{-}N^{+}(CH_3)_3$–$H_2O$ constructed (*a*) from PMR spectra and (*b*) from results obtained by other techniques [32].

width and second moment. Recent work with pulse methods (see below) strongly indicates, however, that the line widths observed in the mesophases may generally be ascribed to direct interactions between the magnetic moments of the nuclei. When an NMR signal for a crystal with a rigid lattice owes its width to this nuclear magnetic dipolar broadening the second moment may be calculated from the positions, magnetic moments and spin quantum numbers of the nuclei [1–3]. Changes in the line width parameters on heating, for example, may be due to changes in the molecular arrangements or to the onset of molecular motion with a frequency at least of the order of the line width in frequency units (normally 10^4–10^5 s^{-1}). The fact that the line widths are smaller in the mesophases than in the corresponding crystals was ascribed to motional narrowing of the dipolar broadenings, and the line widths in different mesophases were interpreted in terms of different degrees and types of motion [13–25, 28, 32, 34, 38–42]. There were two observations which seemed to indicate that dipole–dipole interaction is not the only mechanism regulating line-broadening. Firstly several investigations showed that at phase transitions accompanied by drastic changes in resonance line width the spin–lattice relaxation time, T_1, did not change abruptly (see section 3). This was first discovered by Chapman and Salsbury [38]. These investigations seemed to indicate that the molecular mobility in two phases may be very similar although the line widths differ by an order of magnitude (see Fig. 8.4). Secondly, Penkett *et al.* [35] observed with two amphiphilic mesophases that the line width varied with the strength of the applied magnetic field, and was approximately proportional to the magnetic field strength. This indicates that dipolar broadening is only a small part of the total observed broadening, since the dipole–dipole coupling should not be field dependent. Penkett *et al.* [35] proposed that anisotropies in the bulk magnetic susceptibility could be an important source of line broadening, and this would give rise to the observed field dependence. However, in more recent studies no corresponding field dependence has been observed [43–47], which tends to show that the magnetic anisotropy effect is of little importance. Instead the observed angular dependence of the spectral shape [48] and free induction decay [46] as well as spin echo NMR studies [44, 45, 49] (discussed below) strongly favour an explanation in terms of residual dipolar interactions. This should apply to the majority of the systems investigated. However, the situation might differ in different cases and in the cases where critical tests have not yet been performed it is uncertain whether residual dipolar interaction is the main line-broadening factor.

As mentioned, to have an effect on line width or second moment a motion must occur at least with a frequency of 10^4 s^{-1}. Therefore, limits for the rates of motion may be set from line width and second moment data but it is more difficult to state unambiguously what type of motion is the cause of line-narrowing. This may be achieved by comparison with line-narrowing effects observed for simpler systems. The structure of the phase is important for the line width and second moment since it determines the degree of anisotropy of the motion. If the motion is isotropic,

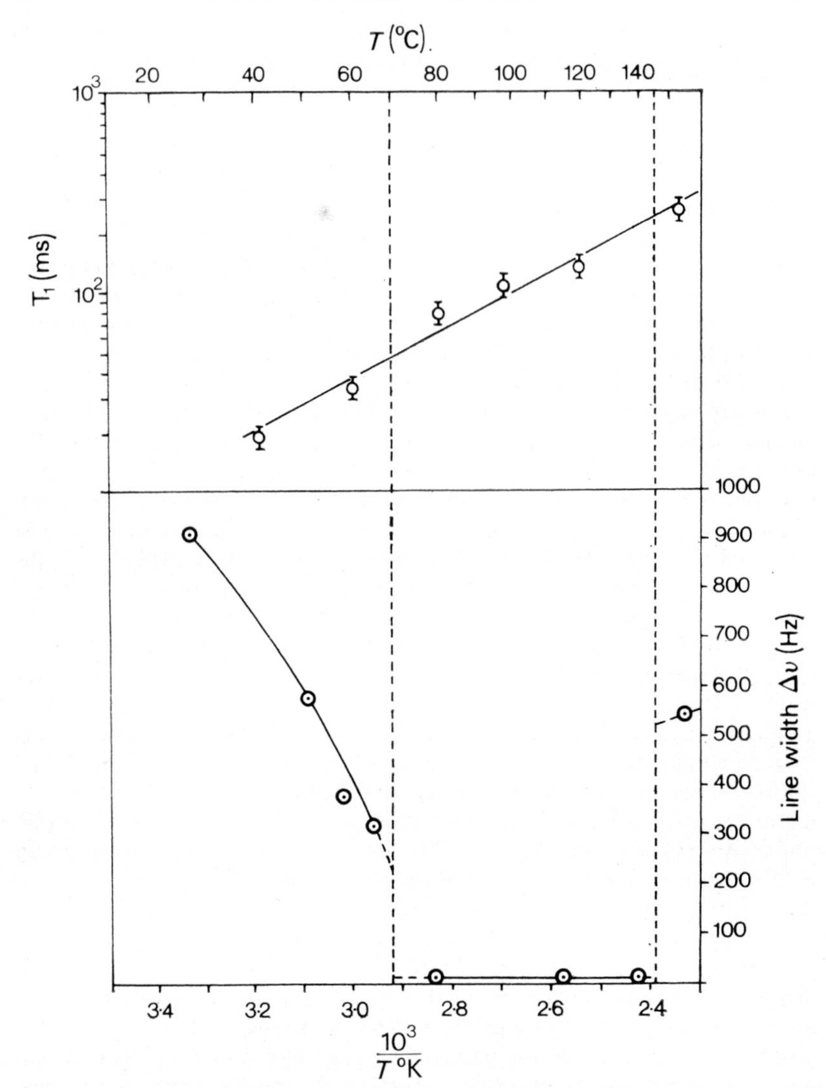

FIG. 8.4. Temperature variation of the PMR spin–lattice relaxation time (T_1) and the PMR line width ($\Delta\nu_{1/2}$) for a mixture of 63% by weight $C_{11}COOK$ and 37% by weight D_2O [35]. Between the dotted lines the mesophase is isotropic; the other phases are anisotropic.

i.e. the inter-spin vectors may occupy all directions with respect to the magnetic field in a time less than about 10^{-4} s, the static dipolar interactions are averaged to zero and the line width is determined by spin-lattice relaxation (see below). Other factors such as unresolved chemical shift differences may also contribute to the line width but in most cases appear to be of little importance in anisotropic mesophases.

In anhydrous sodium soaps, the line width and second moment for the different phases are in the order sub-waxy > waxy > super-waxy > sub-neat [5], i.e. line-narrowing increases with increasing temperature. This may be explained as a decrease in the widths of the polar layers which is expected to permit the hydrocarbon chains to move more freely [5]. For a given temperature there is mostly a decrease in line width with increasing hydrocarbon chain length which would reflect a greater mobility of the longer chains (the second moment, however, changes more irregularly) [5]. The "super-Lorentzian" line shapes in these systems were proposed to be due to a mobility distribution along the chains, with the groups near the polar end possessing little motion and those near the free end of the hydrocarbon chain being relatively mobile [25]. The motional properties of mesophases formed by pure amphiphiles are also discussed by Dunell *et al.* [13–20, 23].

For amphiphile–water systems a striking difference in line width parameters between anisotropic mesophases and optically isotropic cubic mesophases is found [25, 29, 32]. The cubic phases (V_1, V_2, S_{1c}) give high-resolution proton spectra of the amphiphile whereas the anisotropic phases (M_1, M_2 and G) show considerable residual line-broadening. The motional narrowing in the isotropic mesophases is easily explained if the phases can be regarded as built up of micelles which are approximately spherical. Then rotation of the micelles or diffusion of molecules over the micellar surface which is rapid enough can average the dipolar interaction over all possible orientations [34]. The presence of mobile micro-structures was also discussed by Gilchrist *et al.* [32], who pointed out the interesting analogy between optically isotropic cubic mesophases and cubic plastic crystals which also give high-resolution PMR spectra. Winsor [50] gives a thorough account of these matters and describes how NMR spectral features can be used to distinguish between structural models of amphiphilic mesophases (see also ref. 32). Although some motional processes within the micro-crystallites in the anisotropic mesophases may be as fast as in a viscous optically isotropic cubic mesophase (see section 3) the molecules are preferentially oriented within the micro-structures which prevents the dipolar interaction from being completely averaged [34]. The degree of motional narrowing in the anisotropic phases appears to vary along the alkyl chain, being greatest at the non-polar end and least at the polar head [5, 49]. From the line shape Flautt and Lawson [5] found that motional narrowing is generally larger in the middle than in the neat phase.

The water PMR signal in amphiphile–H_2O mesophases is generally relatively narrow corresponding to a considerable mobility [29, 32, 41, 42]. However, in the G and M_1 phases of the $C_{11}CON^-N^+(CH_3)_3$–H_2O system a significant restriction in water mobility was found [32].

Line width measurements have also been made on lamellar phase dispursions, and membrane dispersions after exposure to ultrasonic vibrations [9, 10, 35, 51–56]. For these systems high resolution spectra are observed, and the widths of the individual lines have been determined in order to study the motion of and interaction between different functional groups in lamellar systems. The high resolution spectra arise from the thermal tumbling motions of the finely dispersed particles of lamellar phase or membrane (cf. [10, 56–59]).

2. Chemical Shifts of Proton Magnetic Resonance Signals

The resonance condition for a given nucleus will vary slightly in different chemical compounds and in different positions in a single molecule, since the magnetic field at the nucleus will differ from the applied field, B_0, due to the screening effect of neighbouring electrons. Since the change in field caused by the electrons is proportional to B_0, this effect may be represented by a dimensionless parameter, δ, called the chemical shift:

$$\delta = \frac{B - B_{ref}}{B_{ref}} . 10^6 \tag{8.2}$$

B is the applied field at the resonance condition for the group studied and B_{ref} that of a suitable reference signal. For convenience the factor 10^6 is included and δ is therefore expressed in parts per million (ppm). In the preceding section it was demonstrated that PMR line widths in mesophases are often quite large and this will effectively mask differences in chemical surroundings between different protons. Accordingly, the chemical shift parameter can only be evaluated in certain special cases which are now discussed. Apart from differences in screening, fine structure may also be due to macroscopic ordering. (See section 8.)

For two anhydrous phospholipids (2,3-dimyristoyl-DL-phosphatidylethanolamine and 2-oleoyl-3-stearoyl-DL-phosphatidylethanolamine), Chapman and Salsbury [38] observed a splitting of the PMR signal at temperatures above 135°C. The small peak situated downfield from the main peak was attributed to the protons adjacent to the polar group, which is consistent with an electron-withdrawing effect of the polar group.

The temperature dependence of the water proton chemical shift ($^{\delta}H_2O$) in mixtures of H_2O and $C_{12}(OCH_2CH_2)_6OH$ was studied by Corkill *et al.* [42], $^{\delta}H_2O$ being measured downfield of tetramethylsilane. For a sample which is not macroscopically aligned there is a nearly constant decrease in $^{\delta}H_2O$ with increasing temperature and no abrupt changes were detected on going from a mesophase (neat or middle) to an isotropic solution. Thus there are no drastic changes in hydrogen bonding when these phase transitions take place.

Zlochower and Schulman [41] recorded the PMR spectra of solutions containing $C_{10}N^+(CH_3)_3Br^-$ and H_2O at a constant ratio and varying concentrations of $CHCl_3$. At intermediate chloroform concentrations the system is liquid crystalline, but at high and low $CHCl_3$ concentrations it consists of isotropic micellar solutions. The frequency separation between the water and chloroform protons varies continuously over the whole concentration range investigated and decreases with increasing chloroform content. This was attributed to an interaction of the $CHCl_3$ molecule with the polar end of the surfactant molecule. At higher concentrations of $CHCl_3$ the chloroform molecules are assumed located between the non-polar chains as well [41].

Paramagnetic species affect the resonance frequency of protons by so-called contact and pseudo-contact shifts [60]. This effect was used by Bergelson *et al.* [61, 62] to discriminate between the internal and external surfaces of vesicles produced by sonication of lecithin–D_2O dispersions. If Eu^{3+} ions are added to such a dispersion two —$N^+(CH_3)_3$ signals appear in the PMR spectrum. Addition of Mn^{2+} ions also gives two —$N^+(CH_3)_3$ components in the spectrum, one broad and one narrow signal. In both cases the two signals were attributed to groups situated on the outer and inner surfaces of the vesicles, respectively. For water, on the other hand, only one signal appears in both cases. These experiments show that the exchange of the paramagnetic ions and the lecithin molecules between the internal and external surfaces is slow (lifetimes of at least 1 s), whereas the exchange of water molecules is a much more rapid process.

Whereas the use of high-resolution PMR spectroscopy is of limited applicability for the study of mesophase structure, the use of liquid crystalline phases as solvents for organic molecules represents an important method in NMR spectroscopy. Nematics formed from non-amphiphilic compounds have been most widely used and these studies are discussed in the preceding article, where the basic principles of the nuclear magnetic resonance of oriented molecules are also presented. Lawson and Flautt [63] have described an amphiphile–water mesophase which aligns in a magnetic field. This phase offers several advantages over non-amphiphilic nematics; the most important is that the sample can be spun around an axis perpendicular to the magnetic field without reducing the molecular orientation. A more detailed description of these studies is given in section 8.

3. Proton Magnetic Relaxation

In a system of magnetic nuclei exposed to an external magnetic field the resultant magnetization vector of the nuclear magnetic moments will at equilibrium be directed parallel to the applied magnetic field. If the magnetization vector is turned from the direction of the applied field, the z direction, there will be processes tending to restore the equilibrium. This restoration of the equilibrium is called nuclear magnetic relaxation and the time constants for changes in magnetization (assuming exponential time dependence) are called relaxation times. The change in the z component is characterized by T_1, the longitudinal or spin–lattice relaxation time, and the changes in the x and y components by T_2, the transverse or spin–spin relaxation time. Longitudinal relaxation involves energy transfer from the spin system to a second system called the lattice and presumes an interaction between these two systems. Transverse relaxation may also be affected by spin exchange processes.

NMR relaxation times are conveniently determined by exposing the sample to intense radiofrequency pulses, the durations of which are short compared to the relaxation times. (For descriptions of pulse methods see refs. 2, 64, 65.) If the rf frequency equals the resonance frequency of the nuclei the magnetization vector will precess about the radiofrequency field vector, B_1, which is rotating in the x–y plane. The lengths of the pulses are normally chosen so that the magnetization vector precesses about B_1 by $\pi/2$ (thus taking the magnetization into the x–y plane) or π (thus reversing the direction of the magnetization). By studying the magnetization after certain sequences of $\pi/2$ and π pulses, T_1 and T_2 can be measured.

The spin–lattice relaxation time is normally measured by (π, $\pi/2$) or ($\pi/2$, $\pi/2$) pulse sequences with variable time spacing between the pulses. If the logarithm of the difference of the magnetization from its equilibrium value varies linearly with the pulse spacing (which is normally observed for amphiphilic mesophases), the slope in the semilog plot gives T_1. For anisotropic amphiphilic mesophases transverse relaxation is normally non-exponential. This has the consequence that, unlike isotropic liquids, T_2 is not uniquely defined but depends on the method employed, as discussed below. If the free induction decay after a $\pi/2$ pulse is exponential the slope in a semilog plot defines T_2. If the free induction decay is non-exponential T_2^{eff} will denote the time for the magnetization to decay to e^{-1} of its original value. In the Carr–Purcell method T_2 is obtained by application of a $\pi/2$ pulse followed by a train of π pulses. We denote the relaxation time obtained in this way by T_2^{CP}. Meiboom and Gill modified the Carr–Purcell method by the introduction of a $\pi/2$ phase shift between the $\pi/2$ pulse and the π pulses and the relaxation time obtained in this way is denoted T_2^{MG}. The relaxation time in the rotating frame ($T_{1\rho}$) is obtained by applying a $\pi/2$ pulse followed immediately by a second pulse of variable length, which is phase-shifted by $\pi/2$.

Longitudinal Relaxation

The necessary condition for spin–lattice relaxation to occur is that the spins are involved in some type of time-dependent interaction. Since the time-dependence is connected with the molecular motion in the sample, spin–lattice relaxation studies are important for elucidating the dynamic properties of a system. In isotropic liquids rotational and translational motions exert very important effects on relaxation. In liquid crystals the situation is more complex and other types of motion, e.g. torsional and oscillatory, have also to be considered. A separation of the effects of different types of motion on the relaxation rate is desirable but often difficult to make.

For nuclei with spin quantum number 1/2 the most important source of relaxation is usually due to interaction between magnetic dipoles. The contribution to the longitudinal relaxation rate due to the dipolar interaction between two nuclei (with spin quantum number I) within the same molecule as modulated by the rotation of that molecule with respect to the magnetic field is for isotropic rotation given [66] by

$$\frac{1}{T_1} = \frac{2\gamma^4\hbar^2}{5r^6} I(I+1)\left[\frac{\tau_c}{1+\omega^2\tau_c^2} + \frac{4\tau_c}{1+4\omega^2\tau_c^2}\right]. \tag{8.3}$$

Here γ is the nuclear magnetogyric factor,

$\hbar = h/2\pi$ (h is Planck's constant),

r the internuclear distance,

τ_c the correlation time characterizing the molecular rotation (τ_c may roughly be taken as the time for rotation of the molecule through one radian)

and ω the nuclear angular precession frequency.

The more general equations for the case of rotational motion and also for translational motion are in ref. 2. The case of oscillatory motion has been treated by van Putte [67]. Whether these cases are sufficient to account for the nuclear magnetic relaxation in real systems or whether other mechanisms have to be considered remains to be investigated. Long-range order fluctuations which have been used to explain spin–lattice relaxation in non-amphiphilic nematics give a frequency dependence of T_1 not consistent with experimental data for amphiphilic mesophases [34].

As demonstrated by McLachlan *et al.* [34] the frequency dependence of T_1 conforms closely to equation (8.3) for some amphiphilic mesophases. It appears, therefore, that equations of this type may be used to extract information of motional processes in the system. Spin–lattice relaxation is most efficient when $\omega\tau_c$ is of the order of unity, i.e. when τ_c is 10^{-8}–10^{-9} s for conventional resonance frequencies. Considerably slower processes, causing rapid transverse relaxation (see below), are unable to affect the longitudinal relaxation appreciably. Even if it is possible to obtain information on the rate of motion from T_1 it is more difficult to specify the nature of the motion.

It can normally be assumed that τ_c decreases with increasing temperature. Since equation (8.3) predicts a minimum in T_1 at $\omega\tau_c = 0{\cdot}616$, the temperature dependence of T_1, ω being constant, will give information concerning the rate of the motion responsible for spin–lattice relaxation (equations similar to equation (8.3) are valid also for the other types of motion mentioned above). In this way, using ordinary resonance frequencies, molecular motion corresponding to a correlation time of the order of 10^{-8} s may be studied. Extensive variation of the resonance frequency may also give information regarding slower motion (cf. ref. 68) but this has not yet been done with amphiphilic liquid crystals. Slower motion (correlation times of the order of 10^{-5} s) can also be studied by measuring the spin–lattice relaxation times in the rotating frame, $T_{1\rho}$ (see e.g. ref. 69). $T_{1\rho}$ is the characteristic time constant for the establishment of equilibrium in the direction of the effective field in a co-ordinate system rotating with the radiofrequency field (B_1). Usually the effective field, at resonance, coincides with B_1. Minima in $T_{1\rho}$ occur for $\gamma B_1 \tau_c \simeq 0{\cdot}5$.

Recent studies by Charvolin and Rigny [44, 49] of T_1 for protons in the lamellar phase of the $C_{11}COOK$–D_2O system are informative regarding the relaxation process and the motions causing relaxation. Firstly, the exponential spin–lattice relaxation indicates the existence of one single T_1 (0·68 s at 90° C) although the motion can be assumed to vary along the chain. This indicates that only a fraction of the alkyl protons are significantly coupled to the lattice. The spin–spin interactions in the rest of the alkyl protons can then transfer energy to the fast relaxing part which acts as an energy sink for the whole system. This so-called spin diffusion mechanism, if effective enough, makes T_1 uniform over the whole system and the relaxation rate can be limited either by the spin diffusion rate or by the relaxation rate of the protons strongly coupled to the lattice. The absence of frequency dependence of T_1 and the increase of T_1 with increasing temperature shows that $\omega\tau_c < 1$, i.e. a motion rapid compared to the resonance frequency is responsible for spin–lattice relaxation. The energy of activation for the relaxation process is 23 kJ/mole consistent with the heights of the barriers hindering internal rotation in alkanes. It is suggested that the motion responsible for T_1 consists of internal rotations about the C–C bonds of the paraffinic chain. The correlation time is estimated at 10^{-9} s.

Proton T_1 measurements for the mesophases formed by two anhydrous phospholipids (2,3-dimyristoyl- and 2-oleoyl-3-stearoyl-DL-phosphatidyl-ethanolamine) at different temperatures have been presented by Chapman and Salsbury [38]. In the temperature range investigated (20–150°C) T_1 increases with increasing temperature. Unlike the line widths and second moments (see section 1), T_1 does not change abruptly at the phase transitions. Consequently there are no abrupt changes in the molecular motion which affect spin–lattice relaxation on going from one phase to another. Since T_1 in these mesophases is much smaller than in the corresponding liquid hydrocarbons, it was concluded that the motion in the liquid crystals is restricted [38]. Continuous changes in proton T_1 at phase transitions have also been observed by van Putte [67] for some anhydrous soaps, by Penkett *et al.* [35] for egg yolk lecithin and for a $C_{11}COOK–D_2O$ mixture (Fig. 8.4), and by Hansen and Lawson [70] for the $C_{12}NO(CH_3)_2$–D_2O system. Van Putte [67] attributes minima in T_1 to phase transitions, but this is not necessarily the case since within one phase a T_1 minimum occurs if $\omega\tau_c \simeq 0{\cdot}616$. T_1 minima are accordingly also observed with amorphous isotropic liquids of high viscosity (see e.g. ref. 1). Furthermore, according to the expressions used the temperature for the T_1 minimum should depend strongly on the magnetic field strength (cf. ref. 21).

McLachlan, Natusch and Newman [34] investigated the proton spin–lattice relaxation in the $C_{10}SO_4Na$–$C_{10}OH$–Na_2SO_4–D_2O system at different temperatures and resonance frequencies. T_1 varies continuously with temperature and no changes occur at the phase transitions. This shows that the motion responsible for longitudinal relaxation is independent of phase structure. The observed frequency dependence of T_1 was separated into one frequency-independent and one frequency-dependent term. The frequency-independent term (τ_c of the order of 10^{-9} s) seems to be due to rotations about the long axis of the alkyl chain. The frequency-dependent term (τ_c about 10^{-8} s) might result from intermolecular relaxation within the micro-crystallites.

The dependence of proton T_1 on temperature and composition for several lecithin–D_2O systems with saturated alkyl chains was investigated by Daycock *et al.* [71] (see also ref. 45). The exponential longitudinal relaxation was attributed to spin diffusion. No discontinuities at phase transitions were observed. In the liquid crystalline range, T_1 increases with increasing temperature and, therefore, sets a limit for the correlation time ($\tau_c < 10^{-8}$ s). The activation energy of the relaxation process varies only moderately with alkyl chain length and water content and is 19 ± 3 kJ/mole. T_1 measurements with a selectively deuteriated analogue of dimyristoyl lecithin indicate that the motion of the choline group determines the longitudinal relaxation in lecithin–water systems. Proton T_1 relaxation studies of lecithin bilayers were presented by Chan *et al.* [47]. The longitudinal relaxation was characterized by a single relaxation time (0·22 s at 30°C) which indicates the occurrence of spin diffusion due to an efficient spin–spin coupling between the protons.

Penkett *et al.* [35] investigated the effect of ultrasonic treatment on the proton T_1 in lecithin–D_2O mixtures. The observed increase in T_1 was interpreted in terms of an alteration in the intermolecular spacing of the lipid resulting in a greater freedom of chain motion.

Ripmeester and Dunell [21] presented variable temperature *vs* T_1 data for anhydrous alkali stearates and oleates. In many cases the crystal to waxy liquid crystalline phase transition gave an abrupt decrease in T_1. For sodium stearate above 360° K, the longitudinal relaxation was non-exponential. No definite explanations of these observations were given.

Addition of $C_{12}(OCH_2CH_2)_6OH$ to water decreases the water proton T_1 [42]. In going from an amorphous isotropic solution to a neat or to a middle phase by variation of the temperature there is no abrupt change in T_1, but T_1 increases continuously with increasing temperature. Comparison of T_1 data for the different phases at a certain temperature indicates that the motion of the water molecules in the middle phase is slower than that in an amorphous isotropic solution while that in the neat phase is still slower. At a certain temperature, T_1 in the neat phase is less than T_1 in water by a factor of only about 10. The motion in the mesophases is thus still relatively fast. Clifford *et al.* [72] determined T_1 for the water protons in the lamellar mesophase of the system C_7COONa–$C_{10}OH$–H_2O and found that T_1 is predominantly determined by the ratio of C_7COONa to H_2O. As this ratio increases (and the thickness of the water layers decreases) an increase in the relaxation rate is observed. The largest changes in relaxation rate occur when the number of water molecules per octanoate ion is reduced below about 20. Since the marked reductions in water molecular mobility occur only when the water layer thickness is less than about 2 nm Clifford *et al.* [72] concluded that no long-range effect of surfaces on water structure appears to be present. In the lamellar mesophase in the $C_{15}COONa$–H_2O system, water proton T_1 seems to be controlled by translational diffusion [73]. T_1 data presented by Clifford *et al.* [6] show that part of the water is tightly bound in erythrocyte membranes.

Tiddy [74] has measured proton T_1 values for various amphiphilic molecules solubilized in the lamellar phase of ammonium perfluorooctanoate–D_2O. It was concluded that the CH_2 group rotation becomes more restricted with increasing length of the hydrocarbon chains whereas no large distribution of correlation times along the chain is found.

Spin–lattice relaxation times in the rotating frame ($T_{1\rho}$) were measured by Salsbury *et al.* [69] for 1,2-dipalmitoyl-L-phosphatidylcholine monohydrate in the temperature range -210 to $+170°C$. A low-frequency motion (34 kHz) observed in the lamellar mesophase at 100°C was attributed to a reorientation of the head group protons. No significant change of $T_{1\rho}$ occurs at the lamellar–cubic phase transition, indicating that the low-frequency motion of the polar head groups is essentially the same in the two mesophases.

Transverse Relaxation

As explained, a unique proton T_1 can usually be defined for amphiphilic liquid crystals and equations analogous to those valid for amorphous isotropic solutions may be employed to extract information on the rate of motional processes causing spin–lattice relaxation. For proton spin–spin relaxation in anisotropic mesophases the situation is more complex; e.g. the transverse relaxation is normally non-exponential and the T_2 value depends on the method of measurement. Furthermore, it has often been observed that T_2 changes abruptly at phase transitions. Several models have been proposed to explain the irregular behaviour of the transverse relaxation. Kaufman *et al.* [75] attributed these observations to inhomogeneity broadening arising from internal magnetic field gradients, whereas Hansen and Lawson [70] ascribed the relaxation behaviour to molecular diffusion through such gradients. However, recent pulse experiments by Charvolin and Rigny [44, 49], Tiddy [46] and others [45, 47] tend to show that other explanations must be found. The angular dependence of spin–spin relaxation in a macroscopically aligned sample [46], the shape of the free induction decay [44, 49] and the occurrence of dipolar echoes [44, 45, 49] strongly favour a model with residual dipolar interactions, i.e. dipolar couplings not averaged out by the motion. The non-exponential transverse relaxation due to residual dipolar interactions is sensitive to motions with correlation times shorter than about 10^{-4} s. For this case the discussion of motional effects on line width and second moment presented in section 1 can be applied, since T_2^{eff} is roughly inversely proportional to the line width or the square root of the second moment. In some cases [44, 49], a slower exponential decay is superimposed on the non-exponential decay. This was interpreted as transverse relaxation due to spin–lattice coupling. The equation corresponding to equation (8.3) for this process is [66]

$$\frac{1}{T_2} = \frac{\gamma^4\hbar^2}{5r^6} I(I+1)\left[3\tau_c + \frac{5\tau_c}{1+\omega^2\tau_c^2} + \frac{2\tau_c}{1+4\omega^2\tau_c^2}\right]. \tag{8.4}$$

In contrast to longitudinal relaxation which is most sensitive to relatively fast motion (τ_c of the order of 10^{-8} s), transverse relaxation resulting from spin–lattice coupling is most sensitive to slower motion. Consequently, different types of motions may determine T_1 and T_2. In amorphous isotropic solutions of low viscosity T_1 and T_2 are usually equal, since according to equations (8.3) and (8.4) T_1 and T_2 become identical for small τ_c.

Charvolin and Rigny [44, 49] recently reported detailed studies of proton spin–spin relaxation for the $C_{11}COOK$–D_2O system at varying compositions, temperatures and resonance frequencies. For the lamellar mesophase the free induction decay could be divided into three individual decays, two of which decay as e^{-kt^2} and the third as $e^{-k't}$ (k and k' are constants and t is the time after the pulse). These decays correspond to the Fourier transforms of Gaussian and Lorentzian lines respectively. (After the completion of the present review, the original interpretation given in refs. 44 and 49 has been modified. Important progress in the understanding of relaxation in general and of the shape of the free induction decay in particular has been achieved predominantly by Charvolin and Rigny and by Wennerström. The references are listed at the end of this Chapter.) Charvolin and Rigny [44, 49] found no frequency dependence of the free induction decay. These authors also found that an echo is produced after a $(\pi/2, \pi/2)$ pulse sequence but not after a $(\pi/2, \pi)$ pulse sequence where in both cases the second pulse is phase shifted by $\pi/2$. This is consistent with a dipolar origin of the non-exponential decay. An effect of inhomogeneous magnetic fields, on the other hand, should give spin echoes in both cases. In the viscous optically isotropic cubic phase (V_1) the free induction decay is purely exponential and $\langle \Delta B^2 \rangle$ is zero for all protons.

Tiddy [46] studied the spin–spin relaxation for samples in the C_7COONa–$C_{10}OH$–D_2O system having a lamellar or reversed hexagonal structure; T_2^{eff} was independent of the resonance frequency. Orientation of a lamellar phase sample between glass slides produced an angular dependence of T_2^{eff} with a maximum value of 300 μs at an angle of 55° between the applied field and the glass slides. T_2^{eff} was considerably smaller in the lamellar mesophase than in the reversed hexagonal mesophase. The observations made by Tiddy [46] and especially the angular dependence of T_2^{eff} strongly favour a model for transverse relaxation based on residual dipolar interactions. This model is also consistent with a study by Chan *et al.* [47], who found the transverse relaxation of lecithin bilayers dispersed in D_2O to be non-exponential and T_2^{eff} to be frequency independent. Further evidence for this dipolar origin of spin–spin relaxation was presented by Oldfield *et al.* [45] in studying lamellar mesophases composed of D_2O and $C_{11}COOK$, dipalmitoyl lecithin, egg lecithin or ghost lipids. T_2^{eff} was independent of the resonance frequency and "dipolar echoes" (cf. above) were observed with the dipalmitoyl lecithin–D_2O system. With the systems containing choline groups spin echoes were observed with long T_2's ($\simeq$6 ms). The signal intensity and results on partially deuteriated lecithins indicate that this slow relaxation is predominantly due to the —$N^+(CH_3)_3$ group which should consequently be relatively mobile in the liquid crystalline phase.

In the lamellar mesophase of the C_7COONa–$C_{10}OH$–H_2O system the T_2^{MG} for water protons is much smaller than T_1 and decreases strongly with increasing $C_7COONa : H_2O$ ratio, the decreases being most pronounced at less than about 20 water molecules per octanoate ion [72].

In the $C_{12}NO(CH_3)_2$–D_2O system the proton T_2^{MG} was much smaller than T_1 in the anisotropic phases (M_1 and G), whereas the two relaxation times are approximately equal in the optically isotropic phases (amorphous isotropic and V_1) [70]. The transverse relaxation rate changed abruptly on going from an optically isotropic to an anisotropic phase, where the relaxation is non-exponential. Furthermore, T_2^{MG} was found to depend on the pulse spacing and T_2^{eff} to be frequency dependent. It was suggested [70] that these effects are due to molecular diffusion in an inhomogeneous magnetic field. Effects of magnetic field gradients on transverse relaxation were proposed by Kaufman *et al.* [75], who found T_2^{eff} and T_2^{MG} to be different for erythrocyte ghosts and for total ghost lipids in D_2O. However, these interpretations seem to be overruled by the more recent relaxation studies described above. As pointed out by Tiddy [46] the dependence of T_2^{MG} on pulse spacing occurs also with solid crystalline systems, when the pulse spacing is of the same order of magnitude or smaller than T_2^{eff}. Furthermore one may have contributions from $T_{1\rho}$. Arguments against the magnetic field gradient model were also presented by others [44, 45, 47, 49, 76]. It should be stressed that the frequency dependence of $\Delta\nu$, taken as an argument for the magnetic field gradient effect, is not present in most cases (see section 1). Of course it is possible that the behaviour may be different for different anisotropic mesophases.

By observing the free induction decay following a $\pi/2$ pulse, thermal phase transitions in biomembranes can be conveniently studied [77].

Although the application of NMR relaxation methods to amphiphilic mesophases is of recent date the work described demonstrates their potential value in elucidating primarily the microdynamic behaviour of these systems. The value of these methods will be increased if different types of motion can be distinguished. Methods of doing this, such as studying the translational motion directly by determination of self-diffusion coefficients, comparing relaxation data for different isotopes, and changing the intermolecular dipolar interaction by isotopic substitution, have been used with success for isotropic solutions [78].

4. Self-diffusion Coefficients

Information on translational motion may be obtained from the intermolecular part of the relaxation rate and also more directly from translational self-diffusion coefficients (D) measured by a pulse NMR method [65]. In this a $\pi/2$ pulse is followed by a π pulse after varying times, while the sample is situated in a magnetic field purposely made very inhomogeneous in a controlled way. This method has been applied to study translational motion in amphiphilic mesophases.

Blinc *et al.* [79] have found (by PMR) that the translational self-diffusion coefficient of the amphiphile parallel to the applied field for the neat mesophase formed by anhydrous $C_{17}COONa$ is about 10^{-6} cm^2 s^{-1}, which corresponds to a relatively fast translational motion. In the sub-neat and neat mesophases of anhydrous $C_{15}COONa$, D is about $6 . 10^{-6}$ cm^2 s^{-1} and virtually independent of temperature [80].

Translational diffusion of water in the lamellar and hexagonal mesophases of the $C_{15}COONa$–H_2O system was also studied by Blinc *et al.* [73, 80]. Here D increases exponentially with increasing temperature and drops considerably at the transition to amorphous isotropic solution. The activation energies were 18 kJ/mole for the hexagonal mesophase and 14 kJ/mole for the lamellar mesophase. Translational motion in the mesophases is quite rapid. For example, in the hexagonal phase with 70% H_2O at 350° K, D_{H_2O} is lowered only by a factor of 3 as compared with pure water. The translational motion of water increases with increasing water content. The lower value of D in the amorphous isotropic solution as compared with that in the mesophases might be attributed to a larger fraction of the water molecules being entangled with the amphiphile in the amorphous phase than in the mesophases [73, 80] in which long water channels are present. This is supported by the observation that $D_{H_2O} \simeq D_{C_{15}COO^-}$ in the amorphous isotropic solution whereas in the mesophases $D_{H_2O} \gg D_{C_{15}COO^-}$. The translational diffusion coefficient of the amphiphile in the viscous isotropic cubic mesophase (V_1) in the $C_{11}COOK$–D_2O system was 2.10^{-6} cm^2 s^{-1} at 90°C [49].

5. ^{13}C and ^{19}F Magnetic Resonance

The majority of NMR studies on amphiphilic liquid crystals have used proton magnetic resonance. The reason for this is the great NMR sensitivity of protons as well as the large natural abundance of protons. However, PMR has some drawbacks which are particularly important when dealing with solutions containing large aggregates or large molecules. Thus the dipolar interactions are especially large for protons due to the large nuclear magnetic moment and secondly the chemical shifts in proton NMR are small. ^{13}C NMR possesses advantages in these respects. This nucleus having $I = \frac{1}{2}$ has thus a comparatively small magnetic moment which makes the dipolar broadening an order of magnitude smaller than in PMR. Furthermore, the chemical shifts in ^{13}C spectra are 10–20 times larger than in proton spectra. Consequently, ^{13}C NMR spectra show much better resolution than ^{1}H spectra. However, the use of ^{13}C NMR was, until recently, extremely difficult due to the low sensitivity (about 1·6% of PMR) and the low natural abundance (1·1%). With the development of a novel NMR technique, Fourier transform NMR [81], it has, however, been possible to obtain ^{13}C spectra of non-enriched samples with a good signal-to-noise ratio. In Fourier transform NMR the whole NMR spectrum is excited by a radiofrequency pulse of short duration. The decay obtained is transferred into the continuous wave spectrum by a Fourier transformation.

^{13}C Fourier transform NMR spectra of the lamellar mesophase of the egg yolk lecithin–water system have been reported by Oldfield and Chapman [82]. Well-resolved signals were obtained for the olefinic groups, the *N*-methyl groups, the terminal methyl groups and the methylene groups. The observation of very narrow *N*-methyl ^{13}C signals indicates that the —$N^+(CH_3)_3$ group is relatively mobile. Similar ^{13}C NMR results were obtained by Metcalfe *et al.* [83] for dipalmitoyl lecithin–D_2O with or without ultrasonic irradiation. Even in the case of human erythrocyte membranes a considerable fine-structure was obtained [83].

Fourier transform NMR in combination with the pulse methods for determination of relaxation times (section 3) enables the determination of relaxation times for different signals in a resolved spectrum [84]. T_1 relaxation times for various peaks in the ^{13}C spectrum of sonicated dipalmitoyl lecithin in D_2O have been reported by [83] and are given in Table 8.1. It is significant that T_1 increases along the alkyl chains towards the methyl group, indicating that the motion (cf. equation 8.3), along the alkyl chains, increases towards the methyl group. Also the $—N^+(CH_3)_3$ group seems to be relatively mobile.

TABLE 8.1. *T_1 relaxation times of ^{13}C nuclei in dipalmitoyl lecithin dispersed in D_2O by sonication* [83]. *Palmitic acid carbons are numbered from the methyl group* (1)

Carbon	T_1, s
1	3·34
2	1·76
3	1·14
4–13	0·55
14	0·52
15	0·10
$—N^+(CH_3)_3$	0·57

It can be expected that ^{13}C Fourier transform nuclear magnetic resonance spectroscopy will become a very important tool for studying not only membranes and membrane models but also amphiphilic liquid crystals in general.

^{19}F nuclei and protons are from the magnetic resonance point of view rather similar, both having spin quantum numbers of 1/2 and approximately equal magnetic moments. Since, in addition, hydrocarbons and fluorocarbons have rather similar chemical properties, ^{19}F resonance has been used as an alternative method of investigation in systems involving hydrocarbon chains. Up to now, little use has been made of this method in studies of amphiphilic mesophases.

Tiddy [74] studied the ^{19}F magnetic relaxation behaviour of the lamellar mesophase for ammonium perfluoro-octanoate–D_2O. As in PMR investigations of similar systems a single T_1 is observed while the free induction decay is non-exponential (cf. section 3). Dipole interaction modulated by molecular rotation around the long axis is considered the most probable relaxation mechanism. From a comparison of ^{19}F and proton relaxation times in similar systems Tiddy [74] concludes that molecular rotation is more rapid in the hydrocarbon chains than in the fluorocarbon chains.

A very interesting "nuclear probe" application of ^{19}F resonance is that of Birdsall *et al.* [85] using monofluorostearic acids solubilized in lecithin vesicles. ^{19}F T_2 values, evaluated from line width data, have been measured as a function of the fluorine substituent position along the hydrocarbon chain and the temperature. The results indicate a marked increase of the methylene group mobility with increasing distance from the carbonyl group.

6. Quadrupole Relaxation

For a nucleus with spin quantum number greater than 1/2 the predominant nuclear magnetic relaxation mechanism in fluid systems is usually the interaction between the nuclear electric quadrupole moment and fluctuating electric field gradients (EFG's) at the nucleus. In the simple case where $\omega\tau_c \ll 1$, equation (8.5) applies for the spin–lattice relaxation time, provided the asymmetry parameter may be neglected [2].

$$\frac{1}{T_1} = \frac{3}{40} \cdot \frac{(2I+3)\langle e^4 q^2 Q^2 \rangle}{I^2(2I-1)\hbar^2} \cdot \tau_c \qquad (8.5)$$

Here eQ is the electric quadrupole moment of the nucleus and eq is the principal EFG tensor component. Whether the condition $\omega\tau_c \ll 1$ applies or not may be decided by studying the magnetic field or temperature dependence of the relaxation rate.

The quadrupole relaxation method is one way of studying the binding of small ions in amphiphilic liquid crystals. As seen from equation (8.5) a relaxation time measurement yields the product of $\langle q^2 \rangle$ and τ_c. A separation of these factors is difficult, but a comparison of data for different phases is still informative. Experimentally T_1 is often equal to T_2 [86, 87], and accordingly the longitudinal relaxation rate may be obtained from the NMR line width (the line width is proportional to the transverse relaxation rate). Lindblom *et al.* [86–88] recorded the ^{81}Br magnetic resonance signals in different phases in the three-component system $C_{16}N(CH_3)_3Br$–C_6OH–H_2O (phase diagram given by Ekwall *et al.* [89]). The line width is essentially the same in the middle phase as in amorphous isotropic concentrated aqueous solutions of the surfactant. This shows that the structural and dynamic properties of the two phases are nearly the same in the neighbourhood of the bromide ions. The line widths in the hexagonal phase are about twenty-five times that of a free bromide ion in aqueous solution. Since the EFG values are certainly not smaller in this mesophase than in an ordinary aqueous solution this figure indicates that the correlation time is of the order of 10^{-10} s or less in the middle phase. In the lamellar phase the bromide ions are more firmly bound and the binding is reinforced upon lowering the water concentration. Keeping the $C_{16}N(CH_3)_3Br$ to H_2O ratio constant but increasing the C_6OH concentration in the lamellar phase leads to a reduction in line width corresponding to a release of bromide ions from the amphiphilic lamellae. Similarly, in the lamellar phase of the system C_7COONa–C_7COOH–D_2O an increase in the ratio of acid to soap at constant water content leads to a decrease in the ^{23}Na line width [90]. The counter-ion binding in the systems C_7COONa–$C_{10}OH$–H_2O, C_7COORb–$C_{10}OH$–H_2O and C_8NH_3Cl–$C_{10}OH$–H_2O has also been studied by this method [91, 92]. A marked difference is that the relative change in line width in going from a dilute aqueous solution to a mesophase is an order of magnitude greater for anions (Br^-, Cl^-) than for cations (Na^+, Rb^+). For the cations in lamellar phases the line width is independent of water content at high water concentrations. As the water concentration falls the line widths start to increase rapidly at a certain water-to-soap ratio. A comparison of the ^{85}Rb NMR line widths in the hexagonal phase of the system $C_7COORb–C_{10}OH–H_2O$ with the line widths for amorphous isotropic aqueous solutions of C_7COORb shows no abrupt change in line width in going from the isotropic solution to the mesophase [92, 93].

The possibility of studying ion binding in biologically occurring mesophases is apparent, and although not primarily concerned with liquid crystalline systems, the work of Cope and others [94–100] should be mentioned. These investigations are, however, not generally concerned with the determination of line widths or relaxation times, which makes a dynamical description of the systems difficult. Recently the interaction of sodium ions with erythrocyte membranes has been studied by determination of the ^{23}Na NMR line widths [101]. Competition between sodium and potassium ions was demonstrated by a decrease of the ^{23}Na line width on addition of potassium ions. In the lamellar mesophase of the egg yolk lecithin–sodium cholate–water system T_1 for ^{23}Na is an order of magnitude smaller than in an aqueous sodium chloride solution, indicating a strong interaction between the sodium ions and the lamellae [102]. Nuclear magnetic relaxation of ^{23}Na and ^{39}K in biological cells has also been reported [103–105].

The deuteron has a quadrupole moment and the quadrupole relaxation method may therefore be applicable for studying the water binding in amphiphile–D_2O systems. The experimental data available to date are limited, however [70, 106, 107]. Hansen and Lawson [70] observed that T_2 for the D_2O molecules in such a system depends on the pulse spacing. Charvolin and Rigny [107] suggested from the saturation behaviour of the deuteron signals in the lamellar phase and in a viscous isotropic phase in the $C_{11}COOK–D_2O$ system that two types of deuteron sites corresponding to different relaxation times are present (cf. also section 7).

7. Quadrupole Coupling

A nucleus with $I > \frac{1}{2}$ has an electric quadrupole moment which interacts with electric field gradients (EFG's) in the nuclear position. These arise from a charge distribution in the nuclear environment, the symmetry of which is lower than cubic. The basic theory of quadrupole interaction is given by Cohen and Reif [108], and is further discussed for the case of liquid crystals by Buckingham and McLauchlan [109] and in the preceding chapter. In the simplest case of nuclei with $I = 1$ (for example deuterons) fixed in a crystal lattice and interacting with an axially symmetric EFG, the first-order interaction gives a splitting of the resonance line into a symmetric doublet, the frequency separation being [108]

$$\nu_Q = \frac{3e^2qQ}{4h}(3\cos^2\theta - 1) \tag{8.6}$$

where θ is the angle between the direction of the principal component of the EFG and the magnetic field. The remaining variables have been defined already. The quadrupole coupling constant, e^2qQ/h, is denoted E_Q in the following paragraphs.

In the case of rapid molecular reorientation, the quadrupole splitting will be proportional to $\langle 3\cos^2\theta - 1\rangle$, which for an isotropic reorientation equals zero. However, in systems such as anisotropic liquid crystals, $\langle 3\cos^2\theta - 1\rangle$ will not be effectively reduced to zero and a quadrupole splitting may be observed, which will in general be much smaller than that observed in solids. The first data on quadrupole splittings of the deuteron resonance from D_2O in anisotropic mesophases were reported by Flautt and Lawson [110].

According to the above expression the quadrupole coupling constant, E_Q, may be calculated from a measured value of the quadrupole splitting only if $\langle 3\cos^2\theta - 1\rangle$, i.e. the degree of orientation, is known. This may in some cases be determined by an independent method, e.g. proton magnetic resonance, as discussed in the preceding chapter. This type of measurement has not yet been carried out on amphiphilic mesophases, although systems such as those used by Black *et al.* [111] (see section 8) may also be suitable for the determination of quadrupole coupling constants.

Another application of quadrupole splitting measurements is obvious from equation (8.6). For a molecule with a known value of E_Q the degree of orientation within the micro-crystallites in an anisotropic mesophase may be evaluated from the quadrupole splitting. In this way the influence of, for example, temperature and composition of mesophases on the degree of molecular orientation can be studied, and information obtained concerning the orientation mechanisms. The first measurements of this type were made by Lawson and Flautt [29] on the system $C_{12}NO(CH_3)_2$–D_2O and have been followed by other investigations [37, 73, 80, 90, 107, 112, 113]. All these, except for ref. 112, were concerned with the quadrupole splitting of deuterium in lamellar or hexagonal amphiphilic mesophases containing deuterium oxide. In ref. 112 the deuteron quadrupole coupling has been investigated in mesophases of water and partially deuterated lecithin. The observed splitting from deuterium oxide in anisotropic mesophases was originally interpreted as due to partial orientation of the water molecules [29], but later work [113] demonstrated that rapid exchange between water deuterons and deuterons bound to the amphiphilic molecules may also contribute to the quadrupole splitting. In this connection it is interesting that rapid exchange of nuclei may average out the intramolecular dipole–dipole interaction between the protons in a H_2O molecule but not in general the deuteron quadrupole interaction in D_2O [114].

Buckingham and McLauchlan [109] gave the following expression for the quadrupole splitting, ν_Q, in partially oriented molecules under rapid reorientation:

$$\nu_Q = \left|\tfrac{3}{2}E_Q S.\tfrac{1}{2}\langle 3\cos^2\Omega - 1\rangle\right| \tag{8.7}$$

where S is a factor characterizing the partial orientation of the molecules and Ω denotes the angle between the direction of the constraint, i.e. the axis of rotational symmetry, and the magnetic field. In a lamellar phase, the constraint direction is perpendicular to the lamellar planes and in a hexagonal phase it is parallel to the rods. (As discussed in the references to this section listed at the end of the Chapter, except for partial orientation, S may also be influenced by the asymmetry parameter.) If all directions have equal probabilities, the most probable value of Ω is 90°, which gives rise to a powder pattern in which the difference between the absorption maxima corresponds [108] to

$$\nu_p = \tfrac{3}{4} E_Q S \tag{8.8}$$

Figure 8.5*b* shows the value of $E_Q S$ measured at different compositions of the system $C_{12}NO(CH_3)_2$–D_2O together with a characteristic "powder pattern" from a randomly oriented lamellar phase (Fig. 8.5a). The fact that such powder patterns in some systems remain constant for more than

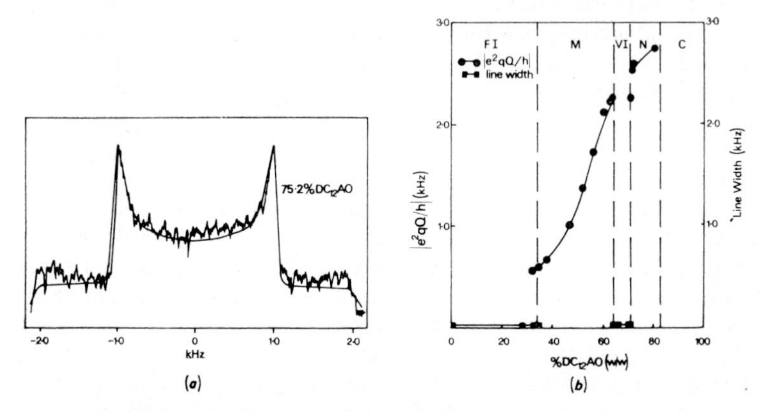

FIG. 8.5. (*a*) Experimental and calculated deuteron spectra of the neat phase of $C_{12}NO(CH_3)_2$ [$DC_{12}AO$]–D_2O.

(*b*) A plot of $E_Q S$ (in the figure denoted by $|e^2qQ/h|$) for anisotropic phases and line width for isotropic phases of the system $DC_{12}AO$–D_2O. The different phases are denoted by FI (fluid isotropic), M (middle), VI (viscous isotropic), N (neat) and C (crystalline) [29].

24 hours in the magnetic field [113] indicates that the reorientation of "micro-crystallites" can be very slow in anisotropic mesophases. Otherwise the diamagnetic anisotropy of the lamellae would lead to a macroscopic orientation within the phase. This is observed in some cases [63, 111, 113] and is discussed in the next section.

The explicit expression for S in the case of partially oriented D_2O molecules is [109, 113]

$$S = S_{11} \cos^2 \beta + S_{22} \sin^2 \beta \tag{8.9}$$

where S_{11} and S_{22} are independent orientation matrix elements, which may have values between $-\tfrac{1}{2}$ and $+1$, and β denotes half the D–O–D angle in the D_2O molecule. Equation (8.9) shows that variations in the S parameter may be attributed either to changes in the *degree* of orientation, i.e. the extent to which the orientation matrix elements deviate from zero or to changes in the *direction* of the preferred orientation, i.e. the ratio between S_{11} and S_{22}.

As indicated by relaxation studies on heavy water solutions [115], the value of E_Q is approximately constant also under different hydrogen bonding conditions. Therefore, variations in the quadrupole splitting may primarily be attributed to changes in the S parameter (cf. equation 8.8). However, since the value of E_Q is not known, in the following paragraphs we use $E_Q S$ for characterizing the partial orientation of the D_2O molecules. The order of magnitude of the S values can be estimated using the assumption that E_Q is approximately equal to the value in solid D_2O, 215 kHz [116] and by this procedure the S values are found to range between $\simeq 0$ and ca. 0·02 [29, 37, 63, 73, 80, 90, 107, 111, 113].

The partial orientation of water molecules may be explained as an interaction (hydrogen bonding, ion–dipole interaction, dipole–dipole interaction and/or van der Waal's interaction) between water and amphiphile molecules resulting in some transfer of partial orientation from the amphiphile to the water molecules [113]. Therefore a change in the degree of preferential orientation of water can result both from an altered amphiphile–water interaction and from a change in the degree of order of the amphiphilic molecules. A separation of these effects is difficult with the experimental results available. However, according to the commonly accepted pictures of lamellar and hexagonal mesophases [117] there is a considerable orientation of the amphiphilic molecules, and thus, the relative variations in their degree of orientation are expected to be small as long as the structure is maintained.

The influence of changes in composition and temperature on $E_Q S$ has been studied for several systems [29, 37, 73, 80, 90, 107, 113]. An increased water content will generally reduce the value of $E_Q S$. This was first interpreted by Lawson and Flautt [29] in terms of a simple two-site model, in which rapid exchange takes place between self-associated water and oriented surfactant-associated water. This model has also been adopted by Charvolin and Rigny for the system $C_{11}COOK$–D_2O [107], but in this case slow exchange is assumed between oriented and unoriented water molecules. However, from the spectra presented [107], one can conclude that the relative amount of unoriented water is extremely small and this may indicate that the unsplit deuteron resonance is due to the presence of small amounts of an isotropic phase (cf. also ref. 29).

An alternative mechanism for the quadrupole splitting in the system $C_7COOCH_2CH(OH)CH_2OH$–D_2O has been discussed by Ellis *et al.* [37] considering rapid deuteron exchange. This interpretation was later abandoned [118].

A later investigation [113] demonstrated that both these effects may contribute to the quadrupole splitting. In a lamellar phase of $C_8ND_3^+Cl^-$–D_2O, separate powder patterns are observed for ammonium and water deuterons while a very small addition of C_8ND_2 is sufficient to increase the rate of deuteron exchange to the extent that the two powder patterns coalesce into one in which the quadrupole splitting $\langle \nu_p \rangle$ is given by the following expression [113].

$$\langle \nu_p \rangle = X_1 \nu_1 + X_2 \nu_2 \tag{8.10}$$

Here X_1 and X_2 are the fractions of the total number of deuterons bound to water molecules and amphiphilic molecules, respectively, and ν_1 and ν_2 are the corresponding quadrupole splittings on slow exchange (cf. also ref. 37). The pH dependence of the rate of amino deuteron exchange observed is in accordance with proton resonance investigations [119].

Recently separate deuteron powder patterns were also observed in lamellar mesophases in the C_7COONa–$C_{10}OH$–D_2O and C_7COOK–$C_{10}OH$–D_2O systems [92]. With increasing temperature the signals broaden and finally coalesce into one powder pattern.

Separate resonance signals can be observed if the lifetime of the deuterons in the distinguishable positions is larger than the inverse difference in quadrupole splitting [113]. The observation of separate resonance signals from deuterons in water and in amphiphilic molecules offers possibilities for studying separately the degree of partial orientation of these molecules.

The averaging of quadrupole splittings by chemical exchange processes offers a new NMR approach to studies of rapid exchange reactions. The conventional NMR method is based upon averaging of chemical shifts and spin–spin splittings in high-resolution spectra and, by application of this method in PMR, average lifetimes in the approximate region of 1–10^{-4} s have been studied [4]. Since the deuteron quadrupole splittings are about two orders of magnitude larger than proton chemical shifts and spin–spin splittings, averaging of quadrupole splittings by chemical exchange allows measurements of average lifetimes down to about 10^{-6} s. It is pointed out that macroscopically aligned systems (see section 8) offer great advantages compared with randomly oriented systems in respect of the spectrum intensity.

The influence of temperature on the value of $E_Q S$ is different in different anisotropic mesophases. In the system $C_7COOCH_2CH(OH)CH_2OH$–D_2O the quadrupole splitting is independent of temperature between 15 and 40°C [37]. In several other lamellar phases the quadrupole splitting decreases with increasing temperature [90, 113], while in the systems $C_{15}COONa$–D_2O [73, 80] and C_8NH_3Cl–D_2O [113] and in some ternary lamellar phases containing aliphatic alcohols [90], the reverse temperature dependence is observed. In the latter cases, the temperature dependence has been explained as partly due to exchange phenomena [90, 113].

The quadrupole splitting in D_2O has also been studied for anisotropic (mostly lamellar) mesophases containing two different types of amphiphilic molecules [90, 113] one being of the ionic type. At constant ratio between the amphiphiles, the value of $E_Q S$ generally decreases with increasing water content as in binary systems, while at constant mole fraction of D_2O, a decrease in the proportion of ionic amphiphile generally gives an increase in the value of $E_Q S$ [90, 113]. The last observation may be due to various effects, e.g. reduced counter-ion binding and higher degree of hydrogen bonding between hydrophilic groups and water [90, 113].

Solubilization of *p*-xylene in lamellar mesophases of

$$C_9H_{19}C_6H_4(OC_2H_4)_nOH^*–D_2O$$

leads to a considerable increase in the value of $E_Q S$ [113]. This is tentatively explained in terms of an increase in the degree of orientation of the amphiphilic molecules [113].

Quadrupole splitting of deuteriated hydrocarbon groups in anisotropic mesophases offers another method for investigation of molecular motion

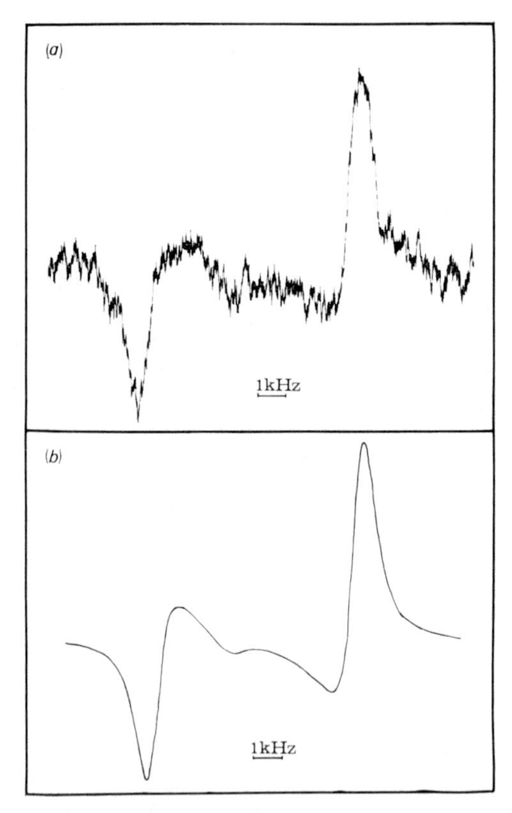

FIG. 8.6. (*a*) Experimental ^{35}Cl spectrum of a lamellar mesomorphic sample having a weight composition of 50% C_8NH_3Cl, 20% $C_{10}OH$ and 30% H_2O. Resonance frequency 5·78 MHz, temperature 29°C [120].

(*b*) The corresponding computed NMR powder pattern under the assumption of second-order quadrupole splitting. The spectrum parameters are given in ref. 120.

* The C_9H_{19} group is branched.

in such phases. This was demonstrated by Oldfield *et al.* [112] for the system di(perdeuterio)myristoyl–L-α-lecithin–H_2O. Below 23°C the system is in the gel phase and no quadrupole splitting is resolved. Above 23°C an anisotropic mesophase is formed in which a quadrupole splitting is observed. The splitting decreases slightly with increasing temperature, probably due to increased molecular mobility. A marked increase of the quadrupole splitting is observed on addition of cholesterol to the mesophase, indicating an increase in the degree of order of the hydrocarbon chains [112].

As mentioned in section 6, line width measurements have been performed on counter-ions in amphiphilic mesophases. Since the nuclei investigated possess a quadrupole moment one could in principle also observe quadrupole splitting in the counter-ion resonance from anisotropic mesophases. This is often not observed, but in a recent investigation [120] it was demonstrated that the ^{35}Cl and ^{37}Cl NMR signals in the

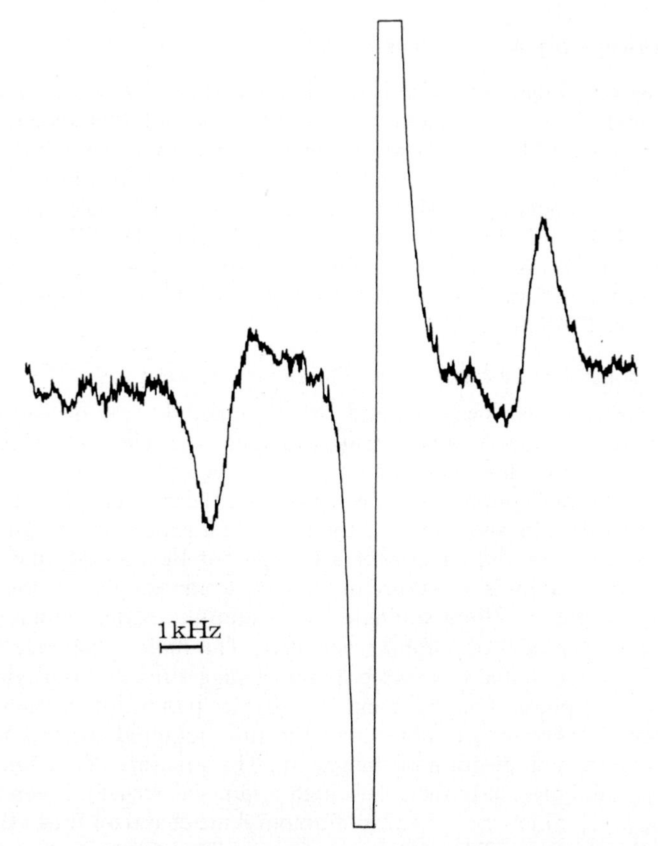

FIG. 8.7. First-order quadrupole coupling in a ^{23}Na NMR spectrum of a lamellar mesomorphic sample composed of 64% lecithin, 16% sodium cholate and 20% water (by weight). Resonance frequency 15·82 MHz, temperature 25°C [102].

lamellar mesophase of the system C_8NH_3Cl–$C_{10}OH$–H_2O show splittings due to second-order quadrupole interaction. Evidence for a second-order splitting was obtained from a line-shape analysis (Fig. 8.6) and the observed inverse dependence of the splitting on the magnetic field strength. The magnitude of the splitting increases with increasing amount of C_8NH_3Cl and with decreasing water content. In the system lecithin–sodium cholate–water the ^{23}Na resonance signal is split by first-order quadrupole interaction as demonstrated by Lindblom [102] (see Fig. 8.7). The ^{14}N NMR signal in the lamellar mesophase of the C_7COONH_4–$C_{10}OH$–H_2O system is also split by quadrupole interactions [92] as well as the NMR signals of ^7Li, ^{39}K, ^{85}Rb, ^{87}Rb, ^{133}Cs and ^{81}Br in lamellar mesophases [92]. The use of quadrupole coupling for simple ions offers a new method in studies of interactions between ions with quadrupolar nuclei and amphiphilic molecules in anisotropic environments, of which membrane systems are especially interesting.

8. Macroscopically Aligned Phases

Macroscopic alignment of anisotropic mesophases by magnetic and/or electric fields is well known and finds extensive use in NMR spectroscopy. However, few NMR investigations have been concerned with macroscopically aligned amphiphilic mesophases. Alignment has been detected by several independent NMR parameters, e.g. quadrupole splitting in D_2O [63, 111, 113], direct dipole coupling [63, 111, 118, 121–130], PMR line width of protons within the amphiphile molecules [48, 113], transverse relaxation time of such protons [46] and width [113] and shape [42, 131] of the water PMR line.

High-Resolution Spectra from Dissolved Molecules

Much interest has been focused on the possibility of measuring the direct magnetic dipole–dipole coupling and the electric quadrupole coupling in molecules dissolved in aligned mesophases. In most such investigations non-amphiphilic nematics have been used. (See the preceding chapter.) However, in 1967 the first investigation of a magnetically aligned amphiphilic liquid crystal was reported by Lawson and Flautt [63, 110]. The system is a mixture of C_8 or C_{10} alkylsulphates, the corresponding alcohols, sodium sulphate and deuterium oxide, and has several advantages over non-amphiphilic nematics. The main advantage is that orientation is facilitated by sample spinning, suggesting an axial symmetry of the aligned phase. On alignment the powder pattern of the deuterium resonance changes to a doublet, and the time required for this process depends on the composition of the system. The structure of the phase has not been definitely established, although a lamellar structure seems most probable [111]. Since the system is of amphiphilic character it can dissolve both hydrophilic and lipophilic compounds. The temperature range of stability (10–75°C) is such that spectra can be recorded at ordinary probe temperature. Furthermore, the sample can be spun perpendicular to the magnetic field direction without destroying the alignment. The line widths observed from molecules dissolved in this phase are generally of the order of 2 Hz, which is less than those commonly observed in non-amphiphilic nematics. Several compounds have been dissolved in this lyotropic phase, e.g. methanol [63], benzene [121], ethanol [111], halogen-substituted benzenes [122], dimethyl sulphone [123], α-L-alanine [123], β-alanine [123], selenophene [124], 2-furan aldehyde [125], 3-furan aldehyde [125], methyl halides [126], *N,N'*-dimethylacetamide-d_3 [127, 128], tetrafluoro-1, 3-dithietane [129], acetone [130] and dimethylsulphoxide [130]. The spectra have been analysed in terms of chemical shifts, δ, indirect dipole coupling constants, J, and direct dipole coupling constants, D, according to the principles outlined by Luckhurst. In the case of halogen-substituted benzenes [122], methyl halides [126] and tetrafluoro-1,3-dithietane [129], chemical shift anisotropies have been evaluated, while for *N,N'*-dimethylacetamide-d_3 [127, 128] the rotation around the amide bond has been studied. Experimental and theoretical spectra for the case of ethanol are in Fig. 8.8. The

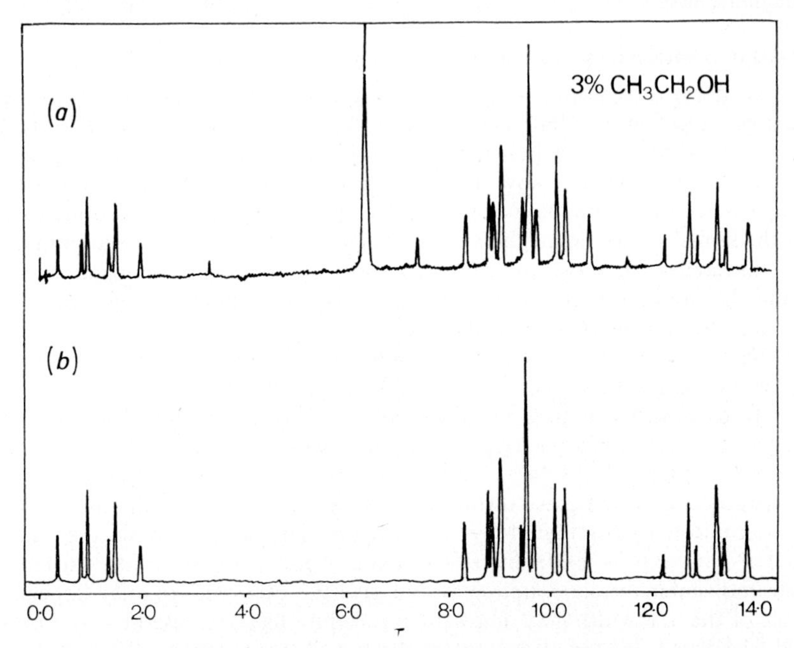

FIG. 8.8. (*a*) Experimental PMR spectrum of C_2H_5OH in a magnetically aligned amphiphilic mesophase. (The intense line is due to HDO.)
(*b*) A computer simulated spectrum with the parameters given in Table 8.2 [111]. (τ is equal to $10 - \delta$, where δ is the chemical shift in ppm referred to tetramethylsilane.)

theoretical spectra were fitted to the experimental by means of a modified version of a LAOCN3 computer program, and the resulting parameters are in Table 8.2

In the alkylsulphate mesophase, benzene molecules are oriented with the molecular plane perpendicular to the magnetic field and ethanol

TABLE 8.2. *Parameters of the spectrum of ethanol oriented in an amphiphilic mesophase.* A *and* B *denote protons of methylene and methyl groups, respectively. A positive sign of* J_{AB} *has been assumed.* (*From ref.* 111.)

Parameter	Value
$\delta_B - \delta_A$	$2{\cdot}462 \pm 0{\cdot}11$ ppm
J_{AB}	$+7{\cdot}05 \pm 0{\cdot}07$ Hz
D_{AA}	$-46{\cdot}38 \pm 0{\cdot}14$ Hz
D_{BB}	$-780{\cdot}69 \pm 0{\cdot}13$ Hz
D_{AB}	$-62{\cdot}59 \pm 0{\cdot}11$ Hz

molecules are aligned with the long molecular axis perpendicular to the magnetic field [111, 121].

ALIGNED MESOPHASE STRUCTURE

In other investigations, only studies of the aligned mesophase itself were carried out. The lamellar mesophase of $C_9H_{19}C_6H_4(OC_2H_4)_{10}OH$–$D_2O$ has been studied [113] in which the alignment was produced by phase separation following a slow temperature decrease of an amorphous isotropic solution in the magnetic field. This phase aligns with the optic axis of the lamellae parallel to the magnetic field direction [113]. The alignment is maintained for more than 24 hours even if the sample is rotated 90° from the original direction. On the other hand, the macroscopic viscosity is such that the phase can easily flow.

The tendency of anisotropic mesophases to orient close to polar surfaces such as glass has been used by de Vries and Berendsen [48] to produce a macroscopically aligned phase of potassium oleate–water. The angular dependence of the proton resonance line width has been shown to be proportional to $(3\cos^2\theta - 1)$, θ being the angle between the direction perpendicular to the glass surface and the magnetic field. If the molecular reorientation is dominated by rotation around the long molecular axis and if this axis is preferentially oriented perpendicular to the surface, intramolecular dipole coupling would give the observed angular dependence of the line width [48]. The same procedure has been used by McDonald and Peel [118] for alignment of the lamellar mesophase of the system $C_7COOCH_2CH(OH)CH_2OH$–H_2O. In this case two sets of doublets have been observed and ascribed to methylene and water protons, respectively. Both doublets show the $(3\cos^2\theta - 1)$ dependence described above and the splittings have been measured at different compositions and temperatures. The results are remarkable for several reasons. Firstly, according to this investigation, all methylene and methyl protons have essentially the same dipolar coupling while one might expect a distribution of dipolar couplings along the hydrocarbon chain. Secondly, in order to observe a dipolar coupling of the order of 1 kHz the average lifetime of the O–H bonds must be significantly longer than that in pure water at neutral pH. A broadening of the water signal at higher temperatures was ascribed to an increased proton exchange rate. The same method of alignment was employed by Tiddy [46] for the system $C_7COONa–C_{10}OH–D_2O$ and in this case the angular dependence of T_2^{eff} was studied (cf. section 3). Corkill *et al.* [42] studied the alignment of a lamellar phase in a glass capillary. A shift of the water proton resonance was observed between an aligned and a randomly oriented phase and discussed in terms of magnetic susceptibility effects. Such effects were also investigated by Drakenberg *et al.* [131].

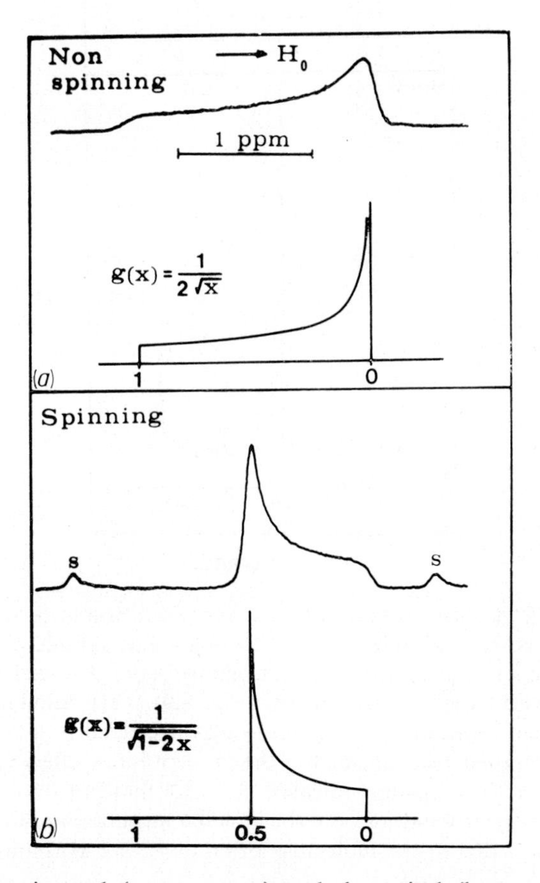

FIG. 8.9. Experimental (upper curves) and theoretical (lower curves) water PMR line shapes for a lamellar phase of $C_8NH_2–C_8NH_3Cl–H_2O$ (mole fractions 0·115, 0·115 and 0·770 respectively) [131]. The scale of the abscissa in the theoretical spectra is defined in ref. 131. $g(x)$ denotes the theoretical line shape and the limits of x are given below.

a. Random orientation, non-spinning sample ($0 \leqslant x \leqslant 1$)
b. Random orientation, spinning sample ($0 \leqslant x \leqslant \frac{1}{2}$)
c. Cylindrical orientation, non-spinning sample ($0 \leqslant x \leqslant 1$)
d. Cylindrical orientation, spinning sample ($x = \frac{1}{2}$)

Spinning sidebands are indicated by s.

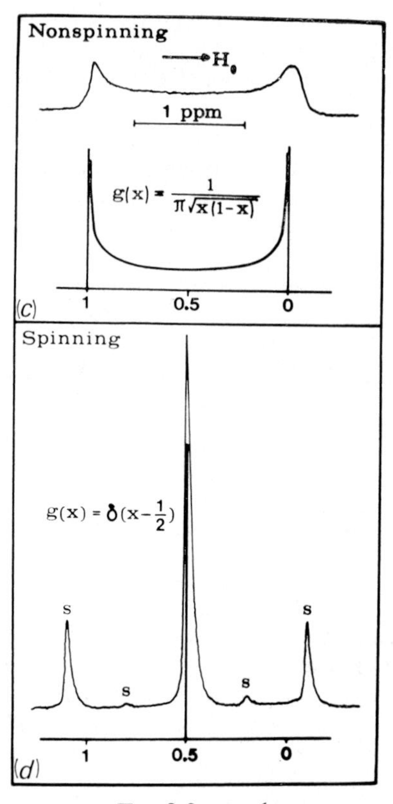

FIG 8.9 contd.

Characteristic proton resonance line shapes were found in some lamellar systems for both randomly oriented samples and samples oriented with an axial symmetry, and theoretical line shapes were deduced in very good agreement with experimental spectra (Fig. 8.9) [131]. Similar line shapes have also been found for various molecules dissolved in a lamellar phase [132]. The detailed mechanism for the susceptibility effect has not been fully explained. The results indicate [131, 132] that not only the susceptibility anisotropy originating from the lamellar structure itself is responsible but that effects due to the molecular orientation are also important.

Recent Work

After the completion of this review, the number of applications of NMR to amphiphilic mesophases has grown strongly. The following references present examples of novel applications or of progress in interpretation.

Section 1. WENNERSTRÖM, H. *Chem. Phys. Lett.* **18,** 41 (1973).
Section 2. KOSTELNIK, R. J. and CASTELLANO, S. M. *J. magn. Resonance* **9,** 291 (1973).

Section 3. CHARVOLIN, J. and RIGNY, P. *J. chem. Phys.* **58,** 3999 (1973) and *Nature, New Biology* **237,** 127 (1972).
WENNERSTRÖM, H. *J. magn. Resonance* (in press).

Section 4. ROBERTS, R. T. *Nature* **242,** 348 (1973).
CONLON, T. and OUTHRED, R. *Biochim. biophys. Acta* **288,** 354 (1972).

Section 5. LEE, A. G., BIRDSALL, N. J. M. and METCALFE, J. C. *Chemistry in Britain* **9,** 116 (1973).

Section 6. CHARVOLIN, J. and RIGNY, P. *Chem. Phys. Lett.* **18,** 575 (1973).

Section 7. LINDBLOM, G. and LINDMAN, B. *Molec. Crystals Liqu. Crystals* (in press).
PERSSON, N.-O., WENNERSTRÖM, H. and LINDMAN, B. *Acta Chem. Scand.* (in press).

Section 8. SAMULSKI, E. T., SMITH, B. A. and WADE, C. G. *Chem. Phys. Lett.* **20,** 167 (1973).

Acknowledgements

The authors are grateful to T. J. Flautt, S. Forsén, J. R. Hansen, K. D. Lawson and P. A. Winsor for their valuable comments on the manuscript.

REFERENCES

[1] CARRINGTON, A. and MCLACHLAN, A. D. *Introduction to Magnetic Resonance with Applications to Chemistry and Chemical Physics*, Harper & Row, New York (1967).

[2] ABRAGAM, A. *The Principles of Nuclear Magnetism*, Clarendon Press, London (1961).

[3] SLICHTER, C. P. *Principles of Magnetic Resonance*, Harper & Row, New York (1963).

[4] EMSLEY, J. W., FEENEY, J. and SUTCLIFFE, L. H. *High Resolution Nuclear Magnetic Resonance Spectroscopy*, Pergamon Press, London (1965).

[5] FLAUTT, T. J. and LAWSON, K. D. *Adv. Chem. Ser. Ordered Fluids and Liquid Crystals, Am. chem. Soc.* No. 63 (1967), p. 26.

[6] CLIFFORD, J., PETHICA, B. A. and SMITH, E. G. *Membrane Models and the Formation of Biological Membranes*, Proc. of the 1967 meeting of the Intern. Conference on Biological Membranes (edited by L. Bolis and B. A. Pethica), North Holland, Amsterdam (1968), p. 19.

[7] CHAPMAN, D. *Biological Membranes, Physical Fact and Function* (edited by D. Chapman), Academic Press, London (1968), p. 125.

[8] CHAPMAN, D. and SALSBURY, N. J. *Recent Progress in Surface Science*, Vol. 3, Academic Press, New York (1970), p. 121.

[9] CHAPMAN, D. *Adv. Chem. Ser. Molecular Association in Biological and Related Systems, Am. chem. Soc.* No. 84 (1968), p. 88.

[10] STEIM, J. M. *Adv. Chem. Ser. Molecular Association in Biological and Related Systems, Am. chem. Soc.* No. 84 (1968), p. 259.

[11] WALTER, J. A. and HOPE, A. B. *Progress in Biophysics and Molecular Biology* (edited by J. A. V. Butler and D. Noble), Pergamon Press, Oxford, **23**, 1 (1971).
[12] GRANT, R. F., HEDGECOCK, N. and DUNELL, B. A. *Can. J. Chem.* **34,** 1514 (1956).
[13] DUNELL, B. A. and JANZEN, W. R. *Wiss. Z. Friedrich Schiller-Univ.* Jena **14,** 191 (1965).
[14] CYR, T. J. R., JANZEN, W. R. and DUNELL, B. A. *Adv. Chem. Ser. Ordered Fluids and Liquid Crystals, Am. chem. Soc.* No. 63 (1967), p. 13.
[15] GRANT, R. F. and DUNELL, B. A. *Can. J. Chem.* **38,** 2395 (1960).
[16] BARR, M. R. and DUNELL, B. A. *Can. J. Chem.* **42,** 1098 (1964).
[17] GRANT, R. F. and DUNELL, B. A. *Can. J. Chem.* **38,** 1951 (1960).
[18] SHAW, D. J. and DUNELL, B. A. *Trans. Faraday Soc.* **58,** 132 (1962).
[19] GRANT, R. F. and DUNELL, B. A. *Can. J. Chem.* **39,** 359 (1961).
[20] JANZEN, W. R. and DUNELL, B. A. *Trans. Faraday Soc.* **59,** 1260 (1963).
[21] RIPMEESTER, J. A. and DUNELL, B. A. *Can. J. Chem.* **49,** 731 (1971).
[22] BARRALL II, E. M. and GUFFY, J. C. *Adv. Chem. Ser. Ordered Fluids and Liquid Crystals, Am. chem. Soc.* No. 63 (1967), p. 1.
[23] JANZEN, W. R. and DUNELL, B. A. *Can. J. Chem.* **47,** 2722 (1969).
[24] LAWSON, K. D. and FLAUTT, T. J. *J. phys. Chem.* **69,** 4256 (1965).
[25] LAWSON, K. D. and FLAUTT, T. J. *Molec. Crystals* **1,** 241 (1966).
[26] WINSOR, P. A. *Chem. Rev.* **68,** 1 (1968).
[27] EKWALL, P., MANDELL, L. and FONTELL, K. *Molec. Crystals Liqu. Crystals* **8,** 157 (1969).
[28] McDONALD, M. P. *Arch. Sci., Genève*, **12,** 141 (1959).
[29] LAWSON, K. D. and FLAUTT, T. J. *J. phys. Chem.* **72,** 2066 (1968).
[30] LUTTON, E. S. *J. Am. Oil Chem. Soc.* **43,** 28 (1966).
[31] LAWSON, K. D., MABIS, A. J. and FLAUTT, T. J. *J. phys. Chem.* **72,** 2058 (1968).
[32] GILCHRIST, C. A., ROGERS, J., STEEL, G., VAAL, E. G. and WINSOR, P. A. *J. Colloid Interface Sci.* **25,** 409 (1967); GILCHRIST, C. A., ROGERS, J., VAAL, E. G. and WINSOR, P. A. *Discuss. Faraday Soc.* **42,** 134 (1967).
[33] BALMBRA, R. R., CLUNIE, J. S. and GOODMAN, J. F. *Nature*, **222,** 1159 (1969).
[34] McLACHLAN, L. A., NATUSCH, D. F. S. and NEWMAN, R. H. *J. magn. Resonance* **4,** 358 (1971).
[35] PENKETT, S. A., FLOOK, A. G. and CHAPMAN, D. *Chem. Phys. Lipids* **2,** 273 (1968).
[36] LAWRENCE, A. S. C. and McDONALD, M. P. *Molec. Crystals* **1,** 205 (1966).
[37] ELLIS, B., LAWRENCE, A. S. C., McDONALD, M. P. and PEEL, W. E. *Liquid Crystals and Ordered Fluids* (edited by J. F. Johnson and R. S. Porter), Plenum Press, New York (1970), p. 277.
[38] CHAPMAN, D. and SALSBURY, N. J. *Trans. Faraday Soc.* **62,** 2607 (1966).

[39] SALSBURY, N. J. and CHAPMAN, D. *Biochim. biophys. Acta* **163,** 314 (1968).
[40] VEKSLI, Z., SALSBURY, N. J. and CHAPMAN, D. *Biochim. biophys. Acta* **183,** 434 (1969).
[41] ZLOCHOWER, I. A. and SCHULMAN, J. H. *J. Colloid Interface Sci.* **24,** 115 (1967).
[42] CORKHILL, J. M., GOODMAN, J. F. and WYER, J. *Trans. Faraday Soc.* **65,** 9 (1969).
[43] LAWSON, K. D. Personal communication.
[44] CHARVOLIN, J. and RIGNY, P. *Molec. Crystals Liqu. Crystals* **15,** 211 (1971).
[45] OLDFIELD, E., MARSDEN, J. and CHAPMAN, D. *Chem. Phys. Lipids* **7,** 1 (1971).
[46] TIDDY, G. J. T. *Nature, Physical Science* **230,** 136 (1971).
[47] CHAN, S. I., FEIGENSON, G. W. and SEITER, C. H. A. *Nature* **231,** 110 (1971).
[48] DE VRIES, J. J. and BERENDSEN, H. J. C. *Nature*, **221,** 1139 (1969).
[49] CHARVOLIN, J. and RIGNY, P. *J. magn. Resonance* **4,** 40 (1971).
[50] WINSOR, P. A. (to be published).
[51] CHAPMAN, D. and PENKETT, S. A. *Nature*, **211,** 1304 (1966).
[52] CHAPMAN, D., KAMAT, V. B., DE GIER, J. and PENKETT, S. A. *Nature*, **212,** 74 (1967).
[53] CHAPMAN, D., KAMAT, V. B., DE GIER, J. and PENKETT, S. A. *J. molec. Biol.* **31,** 101 (1968).
[54] JENKINSON, T. J., KAMAT, V. B. and CHAPMAN, D. *Biochim. biophys. Acta* **183,** 427 (1969).
[55] STEIM, J. M., EDNER, O. J. and BARGOOT, F. G. *Science, N.Y.* **162,** 909 (1968).
[56] FINER, E. G., FLOOK, A. G. and HAUSER, H. *FEBS Letters* **18,** 331 (1971); *Biochim. biophys. Acta* **260,** 59 (1972).
[57] SHEARD, B. *Nature*, **223,** 1057 (1969).
[58] SAUNDERS, L. *Biochim. biophys. Acta* **125,** 70 (1966).
[59] HAUSER, H. *Biochem. biophys. Res. Comm.* **45,** 1049 (1971).
[60] See e.g. ref. 1, chap. 13.
[61] BERGELSON, L. D., BARSUKOV, L. I., DUBROVINA, N. I. and BYSTROV, V. F. *Dokl. Akad. Nauk SSSR* **194,** 222 (1970).
[62] BYSTROV, V. F., DUBROVINA, N. I., BARSUKOV, L. I. and BERGELSON, L. D. *Chem. Phys. Lipids* **6,** 343 (1971).
[63] LAWSON, K. D. and FLAUTT, T. J. *J. Am. chem. Soc.* **89,** 5489 (1967).
[64] HAHN, E. L. *Physics Today* **6,** 4 (1953).
[65] PFEIFER, H. *Hochfrequenzspektroskopie* (edited by A. Lösche and W. Schütz), Akademie-Verlag, Berlin (1961), p. 30.
[66] KUBO, R. and TOMITA, K. *J. phys. Soc. Japan* **9,** 888 (1954).
[67] VAN PUTTE, K. *J. magn. Resonance* **2,** 23 (1970).
[68] FABRY, M. E., KOENIG, S. H. and SCHILLINGER, W. E. *J. biol. Chem.* **245,** 4256 (1970).
[69] SALSBURY, N. J., CHAPMAN, D. and JONES, G. P. *Trans. Faraday Soc.* **66,** 1554 (1970).

[70] HANSEN, J. R. and LAWSON, K. D. *Nature* **225,** 542 (1970).
[71] DAYCOCK, J. T., DARKE, A. and CHAPMAN, D. *Chem. Phys. Lipids* **6,** 205 (1971).
[72] CLIFFORD, J., OAKES, J. and TIDDY, G. J. T. *Spec. Discuss. Faraday Soc.* 1970, No. 1, p. 175.
[73] BLINC, R., EASWARAN, K., PIRŠ, J., VOLFAN, M. and ZUPANČIČ, I. *Phys. Rev. Lett.* **25,** 1327 (1970).
[74] TIDDY, G. J. T. *Faraday Soc. Symposium* **5,** 150 (1971).
[75] KAUFMAN, S., STEIM, J. M. and GIBBS, J. H. *Nature* **225,** 743 (1970).
[76] GLASEL, J. A. *Nature*, **227,** 704 (1970).
[77] STEIM, J. M. *Liquid Crystals and Ordered Fluids* (edited by J. F. Johnson and R. S. Porter), Plenum Press, New York (1970), p. 1.
[78] HERTZ, H. G. *Progress in NMR Spectroscopy,* Vol. III, Pergamon Press, Oxford (1967), p. 159.
[79] BLINC, R., HOGENBOOM, D. L., O'REILLY, D. E. and PETERSON, E. M. *Phys. Rev. Lett.* **23,** 969 (1969).
[80] BLINC, R., DIMIC, V., PIRŠ, J., VILFAN, M. and ZUPANČIČ, I. *Molec. Crystals Liqu. Crystals* **14,** 97 (1971).
[81] ERNST, R. R. and ANDERSON, W. A. *Rev. Sci. Instr.* **37,** 93 (1966).
[82] OLDFIELD, E. and CHAPMAN, D. *Biochem. biophys. Res. Comm.* **43,** 949 (1971).
[83] METCALFE, J. C., BIRDSALL, N. J. M., FEENEY, J., LEE, A. G., LEVINE, Y. K. and PARTINGTON, P. *Nature* **233,** 199 (1971).
[84] VOLD, R. L., WAUGH, J. S., KLEIN, M. P. and PHELPS, D. E. *J. chem. Phys.* **48,** 3831 (1968).
[85] BIRDSALL, N. J. M., LEE, A. G., LEVINE, Y. K. and METCALFE, J. C. *Biochim. biophys. Acta* **241,** 693 (1971).
[86] LINDBLOM, G., LINDMAN, B. and MANDELL, L. *J. Colloid Interface Sci.* **34,** 262 (1970).
[87] LINDBLOM, G. and LINDMAN, B. *Molec. Crystals Liqu. Crystals* **14,** 49 (1971).
[88] LINDBLOM, G., LINDMAN, B. and MANDELL, L. (to be published).
[89] EKWALL, P., MANDELL, L. and FONTELL, K. *J. Colloid Interf. Sci.* **29,** 639 (1969).
[90] PERSSON, N.-O. and JOHANSSON, Å. *Acta Chem. Scand.* **25,** 2118 (1971).
[91] LINDMAN, B. and EKWALL, P. *Molec. Crystals* **5,** 79 (1968).
[92] LINDBLOM, G. and LINDMAN, B. *Proc. Internat. Congr. Surface Active Agents*, 6th, Zürich, 1972 (in press); LINDBLOM, G., PERSSON, N.-O. and LINDMAN, B., *Proc. Internat. Congr. Surface Active Agents*, 6th, Zürich, 1972 (in press).
[93] LINDMAN, B. and DANIELSSON, I. *J. Colloid Interface Sci.* **39,** 349 (1972).
[94] COPE, F. W. *Proc. Nat. Acad. Sci. U.S.A.* **54,** 225 (1965).
[95] COPE, F. W. *J. gen. Physiol.* **50,** 1353 (1967).
[96] COPE, F. W. *Molec. Crystals* **2,** 45 (1966).
[97] COPE, F. W. Paper presented at the Annual NMR Conference, Pittsburgh, Pa., April 1970.

[98] Ling, G. N. and Cope, F. W. *Science, N.Y.* **163,** 1335 (1969).
[99] Rotunno, C. A., Kovalewski, V. and Cereijido, M. *Biochim. biophys. Acta* **135,** 170 (1967).
[100] Czeisler, J. L., Fritz, O. G. and Swift, T. J. *Biophys. J.* **10,** 260 (1970).
[101] Magnuson, J. A., Shelton, D. S. and Magnuson, N. S. *Biochem. biophys. Res. Comm.* **39,** 279 (1970).
[102] Lindblom, G. *Acta Chem. Scand.* **25,** 2767 (1971).
[103] Cope, F. W. *Biophys. J.* **10,** 843 (1970).
[104] Cope, F. W. *Physiol. Chem. Physics* **2,** 545 (1970).
[105] Cope, F. W. and Damadian, R. D. *Nature* **228,** 76 (1970).
[106] Charvolin, J. and Rigny, P. *C. r. hebd. Séanc. Acad. Sci., Paris,* **B269,** 224 (1969).
[107] Charvolin, J. and Rigny, P. *J. Phys. Radium, Paris* **30,** Suppl. C4, No. 11–12, 76 (1969).
[108] Cohen, M. H. and Reif, F. *Solid State Physics,* Vol. V (edited by F. Seitz and D. Turnbull), Academic Press Inc., New York (1962), p. 321.
[109] Buckingham, A. D. and McLauchlan, K. A. *Progress in NMR Spectroscopy* Vol. II, Pergamon Press, Oxford (1967), p. 63.
[110] Flautt, T. J. and Lawson, K. D. *Magnetic Resonance and Relaxation,* Proc. of the XIVth Colloque AMPERE, Ljubljana, Yugoslavia, Sept. 6–11, 1966 (edited by R. Blinc), North Holland, Amsterdam (1967), p. 759.
[111] Black, P. J., Lawson, K. D. and Flautt, T. J. *Molec. Crystals Liqu. Crystals* **7,** 201 (1969).
[112] Oldfield, E., Chapman, D. and Derbyshire, W. *FEBS Letters* **16,** 102 (1971).
[113] Johansson, Å. and Drakenberg, T. *Molec. Crystals Liqu. Crystals* **14,** 23 (1971).
[114] Woessner, D. E. and Snowden, B. S., Jr. *J. chem. Phys.* **50,** 1516 (1969).
[115] Deverell, C. *Progress in NMR Spectroscopy,* Vol. IV, Pergamon Press, Oxford (1969), p. 268.
[116] Waldstein, P., Rabideau, S. W. and Jackson, J. A. *J. chem. Phys.* **41,** 3407 (1964).
[117] See e.g. Luzzati, V., Mustacchi, H., Skoulios, A. and Husson, F. *Acta Crystallogr.* **13,** 660 (1960).
[118] McDonald, M. P. and Peel, W. E. *Trans. Faraday Soc.* **67,** 890 (1971).
[119] Grunwald, E., Löwenstein, A. and Meiboom, S. *J. chem. Phys.* **27,** 630 (1957).
[120] Lindblom, G., Wennerström, H. and Lindman, B. *Chem. Phys. Lett.* **8,** 489 (1971).
[121] Black, P. J., Lawson, K. D. and Flautt, T. J. *J. chem. Phys.* **50,** 542 (1969).
[122] Hayamizu, K. and Yamamoto, O. *J. magn. Resonance* **2,** 377 (1970).
[123] Black, P. J. Paper presented at the Sixth NMR Spectroscopy Colloquium, Aachen, West Germany, March 24–29, 1969.

[124] DAHLQVIST, K.-I. and HÖRNFELDT, A.-B. *Chemica Scripta* **1,** 125 (1971).
[125] DAHLQVIST, K.-I. and HÖRNFELDT, A.-B. Personal communication.
[126] HAYAMIZU, K. and YAMAMOTO, O. *J. magn. Resonance* **5,** 94 (1971).
[127] ANDERSON, J. M. and LEE, A. C.-F. *J. magn. Resonance* **3,** 427 (1970).
[128] ANDERSON, J. M. and LEE, A. C.-F. *J. magn. Resonance* **4,** 160 (1971).
[129] LONG, R. C., JR., and GOLDSTEIN, J. H. *J. chem. Phys.* **54,** 1563 (1971).
[130] LINDON, J. and DAILEY, B. P. *Molec. Phys.* **20,** 937 (1971).
[131] DRAKENBERG, T., JOHANSSON, Å. and FORSÉN, S. *J. phys. Chem.* **74,** 4528 (1970).
[132] DRAKENBERG, T. (to be published).

9

Infrared, Raman, Visible and Ultraviolet Spectroscopy of Liquid Crystals

V. D. NEFF

Introduction

The classical methods of molecular spectroscopy are well suited to the study of the structure and physical properties of liquid crystals. In spite of this, the number of reported investigations of infrared, Raman, ultraviolet and visible spectra has been small in comparison with magnetic resonance studies [1]. Furthermore, most of the work has been confined to non-amphiphilic systems. For this reason we will consider only these systems, bearing in mind that most of the techniques can be applied to amphiphilic systems as well.

One might ask what information is expected from a study of the molecular absorption spectra of liquid crystals. The main definitive feature of a mesophase is the existence of a degree of long-range translational or orientational order intermediate between that of the completely ordered solid and the disordered amorphous liquid. For spectroscopic studies, the crystalline solid and the amorphous isotropic liquid serve as points of reference for the comparison of the wave numbers, intensities, and shapes of absorption bands. Such comparisons can lead to more detailed information concerning the dynamics of molecular motion in a mesophase. In addition, the long-range order usually leads to anisotropies in the physical properties including absorption intensities measured in different directions with polarized radiation. These effects may be used to determine the degree of order. Finally some liquid crystals, such as a well-oriented nematic, can be used as an orienting solvent for absorption polarization studies of dissolved molecules.

I Interpretation of Mesophase Spectra

The complete description of the dynamics of molecular motion in a liquid crystal is, like that in a liquid, a difficult problem. For the interpretation of infrared and Raman spectra we can separate the atomic displacements approximately into internal and external vibrations. The internal vibrations are based on those of the isolated molecule perturbed by the molecular field of neighbouring molecules. The external vibrations are approximately the translational and rotational displacements of the rigid molecule about the centre of mass. In the crystal, the external vibrations become the lattice modes, but for the amorphous liquid, the equilibrium positions about which these motions take place are subject to random fluctuations so that the motions of a given molecule are not well correlated in phase with those several molecular distances away. The situation in a liquid crystal is expected to be intermediate between these two extremes.

A Internal Vibrations

For the assignment of the internal molecular vibrations one may consider that most compounds which form nematics or smectics are derived from the basic *para*-disubstituted structures

A—⟨benzene⟩—A (D_{2h}) or A—⟨benzene⟩—B (C_{2v})

The symmetrical D_{2h} structure has been analysed in detail by Schmid and Brandmuller [2]. The A groups are treated as rigid mass units. The expected wave number ranges of most of the thirty normal vibration modes have been determined from the analysis of a large number of *para*-disubstituted benzenes by Schmid and Brandmuller and by Katritzky and Simmons [3]. Actually, for many nematogens, the ring structure will have approximate C_{2v} symmetry. In this case all vibrations will be Raman active and all but three will be infrared active. In addition to these basic motions of the aromatic ring, there will be the vibrations associated with the *para*-substituents and the ring bridging groups such as azo- or azoxy-.

If we examine the wave numbers for the internal vibrations for different mesophases, and compare them with those for the isotropic liquid, we expect that structural differences between phases will show up in terms of wave number shifts, changes in intensity, and changes in band shape. These are discussed separately below.

1. *Wave Number Shifts*

A useful generalization can be made concerning the internal infrared and Raman bands for the nematic phase. For all compounds examined, no significant wave number shifts have been observed compared with the bands for the amorphous isotropic liquid [4]. As an example, the infrared spectra of 4-methoxybenzylidene-4′-cyanoaniline in the nematic and isotropic phases are shown in Fig. 9.1. The same general observation is true for the cholesteric phase, although only a few cholesteric compounds have been studied. This leads one to the conclusion that the short-range intermolecular interactions in a nematic or cholesteric are essentially the same as those in the amorphous isotropic liquid. The situation is not as clear

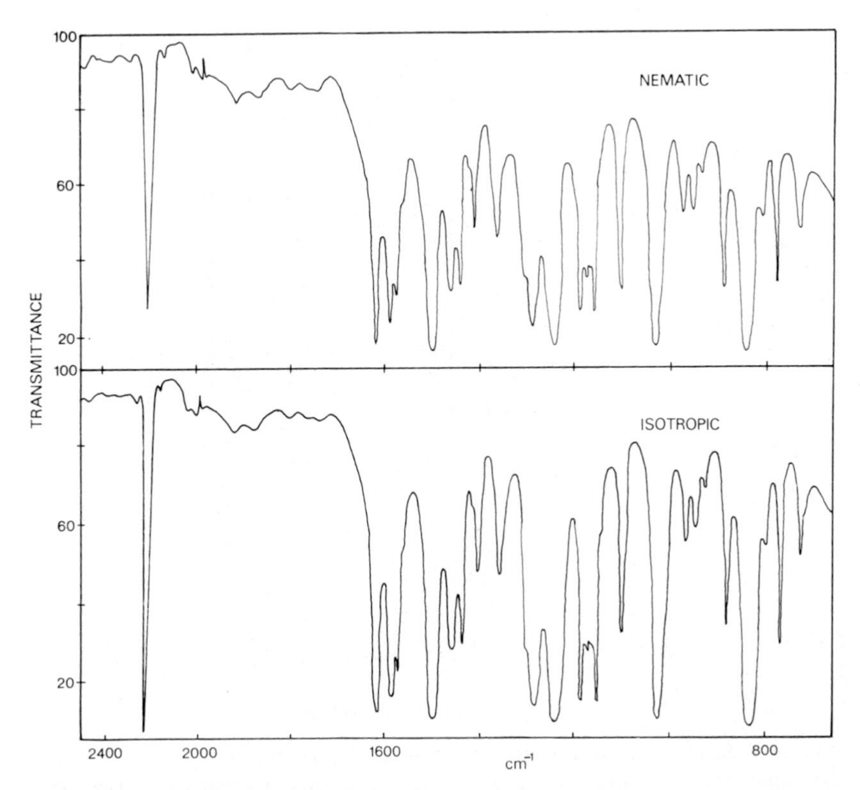

FIG. 9.1. The infrared spectrum of 4-methoxybenzylidene-4′-cyanoaniline at temperatures 4°C above (lower curve) and 4°C below (upper curve) the nematic–amorphous isotropic liquid clearing point.

for the smectic phase. First of all there are a variety of well-authenticated, polymorphic smectic phases which have been classified by Sackmann [5]. No infrared or Raman studies have been reported for most of these structures. Maier and Englert [4] report no significant wave number shifts for the smectic phase of 4-4′-di-n-hexyloxyazoxybenzene. On the other hand, we have observed small shifts for some smectic A compounds. One cannot however make any general statements until more spectroscopic studies have been reported.

2. *Intensity Changes*

The interpretation of intensity changes is not so obvious and this has led to some erroneous conclusions about the structural features of the nematic phase. That is, it is possible to observe intensity changes without wave number shifts. The intensity differences are due to the uniform molecular orientation induced by the long-range order in the nematic. Suppose a well-collimated beam of unpolarized infrared radiation is incident on a sample, as shown in Fig. 9.2 with electric vector in the *xz*

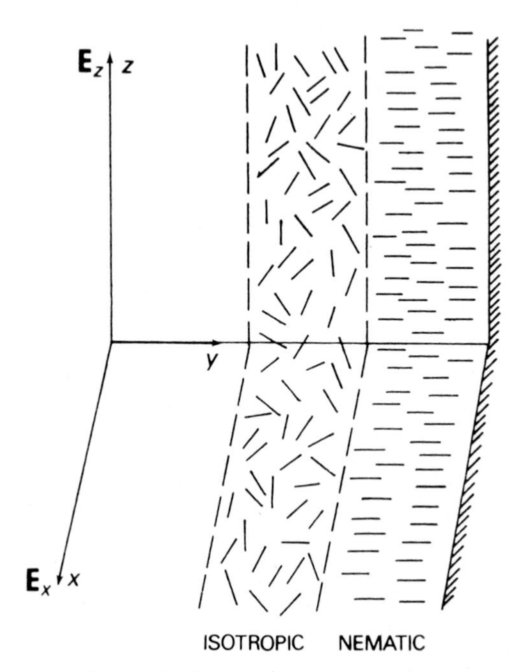

FIG. 9.2. Schematic illustration of intensity change through the amorphous isotropic–nematic transition due to change in molecular orientation. The light wave is travelling in the *y* direction. The direction of the vibrational transition moment is taken parallel to the long axis of the molecule. When the molecules are aligned preferentially perpendicular to the *xz* plane in the nematic phase, the intensity will decrease.

plane. If the vibrational transition moment is directed along the molecular axis and if the molecules tend to align preferentially perpendicular to the surface of the cell in the nematic, the intensity of the absorption band will decrease at the transition from the amorphous isotropic to the nematic. The intensity will decrease because the average number of molecules with transition moment perpendicular to the plane of the electric vector has increased. For example, in Fig. 9.1. the intensity of the C≡N band at 2230 cm^{-1} has clearly decreased in the nematic. The change in intensity at the transition is consistent with the fact that the molecules of this type of Schiff's base orient preferentially perpendicular to the surface of the sodium chloride cell.

3. *Band Shapes*

In passing from the amorphous isotropic liquid through the various mesophases to the crystal one might expect that the molecules would become gradually more rigidly confined to some definite equilibrium configuration. This increase in molecular rigidity, if very pronounced, should be reflected in changes of shape of infrared or Raman absorption bands. In general, the narrower the absorption band, the more immobile is the transition dipole giving rise to the absorption. A more quantitative statement of this fact is found in the work of Gordon [6], who noted that the shape of an infrared or Raman absorption band gives information concerning the time evolution of the motion of the system (molecule) undergoing the transition. We define a unit vector $\mathbf{U}(0)$ in the direction of a molecular vibrational transition moment at time $t = 0$. Then we follow the thermal motion of the molecules and at a later time measure the projection of $\mathbf{U}(t)$ on the original direction: $\mathbf{U}(0)\cdot\mathbf{U}(t)$. Next we take the average of all these trajectories starting at different initial reference times. The average $\langle\mathbf{U}(0)\cdot\mathbf{U}(t)\rangle$ is called the time correlation function for the transition moment. The time correlation function is related to the Fourier transform of the normalized band intensity according to

$$\langle\mathbf{U}(0)\cdot\mathbf{U}(t)\rangle = \int_{\text{band}} I(\omega)\, e^{i\omega t}\, d\omega$$

where $I(\omega)$ is the intensity at frequency ω, and the integral is taken over the absorption band. Gordon makes certain, rather restrictive, assumptions in deriving this expression, one of which involves no interaction of the internal vibration with other molecular degrees of freedom. Because of these restrictions one cannot apply the equation quantitatively to molecular motions in liquid crystals where one expects some vibrational coupling with pseudo-lattice modes. It can, however, be used for qualitative comparisons. A rapidly decaying time correlation function implies a loosely bound system and *vice versa.* Also because of the general inverse relation between the width of a symmetrical function and the width of its Fourier transform, a rapidly decaying correlation function implies a broad absorption band. We have examined the normalized intensities of the internal absorption bands for a large number of nematics and found that invariably the band contours are the same as those for the amorphous isotropic liquid within the limit of experimental error. As an example, in Fig. 9.3 we show the band contour for the C≡N absorption in 4-methoxybenzylidene-4′-cyanoaniline at temperatures above and below the clearing point. These bands were analysed under grating resolution and an expanded scale with a Perkin Elmer Model 621 spectrometer [7]. It will be seen that they are essentially identical in shape. This particular band also has an almost perfectly Lorentzian shape, so that the time correlation function is a simple exponential decay of the form e^{-at} where a, the band half-width, is $2{\cdot}97 \times 10^{11}\ \text{s}^{-1}$ for the C≡N band. These results again underline the

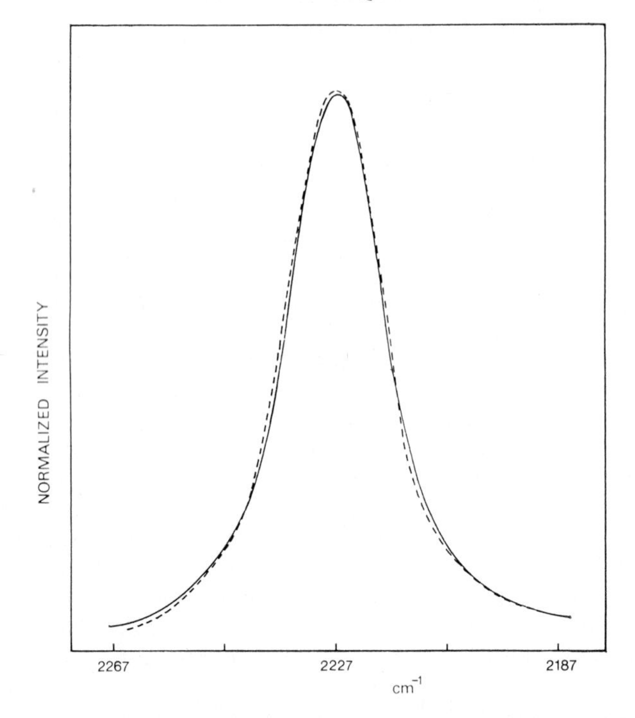

FIG. 9.3. Normalized intensity (arbitrary scale) of the C≡N band of 4-methoxybenzylidene-4′-cyanoaniline at 2227 cm^{-1}. The continuous curve represents the band shape 5°C above, and the dotted curve 5°C below, the transition temperature.

similarity of the nematic and amorphous isotropic phases as regards the interaction of a molecule with its immediate neighbours on a time scale comparable with the period of vibrational motions. They do not give much information concerning co-operative molecular motions on a long time scale, because this information is derived from the far wings of the absorption bands. Such long time co-operative motions are known to occur in nematics in the form of fluctuations about the nematic director [8]. In fact these fluctuations give the light-scattering characteristics of the nematic phase [9].

Again the results are not as clear for smectics, and no detailed studies of the shapes of absorption bands have been reported. We have observed slight narrowing of some of the absorption bands in smectic A of di-4-chlorophenyl azelate as compared with the amorphous isotropic liquid. A comparison of absorption bands is shown in Fig. 9.4. Although generalizations would be premature, it seems that the band shapes in smectics indicate a more rigid structure. This is consistent with the increased viscosity of the smectic compared with that of the nematic phase [10].

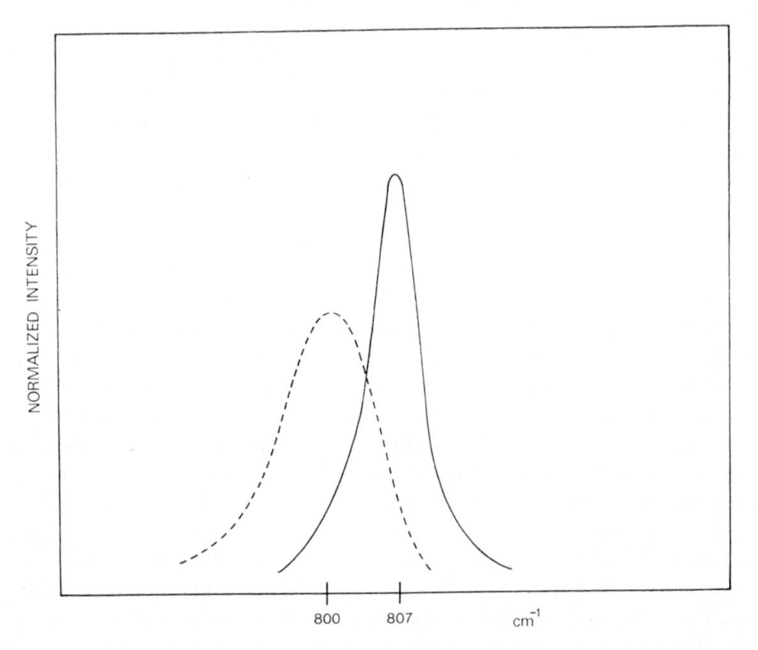

FIG. 9.4. Normalized intensity (arbitrary scale) for the 800 cm^{-1} band of di-4-chlorophenyl azelate. The continuous curve represents the band shape in the isotropic phase at 100°C. The dotted curve is taken at 93°C in the smectic A phase. The wave number is shifted by 7 cm^{-1}.

B EXTERNAL VIBRATIONS

For compounds which form liquid crystals, the external vibrations are expected to occur below 150 cm^{-1}. They can be observed most explicitly in the crystalline solid where they become the lattice modes. We do not discuss the vibrational analysis of molecular crystals in detail, since the methods are well known; e.g., see the review by Mitra [11]. The number of translational and rotational external vibrations is determined by the crystal geometry. If there are m molecules per unit cell, there will be $m(3n - 6)$ internal frequency branches where n is the number of atoms per molecule. In the harmonic approximation, only the $\mathbf{k} = 0$ frequency is observed in each branch, where $\mathbf{k}$ is the wave number vector. There will be $6m - 3$ optical external branches and three acoustical branches. The acoustical frequencies are zero when $\mathbf{k} = 0$. The symmetry of the vibrational modes is determined from the factor group analysis of the unit cell. As an example of this kind of analysis, we choose the nematogen p-azoxyanisole because the crystal structure is known [12] and some low-frequency bands have been recorded in the Raman spectrum [13]. The space group is $P2_1/c(C_{2h}^5))$ with four molecules in the unit cell. The twenty-one optical lattice vibrations separate into $6A_g + 5A_u + 4B_u + 6B_g$. The unit cell is centrally symmetric so that the rule of mutual exclusion holds. There are six rotational and six translational Raman active modes. These may not all be well separated because the A_g and B_g species are simply different phase combinations of the basic three rotations and translations distributed among the four molecules in the unit cell. The low-frequency vibrations which have been observed to date will be discussed in Part II.

The low-frequency collective motions in amorphous liquids are not as easy to characterize as in the crystalline solid. The long-range order is lost, and it is no longer possible to identify a unit cell which simplifies the discussion to the $\mathbf{k} = 0$ modes. The optical modes extend over a much wider range of frequencies than in a corresponding crystal and consequently the absorption bands become much more diffuse. In addition, the intensities may be reduced to the extent that no bands appear. The situation in the case of liquid crystals is expected to be of similar complexity and no detailed theory of the low-frequency external collective motions in liquid crystals has been attempted. In the amorphous liquid phase, the equilibrium positions about which the molecular rotations and translations take place undergo random fluctuations so that the motions of a given molecule are not coherently related to distant molecules. In the liquid crystal there is some long-range order, so the low-frequency vibrations should retain partial coherence in some degrees of freedom. In general we might expect that as the range of coherence in the collective motion increases, the width of the absorption band would decrease, and this may be used to study the onset of long-range order in different degrees of freedom in a mesophase—see Part II.

C Dichroism Studies and Degree of Order

Dichroism studies on any type of spectroscopic transition in a mesophase may be used to determine the long-range orientational order. The problem of properly describing the long-range order for the different types of mesophase is not completely straightforward. For example, the order is rarely as complete as that in a crystal, and we have to consider a degree of order which, at least in a nematic, may vary considerably with temperature over a single phase [14].

For the purpose of classification, it is useful to visualize a kind of inverse relationship between the types of long-range order and the existence of certain continuous symmetry groups in a mesophase. With respect to the macroscopic physical properties, an amorphous liquid possesses the symmetry of isotropic and homogeneous space, i.e. the full symmetry of the continuous group of translations and rotations. Breaking the rotational symmetry in one dimension, but retaining the translational symmetry, produces the nematic. The continuous rotational symmetry is reduced to either $C_{\infty v}$ or $D_{\infty h}$ depending on whether the order is polar or apolar. The latter is the only one known at this time. If the continuous rotational symmetry is broken further, the orientational symmetry is reduced to that of a discrete point group. For example, suppose the molecules have the symmetry of rectangular parallelepipeds and are uniformly aligned with respect to the symmetry axes. If the continuous translational symmetry is retained, we have $T_3 X D_{2h}$ which would describe a biaxial nematic. Breaking the translational and rotational symmetry in one dimension introduces translational order and produces the symmetry of the uniaxial smectic etc. If all continuous symmetry is broken, the system is left with the full translational and orientational order of the crystalline phase. As the order in the different degrees of freedom increases, the continuous symmetry is reduced.

1. *Degree of Order*

The physical properties of a mesophase which depend on the molecular orientation are anisotropic.* Measurement of some selected physical property in different directions can be used to characterize the state of orientational order. We mention particularly the investigations of Saupe and Maier [15] and of Saupe [16]. It is useful to define a quantity called the degree of order which can be calculated from measurement of any of the anisotropic physical properties. The most general definition is formulated to take advantage of the full freedom of measurement inherent in the spectroscopic determination of dichroism ratios. This should include the possibility of determining the degree of order of molecules dissolved in a liquid crystal solvent. The definition is a simple generalization of that used by Saupe [16]. The orientational order is an intrinsic property of a mesophase independent of co-ordinate systems. It is obvious that the quantity in some way characterizing the degree of order must be an invariant tensor.

We specify a cartesian co-ordinate system fixed in the molecule ($x_1x_2x_3$) and a second system fixed in the liquid crystal ($y_1y_2y_3$). The molecular orientation is defined in terms of the nine direction cosines of the molecule fixed relative to the space fixed system. We define the matrix $\langle \mathbf{T} \rangle$ with elements given by

$$\langle T_{ij} \rangle = \sqrt{3} \langle \cos \theta_{ij} \rangle \tag{9.1.1}$$

where i refers to the space fixed and j to the molecule fixed system, and the average is taken over all molecules in a uniformly oriented macroscopic region of the liquid crystal. If the liquid crystal is apolar, that is, if the state of orientation is invariant under the operation $\mathbf{L} \rightarrow -\mathbf{L}$, where $\mathbf{L}$ is a vector in the molecule, all of the elements of $\langle \mathbf{T} \rangle$ will be zero. This result is easily visualized if we recognize that, for all molecules with orientation $\cos \theta_{ij}$, there will be an equal number with orientation $\cos (\theta_{ij} + \pi)$ for all i and j. This is the situation with all known examples of non-amphiphilic liquid crystals, although it should be pointed out that, for a polar liquid, such as might be obtained in a very strong electric field, the matrix $\mathbf{T}$ should be included in the definition of the degree of order. Next one forms the direct product of $\mathbf{T}$ with itself and again takes the average over all molecules. A general element of the matrix $\langle \mathbf{T}^2 \rangle$ is then

$$\langle T_{ij} T_{i'j'} \rangle = 3 \langle \cos \theta_{ij} \cos \theta_{i'j'} \rangle. \tag{9.1.2}$$

We can refer to $\langle \mathbf{T}^2 \rangle$ as the orientation matrix. The elements transform as a fourth-rank tensor under orthogonal transformation of the co-ordinate system. The tensor character of $\mathbf{T}^2$ is most easily established by

* We do not consider here any optically isotropic mesophases.

recognizing that **T** is obviously a second-rank tensor, because it can be regarded as the matrix of a linear transformation. Since **T** is a tensor, $\mathbf{T}^2$ is a symmetric fourth-rank tensor. Also, since the operation of taking the average values is linear, $\langle \mathbf{T}^2 \rangle$ is also a fourth-rank tensor. Of course, the elements of $\langle \mathbf{T}^2 \rangle$ are highly redundant bearing in mind that we need only three independent parameters to specify completely the average orientation in a uniformly oriented region of a mesophase. The relations between elements can be obtained from the orthogonality conditions on **T**. This definition is still quite useful, however, because one can explicitly keep track of the average orientation of a particular axis in the molecule with respect to a particular axis in the laboratory co-ordinate system. This is not the case with, for example, the Euler angles. We have formulated the definition of $\langle \mathbf{T}^2 \rangle$ in such a way that for the completely disordered isotropic system $\langle \mathbf{T}^2_{iso} \rangle$ is simply the unit matrix. That is

$$\langle \mathbf{T}^2_{iso} \rangle = \delta_{ii'}\delta_{jj'}. \tag{9.1.3}$$

We are now ready to define the order tensor **S**. It is a quantity defined by difference relative to the completely isotropic phase. That is

$$\mathbf{S} = \tfrac{1}{2}(\langle \mathbf{T}^2 \rangle - \langle \mathbf{T}^2_{iso} \rangle). \tag{9.1.4}$$

The factor $\frac{1}{2}$ is introduced simply to scale S so that the diagonal elements will vary from $+1$ to $-\frac{1}{2}$ depending on the degree of order. An arbitrary element of S is given by

$$S_{iji'j'} = \tfrac{1}{2}\langle 3 \cos \theta_{ij} \cos \theta_{i'j'} - \delta_{ii'}\delta_{jj'} \rangle. \tag{9.1.5}$$

The trace of $\langle \mathbf{T}^2 \rangle$ is equal to that of $\langle \mathbf{T}^2_{iso} \rangle$, namely nine, so that **S** is traceless. It is, of course, also symmetric. The diagonal elements of **S** are called the degree of order. We can define the degree of order of an arbitrary axis in the molecule with respect to an arbitrary axis in the liquid crystal. This definition is sufficiently general to include, for example, a biaxial smectic or nematic. If the liquid crystal has axial symmetry, we may choose a representation for **S** in terms of space fixed axes defined with respect to the optic axis of the liquid. In this representation, the **S** matrix is reduced to three 3×3 blocks along the diagonal, two of which are identical. That is all integrals of the type $\langle \cos \theta_{ij} \cos \theta_{i'j'} \rangle$ are zero unless $i = i'$ (recall that i refers to the space fixed axes). The unique 3×3 matrix is a representation of the second-rank tensor defined as the order tensor by Saupe [15]. It measures the degree of order relative to the optic axis of the liquid and, of course, it is traceless. Finally, if the individual molecules have a three or more fold symmetry axis one may also transform to the principal axes of the molecule. In this representation, the order tensor is reduced completely to diagonal form. The diagonal elements are related to one another according to

$$S_{i1i1} = -2S_{i2i2}, \quad S_{i2i2} = S_{i3i3} \qquad i = 1, 2, 3.$$
$$S_{1j1j} = -2S_{2j2j}, \quad S_{2j2j} = S_{3j3j} \qquad j = 1, 2, 3.$$

Hence, in the completely reduced representation, there is only one independent parameter. This is usually measured in terms of the symmetry axis of the molecule relative to the symmetry axis of the liquid crystal and we are left with

$$S_{11} = \tfrac{1}{2}\langle 3\cos^2\theta_{11} - 1\rangle \tag{9.1.6}$$

as the single quantity necessary to specify completely the degree of order.

2. *Dichroism Studies*

The dichroism ratio is defined as the ratio of the absorption coefficients of linearly polarized light measured in perpendicular directions. If the system has uniaxial symmetry, such as a nematic, the absorption coefficients are conveniently measured parallel and perpendicular to the optic axis. In this case, the dichroism ratio R is

$$R = \frac{\varepsilon_{\parallel}}{\varepsilon_{\perp}} \tag{9.2.1}$$

where ε is the absorption coefficient. For the purpose of infrared orientation studies, we can assume the classical relation that the molecular absorption coefficient is proportional to the square of the scalar product of the electric vector with the vibrational transition moment **M**. For Raman studies, we have to replace the dipole vector with the polarizability tensor. We will consider in detail only the infrared dichroism expression. We assume a co-ordinate system as shown in Fig. 9.5, where the optic axis

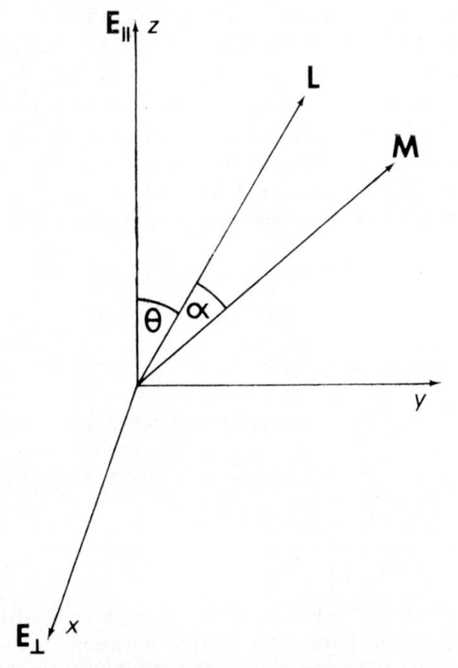

FIG. 9.5. Co-ordinate system for plane polarized light propagating in the y direction. The z-axis is parallel to the optic axis. The vector **L** is in the direction of the long axis of the molecule. The vector **M** represents the direction of the vibrational transition moment.

of a uniaxial liquid crystal is taken along z. The electric vector $\mathbf{E}_{\parallel}$ is directed along the z-axis and $\mathbf{E}_{\perp}$ along the x-axis. The vector $\mathbf{L}$ is taken along the long axis of the molecule. The angle θ is measured between $\mathbf{L}$ and the optic axis, and α is the angle between the long axis of the molecule and the direction of the vibrational transition moment. The angle ϕ is measured between the projection of the $\mathbf{L}$-axis on the xy plane and the x-axis. We define a distribution function $F(\theta)$ which is the probability that a molecule will have its long axis in the direction θ.

The dichroism ratio is then given by the expression

$$R = \frac{\int_0^{2\pi}\int_0^{\pi} F(\theta)[\cos^2\alpha\cos^2\theta + \sin^2\alpha\sin^2\theta\sin^2\phi]\sin\alpha\sin\theta\,\mathrm{d}\theta\,\mathrm{d}\phi}{\int_0^{2\pi}\int_0^{\pi} F(\theta)[\cos^2\alpha\sin^2\theta\sin^2\phi + \sin^2\alpha(\cos\phi + \sin\phi\cos\theta)^2]\sin\alpha\sin\theta\,\mathrm{d}\theta\,\mathrm{d}\phi} \tag{9.2.2}$$

which can be put in the form

$$R = \frac{\cos^2\alpha\langle\cos^2\theta\rangle + \frac{1}{2}\sin^2\alpha\langle\sin^2\theta\rangle}{\frac{1}{2}\cos^2\alpha\langle\sin^2\theta\rangle + \frac{1}{4}\sin^2\alpha\langle 1 + \cos^2\theta\rangle} \tag{9.2.3}$$

where $\langle\cos^2\theta\rangle = \int_0^{\pi} F(\theta)\cos^2\theta\sin\theta\,\mathrm{d}\theta.$

We note that the measurement of R is not sufficient to determine the orientation distribution function but, because of the occurrence of the square of the direction cosine, it does determine the degree of order of the molecular axis relative to the optic axis. That is, knowing the direction of the vibrational transition moment, we can solve for the degree of order S in terms of R. Here S refers to the degree of order measured relative to the long axis of the molecule as defined in section C-1. If the transition moment is directed along the long axis of the molecule, equation (9.2.2) reduces to

$$R = \frac{2\langle\cos^2\theta\rangle}{\langle\sin^2\theta\rangle} = \frac{1 + 2S}{1 - S}$$

and

$$S = \frac{R - 1}{R + 2} \tag{9.2.4}$$

If the vibrational transition moment is perpendicular to the long axis, we have

$$R = \frac{2\langle\sin^2\theta\rangle}{\langle 1 + \cos^2\theta\rangle} = \frac{(1 - S)}{2 + S}$$

and

$$S = \frac{1 - R}{1 + 2R} \tag{9.2.5}$$

In terms of the co-ordinate system we have defined, R varies between zero and infinity. The order parameter varies between $-\frac{1}{2}$ and $+1$. If the vibrational transition moment is parallel to the long axis of a uniaxial liquid crystal, and if we have complete parallel order, R would be infinite and $S = 1$. If the transition moment is perpendicular to the molecular axis, R would be zero and $S = 1$ for the same system.

If $S = 0$, that is $R = 1$, there is no orientational order. If the vibrational transition moment is set at some other angle α with respect to the long axis, we have to use equation (9.2.3) to determine R. In principle, one could invert this process and, knowing the degree of order and the direction of orientation for a nematic, determine the direction of the vibrational transition moment relative to the long axis of the molecule. Similar, but more complicated, expressions can be derived for the dichroism ratio measured in different directions in a biaxial liquid crystal, making use of the definition of the degree of order outlined in section C-1.

D Experimental Pitfalls and Limitations

At this point some of the experimental problems and limitations of spectroscopic methods are mentioned. The state of a liquid crystal is never defined completely unless the purity of the sample and the nature of the confining surface are specified. Minute perturbations in the form of dust particles or surface imperfections can profoundly affect the uniformity of orientation. In the case of infrared or ultraviolet studies, one usually uses samples in the form of thin films which are particularly sensitive to surface effects, and the direction of preferred orientation may be different for cell materials such as sodium chloride or silicon. Alignment of thin films with electric or magnetic fields leads to the formation of domains. The domain boundaries are formed by inversion walls which are regions where the direction of preferred orientation undergoes a directional change through 180° [17].

Comparative studies involving the solid phase introduce additional difficulties. If one allows a mesophase to crystallize, the crystals formed may be highly imperfect or differ from those obtained by crystallization from a solvent. For example, 4-ethoxy-4′-n-heptanoyloxyazobenzene obtained by cooling a thin film from the nematic gives a spectrum different to that obtained with crystals from ethanol solution. In this instance, the differences in spectra are due to the fact that this compound has two crystalline forms, one of which is stable above room temperature [7]. As the nematic is cooled, a non-equilibrium mixture of the two solids is formed. The crystals obtained by recrystallization contain only the low-temperature form. Also, in crystallizing some mesogenic compounds having alkyl chains as *para*-substituents, it is possible for crystals to form with the chains in different rotational conformations [4].

These questions of sample purity and the history of preparation are not trivial; they can lead to serious misinterpretations of the spectra.

II Review of Specific Investigations

A Wave Number Assignments and Interpretation of Spectra

In the first detailed investigation of the infrared spectra of liquid crystals, Maier and Englert [4] studied seven 4,4′-di-n-alkoxyazoxybenzenes (methoxy to n-heptyloxy), 4,4′-di-n-hexyloxyazobenzene, 4-nitrocinnamylidene-4′-ethoxyaniline and 4-n-butyloxybenzoic acid in the solid, liquid crystal and amorphous isotropic liquid phases from 1 to 25 μ. Extensive tables were provided of the wave numbers of the bands observed and many absorption bands were assigned. For all compounds studied, the spectra of the nematic and amorphous isotropic liquids were essentially equivalent except for some changes in relative intensity attributed to the orientation in the nematic. No marked frequency shifts or changes in band shape were observed. We reiterate the general conclusion (Part I) that there is no indication of a significant change in nearest-neighbour intermolecular interactions in the nematic as compared with the amorphous liquid phase. Maier and Englert also reported interesting results concerning the spectra of the crystalline solid phases of the azoxyphenol ethers. The pentyl, hexyl and heptyl ethers crystallized with a single conformation believed to contain the extended zig-zag chain. Other twisted rotational conformations appear to be accommodated in the crystal lattices of the propyl and butyl compounds dependent on the growth rate of the crystals from solution.

The infrared spectra of 4-ethoxy-4′-methoxyazoxybenzene, cholesteryl propionate and *p*-azoxyanisole have been studied for the solid, liquid crystal and amorphous isotropic liquid phases from 1 to 15 μ by L'vova and Sushchinskii [18]. Interest was centred on the temperature dependence of the intensity of the absorption bands. The crystal–mesophase transition produced an abrupt change in the spectrum with the appearance of new absorption bands and changes in widths and wave numbers of existing bands. In contrast, the mesophase–amorphous isotropic liquid transition occurred without qualitative alteration of the spectrum; the intensity changes throughout the mesophase region were explained in terms of the change in the degree of order with temperature (see section A-2).

The first Raman study of the nematic phase was reported by Koller, Lorentzen and Schwab [19] for *p*-n-butyloxybenzoic acid in the region 400–1700 cm^{-1}. The intensities of the absorption bands were determined for the unoriented liquid crystal and compared with those for the system oriented by means of a magnetic field. These studies enabled the authors to assign most of the ring vibrations. This is the only Raman investigation which has made use of the uniform orientation which can be attained by a moderate (ca. 5 kG) magnetic field.

Bulkin, Grunbaum and Santoro [20] have made a detailed study of some of the internal vibrations of the 4,4′-di-n-alkoxyazoxybenzenes (methyl through n-hexyl) under grating resolution. The primary concern was the study of intensity and wave number changes in the region of the crystal–nematic transition. Some sharp bands in the solid state spectra disappear at the phase transition—see Fig. 9.6. The intensity of these bands begins to decrease at temperatures well below the transition temperature. The decrease of intensity with temperature is shown in Fig. 9.7, and is attributed to the gradual introduction of defects in the solid. The solid state spectra were obtained with solids formed by cooling from the nematic. It would be interesting to compare these spectra with those of samples obtained by crystallization from solution. No far infrared spectra have been reported for these compounds.

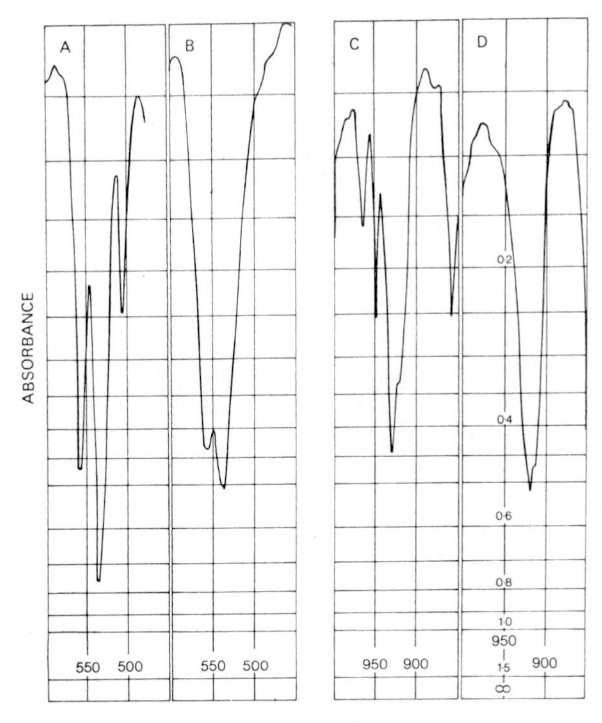

FIG. 9.6. Typical changes in the infrared spectra of 4,4′-di-n-alkoxyazoxybenzenes at the crystal–nematic transition. A and B are for crystalline and nematic 4,4′-di-methoxyazoxybenzene, respectively. C and D are for crystalline and nematic 4,4′-di-ethoxyazoxybenzene, respectively. Nematic melt spectra are for approx. 2°C above the C–N transition temperature [20].

The Raman spectra of two nematic compounds (*p*-azoxyanisole and anisaldazine) and one smectic compound (4,4′-azoxybenzoic acid) have been reported by Zhdanova *et al.* [21]. The spectra of the internal vibrations are recorded only to 500 cm^{-1}.

Few studies have been made on the low-frequency external vibrations of liquid crystal compounds. The Raman spectrum of *p*-azoxyanisole has been reported by Amer, Shen and Rosen [13]. Only three of the expected twelve Raman bands were observed (see Part I). No attempt was made at assignment of the observed bands. Of the three bands at 40, 50 and 72 cm^{-1} in the solid, the 72 cm^{-1} band vanished completely and the other two persisted with much reduced intensities on going from the solid to the nematic. The authors also investigated the region of the internal vibrations, obtaining about 30 lines. Neither was a detectable wave number shift of the low-frequency modes observed at the phase transition nor did the intensity begin to change well below the transition temperature. The intensities of these bands undergo a sharp change at the crystal–nematic transition.

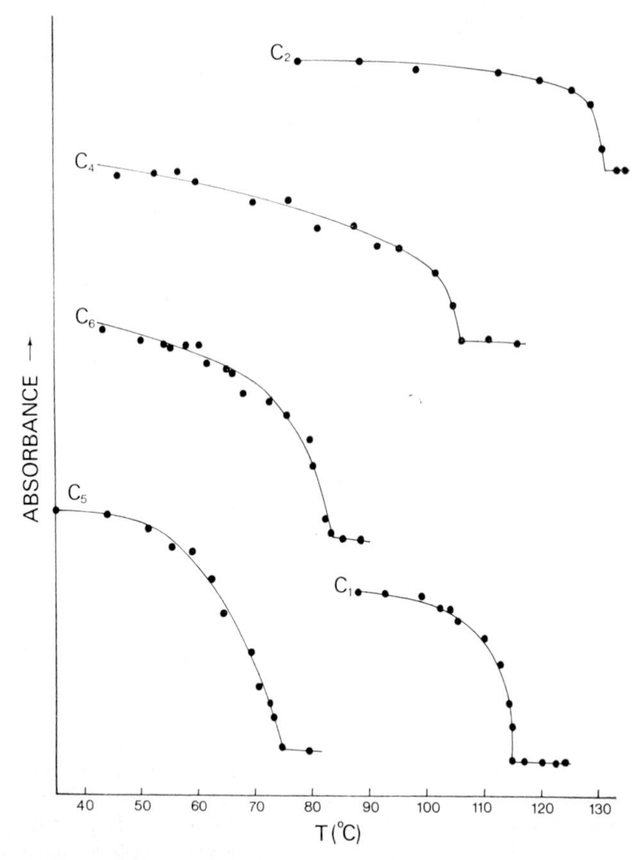

FIG. 9.7. Temperature dependence of the absorbance of the disappearing infrared bands of 4,4′-di-n-alkoxyazoxybenzenes. The curves are labelled according to the length of the alkyl chain, e.g. C_4 = butyl, etc. [20].

The low-frequency infrared spectrum of *p*-azoxyphenetole has been examined by L'vova *et al.* [22]. A very broad band was observed at ca. 100 cm^{-1} with the solid and this persisted with the nematic and amorphous isotropic liquid phases. In the solid and nematic phases, the band contours were similar, but the intensity was somewhat reduced, and the frequency shifted for the liquid phase. Here we have an example of a distinct difference in spectrum between the mesophase and the amorphous isotropic liquid. This does not invalidate previous conclusions, because it is in this low-frequency range that we expect to begin to see the effects of the long-range order. The spectrum was not examined below 70 cm^{-1}.

The infrared spectra of cholesterics show, in general, the same kind of behaviour as the spectra of nematics. We have examined spectra in the internal vibration region for a number of compounds in our laboratory, and again one finds no marked differences between the spectra of the amorphous isotropic liquid and the cholesteric. The low-frequency Raman spectra of cholesteryl propionate, nonanoate and palmitate have been reported by Chang [23]. All three compounds have four low-frequency bands with approximately the same wave numbers in the solid, as shown in Table 9.1. The two bands designated A and B have been tentatively

TABLE 9.1. *The four low wave number vibrations observed in the Raman spectra of cholesteryl propionate, nonanoate and palmitate according to Chang* [23]

Band identification	Cholesteryl alkanoate					
	propionate		nonanoate		palmitate	
	Wave No. cm^{-1}	Intensity ratio	Wave No. cm^{-1}	Intensity ratio	Wave No. cm^{-1}	Intensity ratio
A	18 ± 1	1·0	18 ± 1	1·0	18 ± 1	1·0
B	26	2·2	26	1·8	26	1·5
C	52	0·5	52	0·4	52	0·3
D	60	0·2	60	0·2	60	0·15

assigned as pertaining to longitudinal and transverse vibrations, respectively. These bands are absent in the spectra of the amorphous isotropic liquids.

B DICHROISM RATIOS AND ORIENTATION STUDIES

The first determination of the degree of order in nematics by infrared measurements of dichroism was reported by Maier and Englert [24]. They measured dichroism ratios for a series of 4,4′-di-n-alkoxyazoxybenzenes and for 4,4′-di-n-hexyloxyazobenzene. Frequencies were assigned according to the approximate C_{2v} symmetry of the molecule, and measurements of dichroism were reported for some of the A_1 and B_2 type modes with transition moment parallel and perpendicular to the molecular axis. The dichroism ratio and degree of order as a function of temperature are shown in Table 9.2. In each case, the order parameter S was calculated with respect to the orientation of the long axis of the molecule relative to the optic axis of the liquid crystal. Uniform orientation was achieved by rubbing the surface of the salt plate in one direction. The degree of order obtained from these measurements compares favourably with values determined from measurements of other physical properties [15]. As seen from Table 9.2, the order parameter can vary considerably with temperature from 0·13 to 0·8. The parallel alignment is almost complete in the smectic, with S values of 0·92 and 0·97 for the two smectic compounds investigated. These calculations were made without a correction for the effect of the molecules on the polarized electric field of the radiation, the so-called Lorentz inner field correction. Such corrections have been made to these data by Saupe and Maier [15], but give rise to only small systematic increases in the S values.

TABLE 9.2. *The infrared dichroism ratio and degree of order* (S) *as a function of temperature for a series of liquid crystals according to Maier and Englert* [24]

Substance	Absorption band	Temp. °C	Dichroism ratio	$\sin^2 \theta$	Degree of order (S)
p-azoxyanisole	ω_7	103·5	3·7	0·35	0·47
		110	3·4	0·37	0·45
	1600 cm^{-1}, A_1	117·5	3·4	0·37	0·45
		127	2·7	0·425	0·36
		131·5	2·3	0·465	0·30
		133	1·45	0·58	0·13
,,	γ_3	97	0·35	0·30	0·55
		102·5	0·42	0·35	0·48
	837 cm^{-1}, B_2	106	0·38	0·32	0·52
		113·5	0·45	0·365	0·45
		117·5	0·47	0·38	0·43
		123	0·48	0·385	0·42
		126·5	0·53	0·42	0·37
		132	0·71	0·52	0·22
,,	ω_1	107	4	0·33	0·50
		109·5	4·16	0·325	0·51
	908 cm^{-1}, A_1	114	4	0·33	0·50
		119·5	3·7	0·35	0·47
		123	3·3	0·38	0·43
		125·5	2·6	0·43	0·35
		131	2·36	0·46	0·31
4,4′-di-n-pentyloxy-azoxybenzene	γ_3	76	0·10	0·10	0·85
		84	0·15	0·14	0·82
	833 cm^{-1}, B_2	94	0·17	0·16	0·76
		103	0·22	0·20	0·70
		111	0·22	0·20	0·70
		117	0·30	0·26	0·60
		119·5	0·42	0·34	0·49
		124	1·00	0·66	0·00
		76	0·12	0·12	0·82
4,4′-di-n-heptyloxy-azoxybenzene	ω_7	95	10·5	0·16	0·76
		104	7·4	0·21	0·69
	1600 cm^{-1}, A_1	107·5	6·75	0·23	0·65
		115·5	4·9	0·29	0·57
		120·5	2·8	0·42	0·37

Substance	Absorption band	Temp. °C	Dichroism ratio	$\sin^2 \theta$	Degree of order (S)
4,4′-di-n-heptyloxy-azoxybenzene	γ_3	68	0·055	0·055	0·92 smectic
		84	0·055	0·055	0·92 smectic
	833 cm^{-1}, B_2	105	0·24	0·22	0·67
		114	0·27	0·235	0·65
		120·5	0·43	0·35	0·475
4,4′-di-n-hexyloxy-azobenzene	γ_3	29	$<$0·01	$<$0·02	$>$0·97 smectic
		95	$<$0·01	$<$0·02	$>$0·97 smectic
	844 cm^{-1}, B_2	102·5	0·23	0·21	0·69
		108	0·25	0·22	0·67
		113	0·36	0·23	0·55

The determination of the degree of orientation of the nematic phase of 4-methoxybenzylidene-4′-cyanoaniline in a d.c. electric field has been reported by Neff, Gulrich and Brown [25]. The long axis of the molecule oriented preferentially parallel to the field. The measured dichroism ratios increased with increasing field strength, reaching a saturation value at approximately 3 kV cm^{-1}. The degree of order was rather low, having a maximum value of 0·35 at 106°C, probably due to the fact that d.c. electric fields induce domain formation in thin films, so that the orientation was not uniform [26]. Also, electric fields cause motion due to the flow of electric current. Magnetic fields are more suitable for achieving uniform orientation in a thin film.

The measurement of the dichroism ratio as a function of temperature for a series of *p*-n-alkoxybenzoic acids (C_3 to C_9) has been reported by Kusakov *et al.* [27]. The dichroism ratio changed abruptly at each of the various phase transitions. The relative degree of order in the nematic and the smectic phases depended on the thermal history, differing for the heating and cooling experiments.

Liquid crystals may also be used as an orienting matrix for determining the polarization and degree of order of infrared or Raman bands of dissolved molecules. They have been used extensively in this way for nuclear magnetic resonance and electron spin resonance studies [1]. Infrared and Raman measurements of this type suffer the limitation that the solvent bands often mask the region of interest. The degree of orientation of 4-bromobenzonitrile (BBN) and 4-cyanobenzoic acid (CBA) have been studied using 4,4′-di-n-hexyloxyazoxybenzene as nematic solvent by Hensen [28]. The solvent was oriented uniformly by a combination of directional surface rubbing and an external magnetic field. The maximum degree of order of the solvent was 0·85 at 80°C. The long axes of the dissolved molecules were aligned parallel to the solvent molecules. The degree of order for BBN and CBA was 0·34 and 0·55, respectively. The larger value for CBA is explained in terms of the compound being a dimer, and therefore more extended, in solution. The polarized infrared spectra of the decacarbonyl–dimetal complexes of Mn and Re have been studied using the nematic phase of 4′-ethoxyphenyl 4-n-butyloxycarbonyloxybenzoate by Ceasar, Levenson and Gray [29]. Again the long axis of the solvent molecule tends to align parallel to the optic axis of the liquid crystal. From these dichroism measurements, it was possible to assign the three C=O stretching fundamentals for the $M_2(CO)_{10}$ (D_{4d}) molecule.

III Ultraviolet and Visible Spectra

Ultraviolet and visible spectroscopic investigations of liquid crystals have been confined mostly to polarization studies of certain π-electronic absorption bands of the liquid or of oriented solute molecules dissolved in the liquid.

One again finds no essential differences in the ultraviolet or visible spectra of mesophases and the corresponding amorphous liquid phases, indicating that there are no special kinds of π-electronic short-range intermolecular interactions which can be distinguished in liquid crystals.

The ultraviolet spectra of the π-electronic states of stilbene, benzylideneaniline, azobenzene, azoxybenzene and of several nematogens of related molecular structure have been discussed and assigned by Saupe [30]. Polarization measurements of the uniformly oriented liquid crystal compounds were used as an aid in the assignments of the absorption bands. The starting point for the assignments was based on the π-states of stilbene, calculated with inclusion of first-order configuration interaction. A complete study of the ultraviolet spectra of stilbene, *p*-substituted stilbenes and azobenzene has been reported by Dyck and McClure based on low-temperature matrix isolation techniques [31]. Although they did not study any liquid crystal compounds, their assignments will be of considerable interest to those interested in the assignments of related nematogenic compounds. One of the important conclusions of this study is that, contrary to previous assumptions, the central bond in the lowest excited triplet state has a substantial barrier to internal rotation. The stilbene, azo-, or azoxy-type compounds which form liquid crystals can all exist in *cis*- or *trans*- forms. Azo- compounds can be converted photochemically to the *cis*- form under certain conditions [32]. Only the *trans*- isomers of compounds have been found to be mesomorphic. The same is true of the stilbene compounds [33].

Several investigations have been reported in which the nematic phase or cholesteric mixtures were used as orienting solvents for polarization studies of the ultraviolet spectra. Sackmann has used an optically compensated mixture of cholesteryl derivatives in a magnetic field to achieve a uniform nematic phase for polarization studies of dissolved molecules [34]. He has also reported polarization measurements on cholesteric mixtures in a d.c. electric field [35]. This method is based on the observation by Wysocki [36] that mixtures of cholesteryl chloride and cholesteryl esters will undergo a transition to the nematic state, i.e. the untwisted cholesteric phase, in a d.c. field. An example of the polarization spectrum is shown in Fig. 9.8 for the compound p-$Me_2N—C_6H_4—N{=}N—C_6H_4—NO_2$-$p'$.

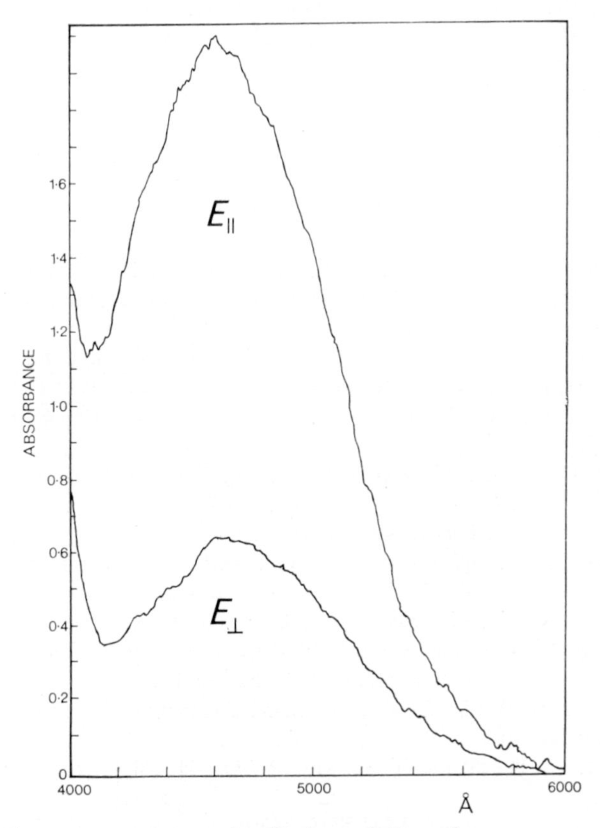

FIG. 9.8. Absorption spectrum of a 10^{-4} M solution of

p-$Me_2N—C_6H_4—N{=}N—C_6H_4—NO_2$-$p'$

in a 1·9 : 1 by weight mixture of cholesteryl chloride and cholesteryl myristate (T_{nem} = 40°C). The spectra were taken at about 45°C with an applied electric field $E = 3{\cdot}5 \times 10^4$ V cm^{-1}. The strong absorption ($E_{\parallel}$) is obtained when the electric vector of the light is parallel to the applied electric field E. The weak absorption ($E_{\perp}$) originates when the polarization is perpendicular to the electric field [35].

Sackmann and Krebs [37] have also reported polarization and electron spin resonance measurements using a sandwich type charge transfer complex of tetrachlorophthalic anhydride with chloranil. It was found that the planes of the complex align parallel with the long axes of the solvent molecules.

An ordinary nematic liquid has been used as a solvent for polarization studies by Ceasar and Gray [38]. The solvent used was 4′-ethoxyphenyl

4-n-butyloxycarbonyloxybenzoate. Heilmeier and Zanoni have used the orientation induced by an electric field to bring about switching in colour of a sample containing dissolved dye molecules [39]. In all these investigations, the long molecular axes of the solute molecules align parallel to the optic axis of the nematic phase.

Despite the obvious limitations, including the high intensity of the solvent bands and the difficulty of achieving uniform alignment, it is apparent that visible and ultraviolet polarization measurements with liquid crystal solvents can be useful.

REFERENCES

[1] Brown, G. H., Doane, J. W. and Neff, V. D. *Critical Reviews in Solid State Sciences*, Chemical Rubber Co., Cleveland, Ohio, **1,** 343 (1970).
[2] Schmid, E. W. and Brandmuller, J. *Z. Elektrochem.* **64,** 940 (1960).
[3] Katritzky, A. R. and Simmons, P. *J. chem. Soc.* 2051 (1959).
[4] Maier, W. and Englert, G. *Z. phys. Chem.* (*Frankfurt*) **19,** 168 (1959).
[5] Sackmann, H. and Demus, D. *Molec. Crystals* **2,** 81 (1966).
[6] Gordon, R. G. *J. chem. Phys.* **43,** 1307 (1965).
[7] Neff, V. D. hitherto unpublished data.
[8] Doane, J. W. and Johnson, D. L. *Chem. Phys. Lett.* **6,** 291 (1970).
[9] Orsay Liquid Crystal Group. *J. chem. Phys.* **51,** 816 (1969).
[10] Helfrich, W. *Phys. Rev. Lett.* **23,** 372 (1969).
[11] *Optical Properties of Solids* (edited by S. Nudelman and S. S. Mitra), Plenum Press, New York (1969), Chapter 14.
[12] Krigbaum, W. R., Chatani, Y. and Barber, P. G. *Acta Crystallogr.* **26B,** 97 (1970).
[13] Amer, N. M., Shen, Y. R. and Rosen, H. *Phys. Rev. Lett.* **24,** 718 (1970).
[14] Saupe, A. *Z. Naturf.* **15A,** 810 (1960).
[15] Saupe, A. and Maier, W. *Z. Naturf.* **16A,** 816 (1961).
[16] Saupe, A. *Z. Naturf.* **19A,** 161 (1964).
[17] Meyer, R. B. *Phys. Rev. Lett.* **22,** 918 (1969).
[18] L'vova, A. S. and Sushchinskii, M. M., *Optics Spectrosc., Wash. suppl. 2*, 139 (1966) [*Optika Spektrosk. suppl. 2*, 266 (1963)].
[19] Koller, K., Lorenzen, K. and Schwab, G. M. *Z. phys. Chem.* (*Frankfurt*) **44,** 101 (1965).
[20] Bulkin, B. J., Grunbaum, D. and Santoro, A. *J. chem. Phys.* **51,** 1602 (1969).
[21] Zhdanova, A. S., Morozova, L. F., Peregudov, G. V., Sushchinskii, M. M. *Optics Spectrosc., Wash.* **26,** 112 (1969) [*Optika Spectrosk.* **26,** 226 (1969)].
[22] L'vova, A. S., Sabirov, L. M., Arefev, I. M. and Sushchinskii, M. M. *Optics Spectrosc., Wash.* **24,** 322 (1968) [*Optika Spektrosk.* **24,** 613 (1968)].

[23] Chang, R. *Molec. Crystals Liqu. Crystals* **12,** 105 (1971).
[24] Maier, W. and Englert, G. *Z. Elektrochem.* **64,** 689 (1960).
[25] Neff, V. D., Gulrich, L. and Brown, G. H. *Molec. Crystals* **1,** 226 (1966).
[26] Williams, R. *J. chem. Phys.* **39,** 384 (1963).
[27] Kusakov, M. M., Khodzhaeva, V. L., Shishkina, M. V. and Konstantinov, I. I. *Soviet Phys. Crystallogr.* **14,** 398 (1969) [*Kristallografiya* **14,** 485 (1969)].
[28] Hansen, T. S. *Z. Naturf.* **24A,** 866 (1969).
[29] Ceasar, G. P., Levenson, R. A. and Gray, H. B. *J. Am. chem. Soc.* **91,** 772 (1961).
[30] Saupe, A. *Z. Naturf.* **18A,** 336 (1963).
[31] Dyck, R. H. and McClure, D. S. *J. chem. Phys.* **36,** 2326 (1962).
[32] Hartley, G. S. *J. chem. Soc.* 633 (1938).
[33] Gray, G. W. *Molecular Structure and the Properties of Liquid Crystals,* Academic Press, London and New York (1962), p. 149.
[34] Sackmann, E. *J. Am. chem. Soc.* **90,** 3569 (1968).
[35] Sackmann, E. *Chem. Phys. Lett.* **3,** 253 (1969).
[36] Wysocki, J. J., Adams, J. and Haas, W. *Phys. Rev. Lett.* **20,** 1024 (1968).
[37] Sackmann, E. and Krebs, P. *Chem. Phys. Lett.* **4,** 65 (1969).
[38] Ceasar, G. P. and Gray, H. B. *J. Am. chem. Soc.* **91,** 191 (1969).
[39] Heilmeier, G. H. and Zanoni, L. A. *Appl. Phys. Lett.* **13,** 91 (1968).

10

Thermal Properties of Liquid Crystals

EDWARD M. BARRALL II and JULIAN F. JOHNSON

Since the middle of the past decade a large body of thermodynamic data on mesophase transitions has become available. Brown and Shaw in their review of the literature to 1957 could find heat of fusion data on only four materials [1]. Thermodynamic data are now known on entire homologous series of mesophase-forming materials.

Knowledge of both the temperature and the heat of transition is necessary if the principles of physical analysis are to be applied to mesophase-forming systems. From this information the transition entropy may be calculated. This acts as the key for evaluating the type and degree of order present in many systems.

In the past it was customary to measure and record only the temperature of mesophase transitions. While this has been very profitable in a number of systems [2], temperature alone cannot furnish a sufficiently encompassing basis for prediction of mesophase behaviour. Attempts to predict mesophase transition temperatures, even in homologous series, using only a knowledge of the transition temperatures of other members, have not met with a high degree of success. A survey of the transition temperatures of any homologous series indicates important trends. However, "exceptions" appear. A list of transition entropies is a somewhat more regular indication of important thermal trends.

The recent proliferation of thermodynamic information is principally due to a pressing need for the information and great improvements in thermal instrumentation. Liquid crystals are currently receiving attention in such varied areas as computer display and biomedical work. [3] Large industrial and governmental research departments are actively engaged in a wider range of liquid crystal research than would have been thought possible a few years ago. The question of thermodynamic properties naturally arises in these research efforts. Recent application of the methods of differential thermal analysis (DTA) and differential scanning calorimetry (DSC) has greatly facilitated the determination of temperatures, heats of transition and heat capacities of various phases. The approach of classical calorimetry has not been neglected. At least one laboratory has made detailed studies of thermodynamic properties by conventional methods [4].

The use of recording differential techniques has greatly emphasized two items of importance. (1) The solid phase polymorphism of many mesophase-forming materials is as complex as the mesophase behaviour. Indeed, this statement may be made in general about organic compounds: solid phase polymorphism is very common. (2) Absolute purity and types of impurity play very important rôles in governing the type, transition temperature and range of mesophase formation. To some extent, these points were recognized by the most careful of the early workers using optical techniques, but these methods did not drive the points home as effectively or dramatically as differential measurements.

The purpose of this chapter is to illustrate the thermodynamic data which are presently available for homologous series of liquid crystal-forming materials. Materials will be listed only when calorimetric data are available. It will become obvious to the reader that certain general statements relate the thermodynamic constants of all liquid crystal-forming materials. Without a doubt these general statements will be revised and grow more powerful as information is added in the future.

Techniques of Measurement

Classical Calorimetry

Arnold and co-workers have made a large number of measurements using classical adiabatic calorimetry [4, 5]. This body of data has the highest degree of estimated precision ($\approx$0·2%). The calorimeter was calibrated using diphenyl ether refined to a purity of 99·984 mole %. The heat of fusion of the diphenyl ether was found to be 4·119 kcal/mole, differing from the value of Furukawa *et al.* by less than $\pm$0·1% [6]. This calorimeter by Arnold is probably one of the most precise ever described for application to liquid crystal systems.

DTA and DSC

The bulk of the thermodynamic data presently available has been obtained by dynamic calorimetry. By the very nature of the instrumentation, dynamic methods are much less accurate than adiabatic calorimetry. The expected accuracy is usually not much better than $\pm$1% and in some cases only $\pm$10%. None the less, the information is useful for general comparisons, so long as the limitations are recognized. Kreutzer and Kast [7] employed dual ice calorimetry which is intermediate in precision between adiabatic and scanning calorimetry.

The techniques of DTA have been described in detail elsewhere [8]. Briefly, a sample and reference material are heated at some linear rate. The absolute temperature and the differential temperature between sample and reference are recorded. The area beneath the differential curve (Fig. 10.1) is related to calories by calibration with a material of known heat of fusion. Since the area is due to a temperature difference,

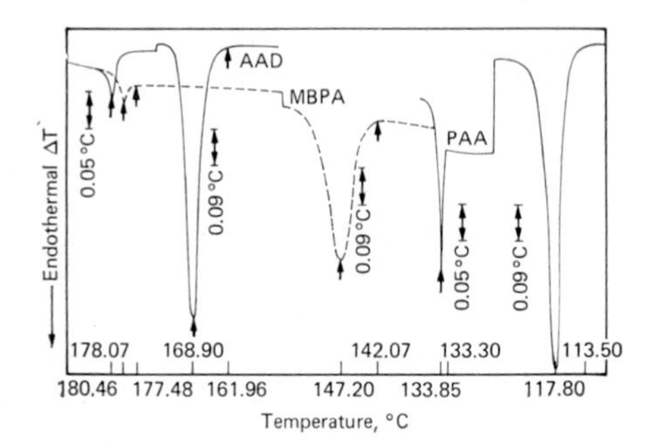

FIG. 10.1. Thermographic traces for three liquid crystalline compounds: Anisaldazine, AAD; *N*-*p*-methoxybenzylidene-*p*-phenylazoaniline,* MBPA; *p*-azoxyanisole, PAA [9].

* $MeO{\cdot}C_6H_4{\cdot}CH{:}N{\cdot}C_6H_4{\cdot}N{:}N{\cdot}C_6H_5$ or 4-(*p*-methoxybenzylidene)aminoazobenzene.

factors such as sample and instrument heat capacity are important. Temperatures, heats of transition and heat capacities may be determined from the curves to about $\pm 1\%$ given adequate calibration. The present authors prefer the DTA cell design shown in Fig. 10.2 for temperature measurement and the design shown in Fig.10.3 for heat measurements. The reasons have been discussed in detail elsewhere [8, 10].

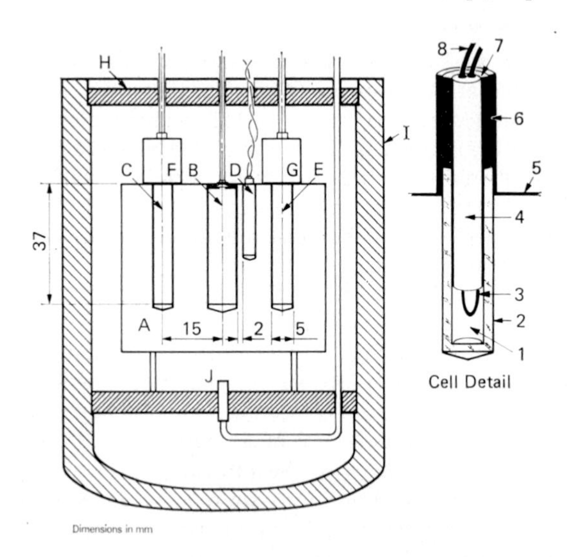

FIG. 10.2. DTA cell suitable for temperature of transition measurement [10].

A, aluminium block (50 mm × 50 mm); B, 500 W cartridge heater; C, sample cell; D, control thermocouple well; E, reference cell; F, G, sample and reference thermocouples; H, lid; I, Dewar flask (90 mm × 200 mm); J, CO_2 cooling jet. Cell detail: 1, sample space; 2, glass cell, 4·5 mm × 45 mm; 3, thermocouple junction; 4, inner ceramic probe, 3 mm; 5, surface of block; 6, outer ceramic cover; 7, sauereisen cement; 8, thermocouple leads.

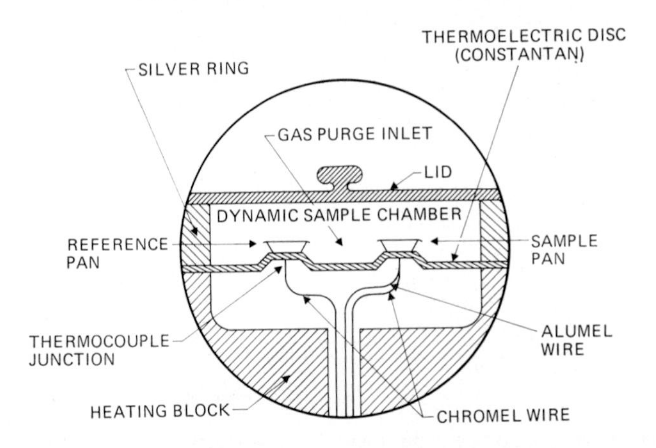

FIG. 10.3. DTA cell suitable for transition heat and heat capacity measurement [11].

DSC also involves the comparison of the sample with an inert reference during a dynamic heating or cooling program. However, instead of a temperature difference being permitted, heat is added *via* the filaments (Fig. 10.4), to keep the sample and reference in balance. To a first approximation, this method is a kind of scanning adiabatic calorimetry.

Obviously, both scanning methods are measuring the same phenomena. However, it is a serious logical error to confuse the two techniques. When adequately calibrated and carefully employed the two methods give comparable results. Indeed, under favourable circumstances DSC data compare closely with data from the highly precise calorimetry of Arnold—see Fig. 10.5—and calculated values. Serious errors may be expected for transitions which do not reach equilibrium quickly. Fortunately, most liquid crystal transitions are rapid.

Other methods for the evaluation of transition heat have been employed occasionally for liquid crystals. These will be mentioned in passing when particular compounds are discussed.

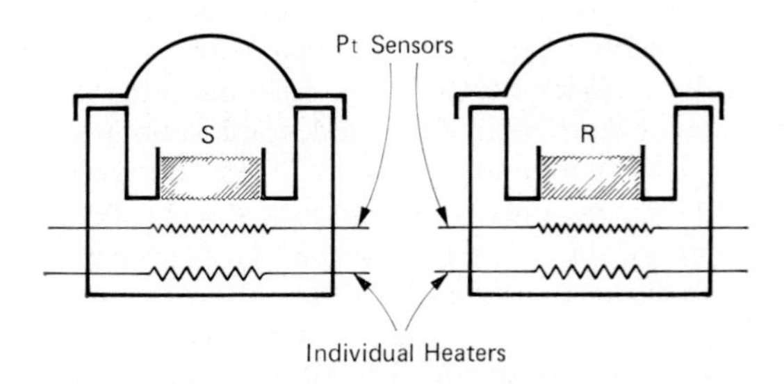

FIG. 10.4. Cell design of a differential scanning calorimeter [12]. Platinum resistance and sample (S) and reference (R) heaters.

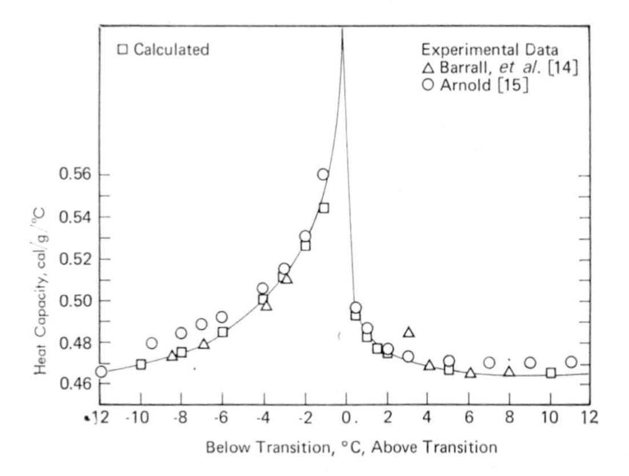

FIG. 10.5. A comparison of experimental (calorimetry and DSC) and calculated heat capacity data [13]. Nematic-isotropic transition for *p*-azoxyanisole, ≈135°C; comparison of calculated and measured specific heats.

Thermodynamic Data

The data in this collection are gathered from various articles in the literature. The original workers reported the results in a variety of units, i.e. cal/g, cal/mole, joules/mole and degrees Kelvin and centigrade. For the sake of unity of presentation all units have been converted to cal/mole for heat of transition and cal/mole/°K for entropy of transition. In such a set of conversions errors are difficult to avoid, but it is hoped that these are few. In some cases, various authors have given several values for a transition heat depending upon the point taken as the start of transition. All values will be cited and the most probable, in the opinion of the present authors indicated. The effect of purity is very large on the heat and temperature of transition [16, 17]. Where possible, the purity of the materials is cited.

When comparing transition heats measured by various methods or by the same method by different authors, it is not uncommon to find variations as large as 10%. This does not imply less precision on the part of one worker over another or necessarily differences in sample purity. The usual source of "error" is how much of the pre-transition effect is included in the measured transition heat. Liquid crystal transitions are certainly not thermodynamically simple. McCullough [18] has identified at least six characteristic non-isothermal melting and solid–solid transitions on the basis of heat capacity–temperature curves of organic compounds. From the available heat capacity curves of liquid crystal-forming materials, it is obvious that several of the above non-isothermal processes are represented in nematic and cholesteric systems [14, 19].

However, it is one thing to recognize a problem and yet another to arrive at a satisfactory solution. For a non-isothermal process, the "pre-transition energy" must be included and properly belongs with the main transition energy—especially if entropy of transition is to be used to judge order changes in a system. Given two sets of measurements on samples of equivalent purity, that involving the larger energy change is more likely to be correct. The statement about equivalent purity is the axis of the argument, for impurities, acting as solvents, can bring about non-isothermal melting in systems which should be isothermal. It is very useful to know not only the temperature of maximum transition rate (T_m), but also the temperature range considered in evaluating the heat of transition. For the present, this sort of information is unavailable in many cases. Hopefully, transition range will be included in more future reports. As for the present study, transition entropies will be reported directly as given in the literature with a minimum of speculation concerning possible pre-transition effects.

4,4′-DI-n-ALKOXYAZOXYBENZENES

Arnold has carried out precision calorimetry on the methyl through dodecyl ethers [5], as well as octadecyl. His values for transition heats are

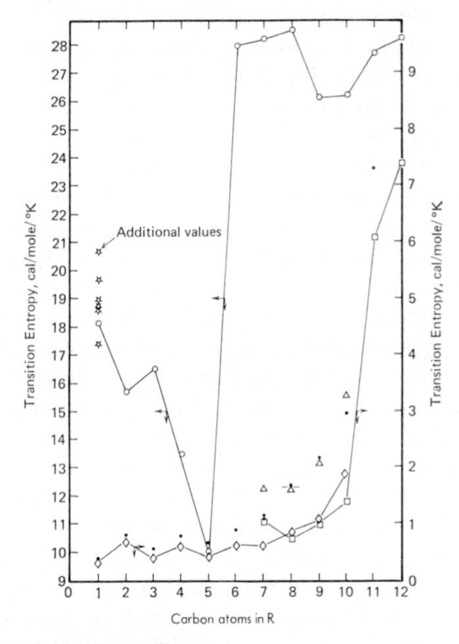

○ Solid → mesophase
◇ Nematic → isotropic
□ Smectic → nematic or isotropic
△ Total mesophase → isotropic liquid
● Transition entropy of last phase by extrapolation

FIG. 10.6. Transition entropies for some 4,4′-di-n-alkoxyazoxybenzenes.

shown in Table 10.1. The first two members of the series, *p*-azoxyanisole (Table 10.2) and *p*-azoxyphenetole, and also 4,4′-di-n-hexyloxyazoxybenzene have been studied by a number of other workers.

Arnold's calorimetric data present a remarkably scattered picture for the transition from solid to mesophase (Fig. 10.6). This is somewhat surprising in consideration of the regular order which is evident for the nematic → isotropic, smectic → nematic and smectic → isotropic transitions in the same study. A possible explanation for such erratic entropy of transition values is offered in Table 10.2 Chow and Martire confirmed the existence of two separate and distinct solid phases of *p*-azoxyanisole (≈99·7 mole %) by scanning calorimetry [23]. Although solid II, the high-temperature form, is metastable at room temperature, it can exist in various mixtures with solid I. The formation of two or more solid phases is very common with mesophase-forming materials (see later sections). In considering the entropy of the transition solid → mesophase it is necessary to use the *total* entropy, i.e. the summed entropy of transition for all crystal forms which are stable between 0°K and the transition temperature. Bondi has adequately demonstrated this for other transitions involving organic compounds [25]. Therefore, the present authors suggest that most of the materials investigated by Arnold were complicated by solid phase polymorphism. The relationship between transition entropy and alkyl chain length should be regular if all crystal forms were accounted for. The data in Table 10.2 for the solid → nematic transition of *p*-azoxyanisole are scattered for probably the same reason. It is interesting to note that the values for the nematic → amorphous isotropic liquid transition as measured by eight workers are in very close agreement. By elimination of the extremes, the value for this transition is between 0·403 and 0·453 cal/mole/°K. Thus, errors due to the small entropy of transition notwithstanding, the nematic → amorphous isotropic transition is better defined than transitions involving the solid phase, i.e. polymorphism does not complicate the nematic mesophase. In Fig. 10.6, the nematic → isotropic transition shows even–odd alternation up to the C_8 side chain.

Chow and Martire [23] measured the thermodynamic properties of the di-n-hexyl ether (99·6 mole %). The entropy of the solid → nematic transition was 29·4 cal/mole/°K (heat of transition, 10·4 kcal/mole). This compares with Arnold's value of 27·92 cal/mole/°K. The nematic → amorphous isotropic liquid transition was found by the same authors to have an entropy of 0·895 cal/mole/°K (heat of transition, 0·359 kcal/mole). This compares closely with Arnold's extrapolated value of 0·898 cal/mole/°K. In addition, Chow and Martire noted the presence of a smectic mesophase very close to the solid transition [23]. They were unable to determine the heat of transition due to poor resolution on the DSC trace.

In general the results of DTA and DSC on the mesophase → amorphous isotropic liquid transition compare more closely with Arnold's "extrapolated" values of transition heat than with the quantities derived by more straightforward measurements. The extrapolated values included pre-transitional heat capacity changes. The method of baseline construction in

TABLE 10.1. *Thermodynamic constants for a series of* 4,4′-*di*-n-*alkoxyazoxybenzenes* [5]

n-Alkyl group	Transition	Trans. temp. (°C)	Trans. heat (cal/mole)	Trans. entropy (cal/mole/°K)
Methyl	solid → nematic	118·2	7067	18·07
	nematic → isotropic	135·3	137	0·336
	extrap. nematic → isotropic	135·3	(165)	(0·404)
Ethyl	solid → nematic	136·6	6422	15·68
	nematic → isotropic	167·5	327	0·741
	extrap. nematic → isotropic	167·5	(358)	(0·814)
Propyl	solid → nematic	115·5	6429	16·55
	nematic → isotropic	123·6	161	0·406
	extrap. nematic → isotropic	123·6	(225)	(0·567)
Butyl	solid → nematic	102·0	5005	13·35
	nematic → isotropic	136·7	247	0·603
	extrap. nematic → isotropic	136·7	(313)	(0·764)
Pentyl	solid → nematic	75·5	3487	10·01
	nematic → isotropic	123·2	173	0·436
	extrap. nematic → isotropic	123·2	(258)	(0·652)
Hexyl	solid → nematic	81·3	9892	27·92
	nematic → isotropic	129·1	250	0·622
	extrap. nematic → isotropic	129·1	(361)	(0·898)
Heptyl	solid → smectic	74·4	9870	28·15
	smectic → nematic	95·4	381	1·03
	nematic → isotropic	124·2	243	0·613
	extrap. nematic → isotropic	124·2	(454)	(1·14)
Octyl	solid → smectic	79·5	10080	28·60
	smectic → nematic	107·7	282	0·741
	nematic → isotropic	126·1	344	0·861
	extrap. nematic → isotropic	126·1	(672)	(1·68)
Nonyl	solid → smectic	75·5	9123	26·18
	smectic → nematic	113·0	395	1·02
	nematic → isotropic	121·5	422	1·07
	extrap. nematic → isotropic	121·5	(851)	(2·16)
Decyl	solid → smectic	78·2	9221	26·26
	smectic → nematic	120·6	554	1·41
	nematic → isotropic	123·4	752	1·90
	extrap. nematic → isotropic	123·4	(1178)	(2·973)
Undecyl	solid → smectic	80·8	9821	27·76
	smectic → isotropic	121·4	2407	6·102
	extrap. smectic → isotropic	121·4	(2894)	(7·339)
Dodecyl	solid → smectic	81·7	10050	28·34
	smectic → isotropic	122·0	2861	7·243
	extrap. smectic → isotropic	122·0	(3325)	(8·417)
Octadecyl	solid → smectic *B*	94·1	18900	51·2
	smectic *B* → smectic *C*	99·0	2526	6·79
	smectic *C* → isotropic	115·3	5410	13·9

Values in ( ) are from extrapolated curves. These values take into consideration pre-transition as well as transition enthalpy change.

TABLE 10.2. *A comparison of thermodynamic data available for the transitions of p-azoxyanisole*

Ref.	Transition type	Trans. temp. (°C)	Trans. temp. (°K)	Trans. heat (cal/mole)	Trans. entropy (cal/mole/°K)
9	solid → nematic nematic → isotropic	117·6 133·9	390·6 406·9	7440 176	19·0 0·432
20	solid → nematic nematic → isotropic	118 132	391 405	7700 176	19·7 0·434
21	solid → nematic nematic → isotropic nematic → isotropic	117 128 onset 132 sharp break	390 401 405	7280 178 178	18·7 0·444 0·440
20	nematic → isotropic	132	405	176	0·434
20	nematic → isotropic	132	405	183	0·453
7, 22	nematic → isotropic	132	404	462	1·14
23	solid → nematic nematic → isotropic solid II → nematic nematic → solid II solid II → solid I	117·5 134·2 104·4 76 61	390·5 407·2 377·4 349 334	7260 181 5630 7230 1030	18·6 0·444 14·9 20·7 3·09
5	solid → nematic nematic → isotropic nematic → isotropic	118·2 135·3 135·3 extrap.	391·2 408·3 408·3	7067 137·2 164·9	18·07 0·3360 0·4039
24	solid → nematic nematic → isotropic	117·5 134·0	390·5 407·0	6800 150	17·4 0·37

DTA and DSC accounts for the pre-transition inclusion. Martin and Müller have discussed this phenomenon [21].

In the homologous series of 4,4′-di-n-alkoxyazoxybenzenes, the alkyl chains do not appear to contribute significantly to the order of the nematic system up to C_5. Below C_6 the alkyl chains are probably not laterally associated and may only determine which of two possible lateral arrangements is adopted by neighbouring azoxybenzene nuclei, in order to account for the odd–even effect. At C_6 the configurational geometry of the side chains becomes such that they are closely associated and contribute to the order of the system, for undoubtedly the appearance at either C_6 or C_7 of a smectic is due to interalkyl chain order. Beyond C_{10} the interchain order is so elaborated that the stability range of the nematic is exceeded when the system has sufficient energy to separate the alkyl chains. Thus, the nematic vanishes from the phase diagram.

Between C_{12} and C_{18} a new type of order appears, and for C_{18} two smectics appear, smectic *B* and smectic *C* of Saupe [26]. It may be characteristic of the smectic *B* → smectic *C* transition that it is smaller in entropy change than the smectic *C* → amorphous isotropic liquid transition. Logically, this is difficult to visualize. If the same chain segments are involved in both transitions, then the transition entropies should be approximately equal—not a factor of two different, 6·79 cal/mole/°K *v.* 13·9 cal/mole/°K. It is possible to speculate that only half the alkyl chains are involved in the smectic *B* → smectic *C* transition. This speculation is predicated on the fact that the smectic *B* modification is not a metastable form.

Arnold has presented a heat capacity *v.* temperature curve for the di-octadecyl ether [27]. The smectic *B* form shows little pre-transition heat capacity anomaly, whereas the smectic *C* modification shows a relatively large pre-transition effect (as does the solid). The heat capacity in the mesophase range is unusually high—well above that of both the amorphous isotropic liquid and the solid phases. This is analogous to the heat capacity phenomena noted for the smectic and cholesteric mesophase of cholesteryl myristate [14].

p-n-Alkyloxybenzoic Acids

Herbert has reported the transition heats of a series of *p*-n-alkyloxybenzoic acids from R = methyl to R = octadecyl [28]. These materials, first described by Gray, are almost surely dimerized in the temperature range studied [29]. The structure would be:

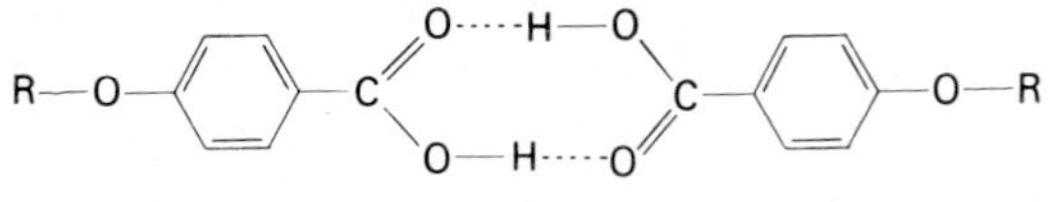

The thermodynamic data were obtained on samples in excess of 99 mole % pure by DSC and are listed in Table 10.3. The first two members of the series, R = methyl and ethyl, do not show a mesophase on heating or cooling. The propyl through hexyl ethers show only a nematic. Solid → solid transitions were noted for the propyl and hexyl materials in this

TABLE 10.3. *Thermodynamic properties of a series of p-n-alkoxybenzoic acids* [28]

Compound	Transition type																		Total ΔS^c
	solid → solid			solid → smectic			solid → nematic			smectic → nematic			smectic → liquid			nematic → liquid			
	T^a	ΔH^b	ΔS^c	T^a	ΔH^b	ΔS^c	T^a	ΔH^b	ΔS^c	T^a	ΔH^b	ΔS^c	T^a	ΔH^b	ΔS^c	T^a	ΔH^b	ΔS^c	
Hydroxy	216 } solid →			—	—	—	—	—	—	—	—	—	—	—	—	—	—	—	—
Methoxy	184 } isotropic liquid			—	—	—	—	—	—	—	—	—	—	—	—	—	—	—	—
Ethoxy	Sublimes			—	—	—	—	—	—	—	—	—	—	—	—	—	—	—	—
Propyloxy	121	1·9	4·8	—	—	—	146·5	4·0	9·5	—	—	—	—	—	—	153·5	0·6	1·4	15·7
Butyloxy	—	—	—	—	—	—	147·5	4·5	10·7	—	—	—	—	—	—	159	0·7	1·6	12·3
Pentyloxy	—	—	—	—	—	—	125	5·2	13·1	—	—	—	—	—	—	149	0·5	1·2	14·3
Hexyloxy	75	1·6	4·6	—	—	—	107	3·3	9·7	—	—	—	—	—	—	153	0·8	1·9	15·2
Heptyloxy	—	—	—	94	4·6	12·5	—	—	—	100	2·6	7·0	—	—	—	147	0·6	1·4	20·9
Octyloxy	75	4·3	12·4	101	2·6	7·0	—	—	—	108	0·3	0·79	—	—	—	146	0·6	1·4	21·6
Nonyloxy	—	—	—	92	8·0	21·9	—	—	—	118	0·4	1·02	—	—	—	145	0·6	1·4	24·3
Decyloxy	86	5·2	14·5	97	2·5	6·8	—	—	—	125	0·4	1·01	—	—	—	143	0·7	1·7	24·0
Undecyloxy	—	—	—	96	9·6	26·0	—	—	—	129	0·5	1·24	—	—	—	140	0·6	1·5	28·7
Dodecyloxy	83	4·3	12·1	90	3·9	10·7	—	—	—	133	0·8	2·0	—	—	—	139	1·0	2·4	27·2
Tridecyloxy	—	—	—	100	11·1	29·8	—	—	—	135	1·0	2·5	—	—	—	137	0·6	1·5	33·8
Tetradecyloxy	—	—	—	96	9·4	26·8	—	—	—	—	—	—	135	2·0	4·9	—	—	—	31·7
Pentadecyloxy	—	—	—	102	12·4	33·1	—	—	—	—	—	—	134	2·3	5·7	—	—	—	38·8
Hexadecyloxy	—	—	—	102	11·6	30·9	—	—	—	—	—	—	133	2·3	5·7	—	—	—	36·6
Heptadecyloxy	—	—	—	104	13·0	34·5	—	—	—	—	—	—	132	2·2	5·4	—	—	—	39·9
Octadecyloxy	—	—	—	105	13·3	35·2	—	—	—	—	—	—	132	2·9	7·2	—	—	—	42·4

a = temperature in °C; b = heat of transition in kcal/mole; c = entropy of transition in cal/mole/°K.

sequence. From heptyl to tridecyl, both a smectic and a nematic are noted, and solid → solid transitions are exhibited by the even carbon numbers of this group. From tetradecyl to octadecyl only a smectic appears.

This series is thermodynamically one of the most regular groups of mesophase-forming materials yet studied. Fig. 10.7 shows the regular increase, with even–odd alternation, of entropy with increasing alkyl

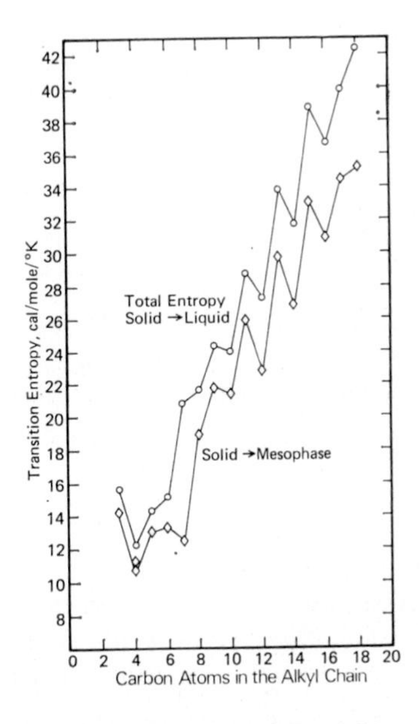

FIG. 10.7. Effect of alkyl chain length on the solid to mesophase transition entropy and total entropy of transition for a series of *p*-n-alkoxybenzoic acids.

chain length for the total entropy change and the entropy change from solid to mesophase. By analogy with normal hydrocarbons, it is possible to postulate that this alternation is due to chain end alignment in a 60° or 90° base plane in the solid state.

The nematic → amorphous isotropic ands mectic → nematic transition entropies are approximately the same order of magnitude up to the C_{13} chain (with the exception of the unexplainably large value of the C_7 smectic → nematic transition). Above C_{13} only the smectic appears and the transition entropy to the amorphous isotropic liquid becomes the *sum* of the previous nematic → amorphous isotropic and smectic → nematic entropies. Throughout the range of the smectic phase the transition entropy either to the nematic or amorphous isotropic liquid phase is dependent on alkyl chain length. Fig. 10.8 indicates the possibility of an even–odd effect. At *p*-n-octyloxybenzoic acid, when the smectic is first noted, the alkyl chain is almost half the length of the whole molecule. Herbert has suggested that in those compounds which exhibit only a

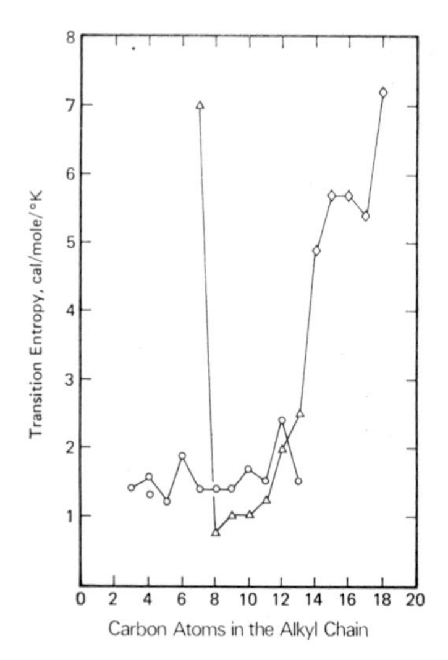

FIG. 10.8. Effect of alkyl chain length on the mesophase transition entropies for a series of *p*-n-alkoxybenzoic acids.

○ Nematic → isotropic liquid
△ Smectic → nematic
◇ Smectic → isotropic liquid

smectic, the solid → smectic transition involves two types of change simultaneously [28]. One type could involve an onset of chain rotation or a change in the angle of tilt *within* planes of molecules. The second type of change would involve the usual formation of smectic layers *between* planes of molecules.

The nematic → amorphous isotropic transition entropy remains essentially constant, and as with the 4,4′-dialkoxyazoxybenzenes, there is evidence of even–odd alternation. This implies that the order involved in the nematic does not depend on alkyl chain length, and the even–odd effect could be due to two stable positions of the aromatic nuclei. The nematic phase predominates until the length of the chain is comparable with the length of the oxybenzoic acid segment of the monomer.

Leclercq *et al.* have examined *p*-n-butoxybenzoic acid by DSC and found that the crystal → nematic transition occurs at 147°C and requires 4·7 kcal/mole (entropy of transition, 11·2 cal/mole/°K) [24]. The nematic → amorphous isotropic liquid transition occurs at 160·5°C and requires 0·57 kcal/mole (entropy of transition, 1·31 cal/mole/°K). These data agree with Herbert's results (Table 10.3) to within 5% for the first transition and 18% on the second transition. The reason for the large difference in mesophase heats is probably the location of the baseline which must be constructed in DSC for integration.

N,α-DI-n-ALKOXYPHENYLNITRONES

Young, Haller and Aviram determined the thermodynamic constants for a series of nitrones, i.e. 4-*p*-n-alkoxybenzylidene-n-alkoxyaniline-*N*-oxides [30]. These have the structure:

O
N— —$O(CH_2)_{n'}H$
$H(CH_2)_nO$— —CH

These materials had not been observed to exhibit mesophases prior to the above work. The authors reasoned that the nitrone series should exhibit mesophases by analogy with known mesophase-forming materials such as Schiff's bases, azoxy- compounds and azo- compounds.

Of the fifteen homologous nitrones (Table 10.4) investigated under the microscope, two showed no mesophase (1 and 3), four were monotropic nematic (2, 4, 14 and 15), and the remainder were either nematic (5–8), smectic (11–13) or both smectic and nematic (9, 10). The corresponding transition temperatures are also listed in Table 10.4. It was not possible actually to form the nematic mesophase of compounds 1 and 3, since the solid phase was not susceptible to supercooling. The transition temperature was estimated in each case by extrapolating the nematic → amorphous isotropic liquid line in a binary mixture phase diagram. This method was employed by Bogojawlensky *et al.* [31] and more recently by Dave and Dewar [32]. Fig. 10.9 is an example of the use of a binary phase diagram.

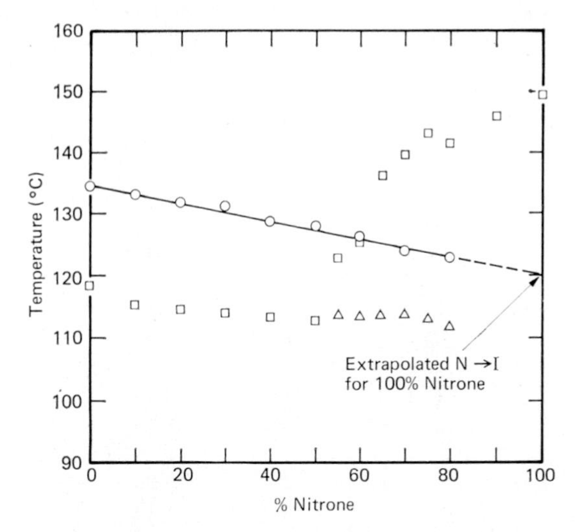

FIG. 10.9. Phase diagram for binary system of nitrone number 1 (Table 10.4) and *p*-azoxyanisole [30].

○ N → I transition observed by microscopy
△ Solid → mesophase or isotropic liquid transition onset, DSC
□ Solid → mesophase or isotropic liquid transition completion, DSC

TABLE 10.4. *Transition temperatures, enthalpies, and entropies for the N-(p-alkoxyphenyl)-α-(p-alkoxyphenyl)nitrones* [30], *i.e. the* 4-*p*-n-*alkoxybenzylidene*-n-*alkoxyaniline-N-oxides*

Graphical code	n,n′[a]	Transition	T (°C)	ΔH (kcal/mole)	ΔS (cal/mole/°K)
1	1,1	C → I N → I[b]	149 120[b]	7·92 —	19·0 —
2	1,2	C → I N → I[c]	146 138	9·00 0·172	21·5 0·42
3	1,3	C → I N → I[b]	155 109[b]	— —	— —
4	1,4	C → I N → I	123 123[c]	8·11 0·150	20·5 0·38
5	1,5	C → N N → I	112 120	7·47 0·168	19·4 0·43
6	1,6	C → N N → I	107 125	6·66 0·197	17·5 0·50
7	1,7	C → N N → I	120 125	8·33 0·219	21·2 0·55
8	1,8	C → N N → I	112 128	11·80 0·268	30·7 0·67
9	1,10	C → S S → N N → I	108 109 129	11·56 *d* 0·373	30·3 *d* 0·93
10	1,12	C → S S → N N → I	110 121 127	13·92 0·045 0·475	36·4 0·12 1·19
11	1,14	C → S S → I	112 124	15·52 0·927	40·3 2·34
12	1,16	C → S S → I	115 128	15·89 1·07	41·0 2·66
13	1,18	C → S S → I	119 127	19·00 1·18	48·5 2·95
14	2,1	C → I N → I[c]	129 128	8·04 0·239	20·0 0·60
15	2,2	C → I N → I[c]	178 158	10·44 0·346	23·1 0·80

References to Table 10.4 at foot of page 269

Fig. 10.9 is interesting from a second viewpoint. The curve formed by the squares (temperature at the vertex of the DSC endotherm for the solid → mesophase or solid → amorphous isotropic liquid transition) has a secondary vertex. This vertex occurs near 75% (wt or mole) dimethoxyphenylnitrone. By inference from non-mesomorphic systems, this indicates that a stable crystal exists with three molecules of nitrone associated with one molecule of *p*-azoxyanisole [33]. A secondary eutectic occurs near 80% nitrone. This is required if the 3 : 1 solid exists. The 3 : 1 solid should act as a pure material with its own heats of transition and mesophase system. Additional thermodynamic studies on phase diagrams of binary mixtures would probably be of great assistance in understanding the rôle of the configuration of the central (non-alkyl) part of the molecule in the formation of the nematic mesophase.

The series of homologous nitrones follows the same pattern as those of the dialkoxyazoxybenzenes and alkoxybenzoic acids. The first members of the series are either non-mesomorphic or monotropic nematic. In the case of the nitrones, when the alkyl group reaches ten carbon atoms, a smectic appears as well as the nematic. For fourteen or more carbon atoms, smectic mesomorphism alone occurs.

Fig. 10.10 shows the effect of alkyl chain length on the solid to mesophase or amorphous isotropic liquid transition entropy. The trend is not as regular as that noted with the alkoxybenzoic acids, but there is a general increase in the transition entropy with increasing alkyl chain length. The irregularities may be due to undetected solid → solid transitions. It is also possible that the descending trend from compounds 2 to 6 is due to a disorder brought about by a "back" folded chain. This alkyl chain would lie at an angle to the long molecular axis. This would obviously decrease the order in the system.

The mesophase transition entropies for the nitrone series are shown graphically in Fig. 10.11. A single smooth curve represents satisfactorily the trend in the entropies of the nematic–amorphous isotropic transitions, with the even–odd alternation falling within the experimental error. For the substances with short alkyl groups, the points level out at about 0·4 cal/mole/°K. This entropy change is of the order of magnitude expected for the replacement of two torsional oscillations in the nematic phase by two nearly free overall rotations in the amorphous isotropic phase. Furthermore, the relative constancy of the observed entropy changes as the chain length varies from 2 to 6 indicates that both the energy difference between alkyl chain conformational isomers and the potential barriers to

---

a n is the number of carbons in the alkoxy group of the α-phenyl ring, while n′ is the corresponding value for that of the *N*-phenyl ring.

b Obtained by extrapolation of the N → I transition temperatures for mixtures, since the supercooled liquid crystallized prior to onset of the nematic phase.

c Monotropic transition.

d Because of the proximity of the C → S and S → N transition temperatures, this value was not determined.

C = crystal; S = smectic; N = nematic; I = amorphous isotropic liquid.

internal rotations change very little between the nematic and isotropic states for short terminal alkyl groups.

For long alkyl chains, on the other hand, the steady, although not quite linear rise in $\Delta S_{N \to I}$ with increasing carbon number above 7 indicates that the probability of certain chain conformations is decreased or the barrier to internal rotations about each carbon is increased in the nematic phase relative to the amorphous isotropic liquid. It is safe to assume that in the

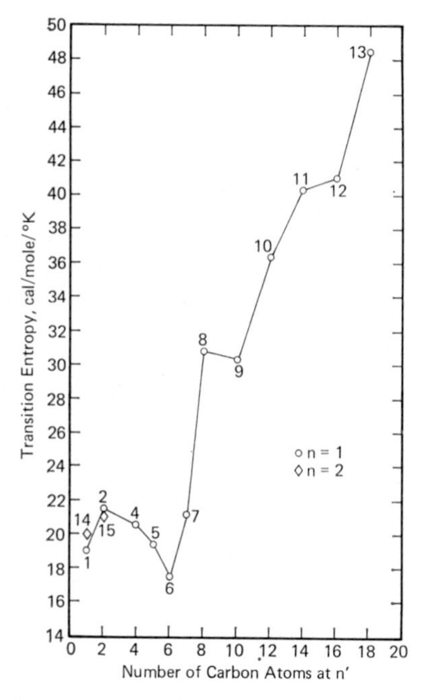

FIG. 10.10. Effect of alkyl chain length on solid → mesophase or isotropic liquid transition entropies for a series of 4-*p*-n-alkoxybenzylidene-n-alkoxyaniline-*N*-oxides.

O
N
$H(CH_2)_n$—O
CH
O—$(CH_2)_{n'}H$

amorphous isotropic liquid, a large number of conformations is virtually equally probable. Since the increment in transition entropy per additional chain carbon atom is only about 0·13 cal/mole/°K, it must be concluded that the terminal groups in the nematic phase *cannot* be restricted to a single (elongated) conformation.

With regard to the overall behaviour of the entropies of the nematic–amorphous isotropic transitions with increasing chain length, the nitrones show a similarity to the homologous series of dialkoxyazoxybenzenes investigated by Arnold.

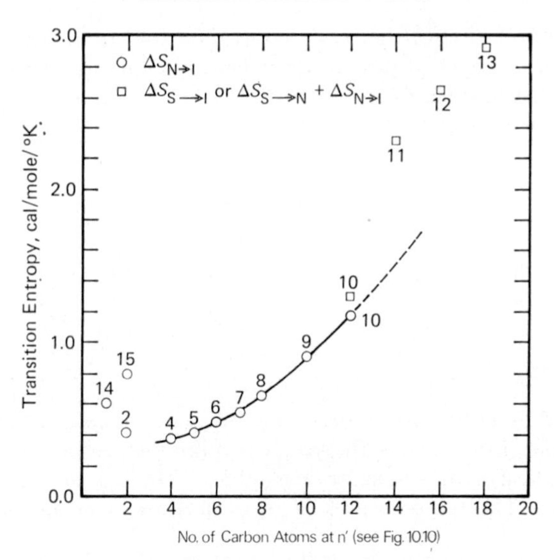

FIG. 10.11. Transition entropies for 4-*p*-methoxybenzylidene-n-alkoxyaniline-*N*-oxides [30].

The entropy difference between the smectic and amorphous isotropic phases as a function of chain length is also shown in Fig. 10.11. Although the point plotted for the 1,12-nitrone represents the sum of the smectic–nematic and nematic–amorphous isotropic entropies of transition, it appears anomalously low. Similar breaks in the $\Delta S_{S \to I}$ curve can be observed in the cholesteryl n-alkanoates, the 4,4′-di-n-alkoxyazoxybenzenes and the *p*-n-alkoxybenzoic acid dimers. In all instances, the break appears at that chain length above which no intermediate nematic exists between the smectic and isotropic phases.

The increment in the smectic–isotropic transition entropy per chain carbon atom is 0·15 cal/mole/°K. This is somewhat smaller than the typical $\Delta S_{S \to I}$ increments in other homologous series. Furthermore, in the nitrone series this increment is barely larger than the increment in $\Delta S_{N \to I}$. This indicates the unusual situation that terminal alkyl groups in nitrones are only slightly more conformationally restricted in the smectic than in the nematic. The 2,2-nitrone, compound 15, is of interest because it is isoelectronic with 4,4′-diethoxyazoxybenzene. The geometry, polarizability, charge distribution, and conformational rigidity of the two compounds are expected to be closely similar. This is indeed reflected in the remarkably close agreement between the nematic–amorphous isotropic transition entropies of the two substances, 0·80 and 0·814 cal/mole/°K, respectively.

Compound 14 differs from compound 2 only in the interchange of the methoxy and ethoxy terminal groups. While it is well known that minor structural changes of this kind influence the nematic–amorphous isotropic transition temperature, it is somewhat surprising that the transition entropies differ by nearly 50%. This points out the need for caution in the detailed interpretation of thermodynamic data for liquid crystalline phase transitions in terms of structural and electronic changes between related molecules. Measurements for numerous mesomorphic compounds are still required before generalizations of acceptable predictive power can be obtained.

Heterocyclic Analogues of 4-Benzylideneamino-4′-methoxybiphenyl

Young, Haller and Williams have prepared a series of compounds of the structure [34]:

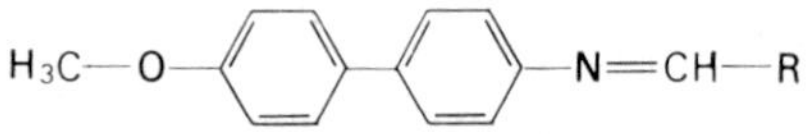

H_3C—O— —N=CH—R

The group R is either aromatic or heterocyclic in nature. Such a series gives valuable insights into the rôle of polar interactions and planarity in the formation of the nematic mesophase. Their data are summarized in Table 10.5. The temperatures of transition are from hot stage micro-

Table 10.5. *Properties of anils of* 4-*amino*-4′-*methoxybiphenyl* [34]

R	No.	Transition	T (°C)	ΔH (kcal/mole)	ΔS (cal/mole/°K)
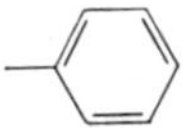	1	C → N N → I	170·0 175·5	8·40 0·094	18·77 0·209
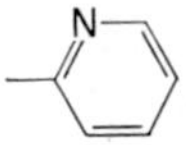 N	2	C → N N → I	116·8 118·8	4·89 0·047	12·56 0·120
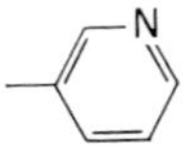 N	3	C → N N → I	175·7 195·2	3·70 0·095	8·25 0·203
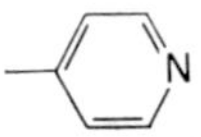 N	4	C → I N → I	193·8 181·3	8·65 0·084	18·53 0·184
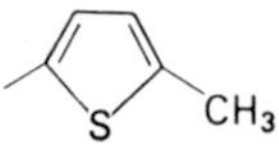 S CH_3	5	C → N N → I	160·1 211·0	7·51 0·153	17·34 0·317

C = crystal; N = nematic; I = amorphous isotropic liquid

scopy and the transition heats from DSC. All mesophases were nematic. The nematic → amorphous isotropic transition temperature is significantly lowered in some cases, e.g. compound 2. In general, compounds with lower nematic stabilities (in the sense implied by Gray [35]) absorb less energy and display less change in gross order on undergoing transition. A substituted thiophen compound (compound 5 in Table 10.5) showed a mesophase, whereas compounds having an unsubstituted thiophen or furan ring as the terminal group (R) exhibited no mesophases.

TABLE 10.6. *Structures and thermodynamic properties of a group of Schiff's bases of p-aminocinnamic acid esters* [37]

Code	R	R′	R″	Transition to smectic			Transition to nematic			Transition to liquid			ΔS total (sum)
				T (°C)	ΔH (kcal/M)	ΔS (cal/M/°K)	T (°C)	ΔH (kcal/M)	ΔS (cal/M/°K)	T (°C)	ΔH (kcal/M)	ΔS (cal/M/°K)	(cal/M/°K)
1	C_2H_5O	H—	—C_2H_5	82[I] 113[II]	6·45 0·41	18·2 1·1	157	1·10	2·56	160	0·17	0·39	22·3
2	—C≡N	H—	—C_3H_7				102·5	5·8	15·5	163	0·16	0·37	15·9
3	CH_3O—	H—	—C_3H_7	[71][I] 79[II]	0·19 5·90	0·55 16·8	101	0·41	1·10	134	0·10	0·25	18·2
4	—C≡N	H—	—C_4H_9				87	5·6	15·6	133	0·090	0·22	15·8
5	CH_3O—	H—	—C_4H_9	66[I] [70][II]*	6·4 0·34	18·9 0·99	91	0·54	1·48	109	0·080	0·21	20·6
6	—C≡N	H—	i-C_5H_{11}				100	6·45	17·3	109	0·090	0·24	17·5
7	CH_3O—	H—	i-C_5H_{11}	47[I] 76[II]	4·10 0·25	12·8 0·72				97	0·80	2·2	15·7
8	—C≡N	H—	—C_6H_{13}				59	5·60	16·9	119	0·090	0·23	17·1
9	CH_3O—	H—	—C_6H_{13}	58·5	7·40	22·3	83·5	0·26	0·73	101	0·11	0·29	23·3
10	—C≡N	H—	—C_8H_{17}				62	5·50	16·4	111	0·22	0·57	17·0
11	—C≡N	—CH_3	—C_3H_7				120	7·50	19·1	122	0·090	0·23	19·3
12	CH_3O—	—CH_3	—C_3H_7				67·5	7·40	21·7	83·5	0·080	0·17	21·9
13	—C≡N	—C_2H_5	—C_3H_7							81·5	?		?
14	CH_3O—	—C_2H_5	—C_3H_7							62	?		?
15	C_2H_5O—	—C_2H_5	—C_3H_7				[63]	0·12	0·36	99	10·8	29·0	29·0

Code refers to Figs. 10.12 and 10.13 and the text. Temperatures in [ ] indicate monotropic transitions. I and II refer to two forms of the smectic mesophase.

* Seen only on rapid *cooling*.

SCHIFF'S BASES OF ESTERS OF *p*-AMINOCINNAMIC ACID

An interesting series of Schiff's bases has been studied by Leclercq, Billard and Jacques [36]. The series has been commented upon by Barrall [37]. Compounds of the type:

$$R-C_6H_4-CH{=}N-C_6H_4-CH{=}C(R')-C(=O)-OR''$$

form variously smectic and nematic mesophases. The calorimetric data (obtained by DSC) in Table 10.6 permit a consideration of some structurally important and general questions:

1. Which part of the molecule controls to a great extent the order/disorder of the transitions solid → smectic, smectic → nematic and nematic → amorphous isotropic liquid?
2. To what extent is molecular geometry important in question 1?
3. What is the extent of the rôle of substituent polarity in mesophase formation?
4. What is the effect of molecular width on mesophase entropy and type?

It is obvious from Table 10.6 and Fig. 10.12 that substituents R, R′ and R″ contribute unique effects. Substituent R produces a unique type of

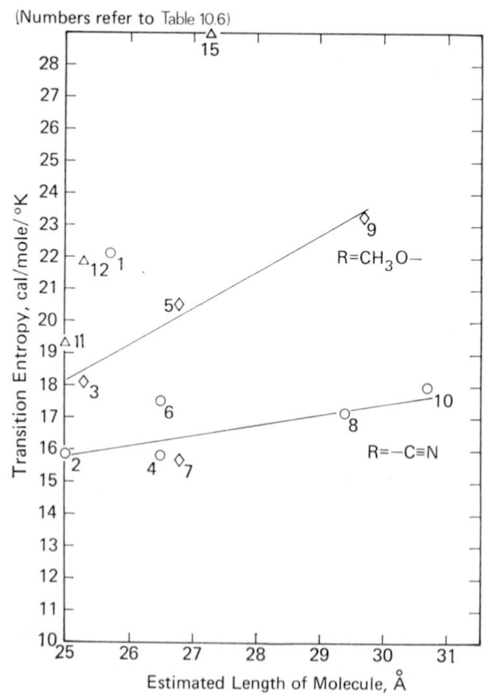

FIG. 10.12. Effect of total molecular length on the solid to isotropic liquid transition entropy [37].

solid order for each R as R″ is extended, and appears to be the primary determinant of solid phase molecular order. R″ extension modifies this order within the limits dictated by R.

The total order of the nematic mesophase depends on the basic internal molecular skeleton. The nematic → amorphous liquid transition entropies shown in Table 10.6 and Fig. 10.13 are almost constant and independent of R, R′ and R″. Physically, this indicates that the groups R, R′ and R″ are almost as completely free to rotate in the nematic state as in the amorphous liquid state.

Ethoxy or methoxy substitution at R alters molecular thickness and permits a smectic order as well as a nematic order. The nitrile substituent at R does not alter the thickness significantly (it is co-planar with the benzene ring), and the nitrile derivatives form only nematic mesophases.

Increasing molecular width by substituting methyl or ethyl for hydrogen

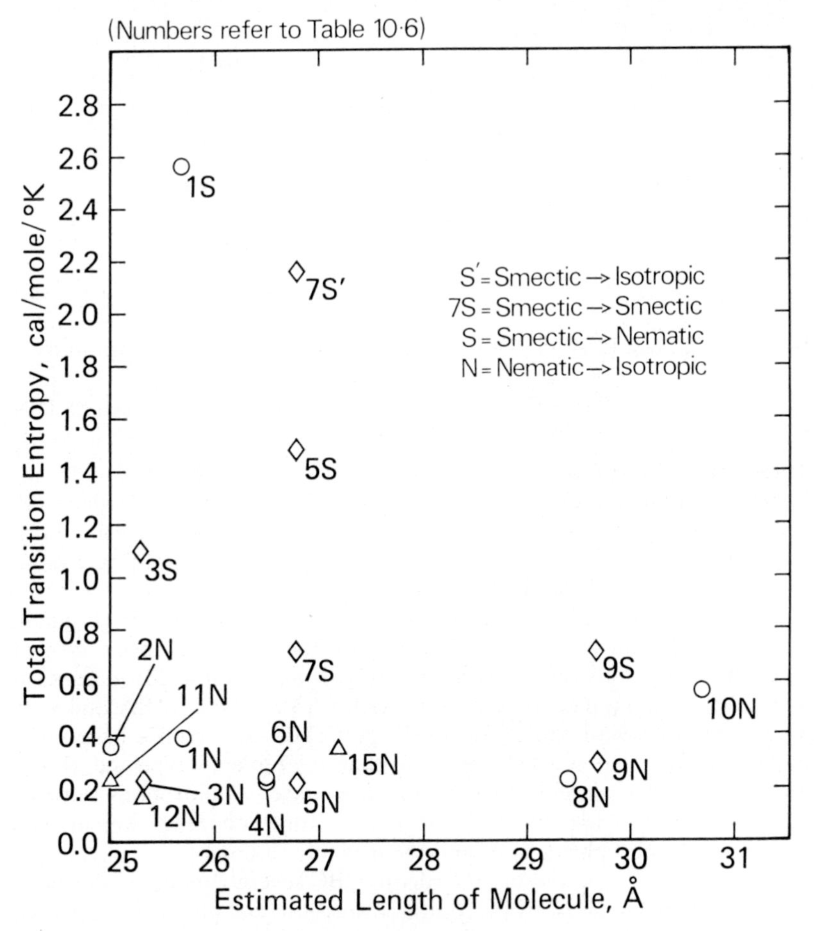

FIG. 10.13. Effect of total molecular length on the mesophase transition entropy [37].

at R′ favours the nematic mesophase over the smectic. However, the substitution at R′ greatly increases the solid order.

This study indicates that simple molecular geometry markedly affects the order of the solid and smectic phases. Nematic order is more complex to define and is a function principally of the nature of the molecular backbone.

Some Other Aromatic Mesophase-forming Materials

Barrall, Porter and Johnson have reported calorimetric values (heats from DSC and temperatures from DTA) on a group of five aromatic mesophase-forming materials [38]. These materials were:

$NC{-}C_6H_4{-}CH{=}N{-}C_6H_4{-}OCH_3$ CBA

$CH_3{-}C_6H_4{-}N{=}CH{-}C_6H_4{-}CH{=}N{-}C_6H_4{-}CH_3$ TBT

$CH_3{-}O{-}C_6H_4{-}CH{=}N{-}C_6H_4{-}O{\cdot}CO{\cdot}CH_3$ MBA

$NC{-}C_6H_4{-}CH{=}N{-}C_6H_4{-}CH{=}CH{\cdot}CO{\cdot}OC_5H_{11}$ CBC*

$CH_3{-}C({=}O){-}O{-}C_6H_4{-}CH{=}CH{\cdot}CO_2H$ ACA

In addition, Arnold *et al.* [39] reported a detailed thermodynamic study of

$C_{12}H_{25}{-}O{-}C_6H_4{-}CH{=}N{-}C_6H_4{-}CH{=}CH{\cdot}CO{\cdot}O{\cdot}C_5H_{11}$ DBC

These materials represent a wide range of variations on the basic molecular backbone. As such they complement the study of Leclercq *et al.* where side chain substitution was varied and the molecular backbone held fairly constant [36]. CBC forms a cholesteric not a nematic mesophase as do the other compounds studied.

The calorimetric data are given in Table 10.7. The authors originally stated cooling and second heating data [38]; the latter have been omitted, since the anomalous results were probably due to sample decomposition. The solid → mesophase transition heats for CBA, TBT, MBA and CBC agree very well with those previously published for Schiff's bases. This molecular type is somewhat more rigid than the biphenyl type and, thus, a greater entropy change is given at the mesophase → amorphous isotropic liquid transition. The introduction of the third benzene ring in TBT increases the solid phase order and the mesophase order.

The existence of "cholesteric" order in CBC severely modified the mesophase → amorphous isotropic liquid transition entropy in comparison with other Schiff's bases with nematic mesophases. CBC is very similar

* optically active amyl ester

TABLE 10.7. *Thermal data on a group of aromatic mesophase-forming compounds*

Compound	Transition	T^* (°C)	ΔH (kcal/mole)	ΔS (cal/mole/°K)
CBA [38]	solid → nematic nematic → isotropic	117·1 124·2	5·98 0·15	15·3 0·38
TBT [38]	solid → nematic nematic → isotropic	186·7 252·6	8·69 0·27	18·9 0·51
MBA [38]	solid → nematic nematic → isotropic	85·2 109·0	5·71 0·22	15·9 0·58
CBC [38]	solid → isotropic "cholesteric" → isotropic	99·6 [99·5	6·52 0·10	17·5 0·27]
ACA [38]	solid → isotropic No mesophase detectable***	208·7	7·61	15·8
DBC [39]	Solid → smectic B	74·6**	6·69	19·2
	smectic B → smectic C	96·3**	1·31	3·55
	smectic C → smectic A	107·3**	0·15	0·39
	smectic A → isotropic liquid	133·1**	2·02	4·98

* DTA endothermal vertex. ** From heat capacity measurements.
*** Although reported to be mesomorphic, the authors found no such transition by DTA or DSC.
[ ] monotropic transition.

to compound 6 reported by Leclercq *et al.* [36] (Table 10.6). They reported data on the i-amyl ester which may not have been completely racemic. The data are in surprisingly good agreement. However, for compound 6 the mesophase is not monotropic. It is possible that the formation of the "twisted" nematic, i.e. cholesteric, mesophase in CBC and possibly in compound 6 (Table 10.6) interrupts polar interactions encountered in the more characteristic nematic and reduces the transition entropy from ≈0·40 to ≈0·25 cal/mole/°K.

The compound ACA is reported to form mesophases [40]. However, Barrall *et al.* found no trace of a nematic mesophase, possibly because their material was purer. The entropy of the solid → amorphous isotropic liquid transition is comparable with that for other mesophase-forming materials studied. However, the high temperature of melting could have caused sufficient decomposition to mask any mesophase formed on cooling.

In close agreement with the predictions made from the work of Leclercq *et al.* [36] by Barrall [37], the compound DBC exhibits a smectic, indeed, three smectic mesophases. Smectic modifications *B* and *A* have large transition entropies similar to those for the previous materials studied by Leclercq *et al.* [36] (Table 10.6, compounds 1, 5 and 7). The long alkyl chain induces two smectic orders of high stability. The transition entropy from smectic *C* to *A* is more nearly of the size expected from a nematic transition. The transition entropies for smectic *B* and *A* indicate that the same structural units must be involved, i.e. 3·55 and 4·98 cal/mole/°K. The 0·39 cal/mole/°K transition entropy may, on the same grounds and by analogy with compounds in Table 10.6, be due to a configurational change in the aromatic portion of the molecule.

Arnold *et al.* have published heat capacity data on DBC [39]. The smectic *B* has a heat capacity similar to that extrapolated from the solid phase. Smectic *C* has a heat capacity which is unusually high; above that of the liquid phase. Smectic *A* has a lower heat capacity than *C*. The smectic *A* heat capacity is very near that which would be obtained by extrapolating the measurements for the solid into the 107° to 133°C temperature range.

DIALKYL *p*-AZOXY-α-METHYLCINNAMATES

Arnold, Demus, Koch, Nelles and Sackmann [41] have prepared and thermally characterized a series of compounds with the general structure:

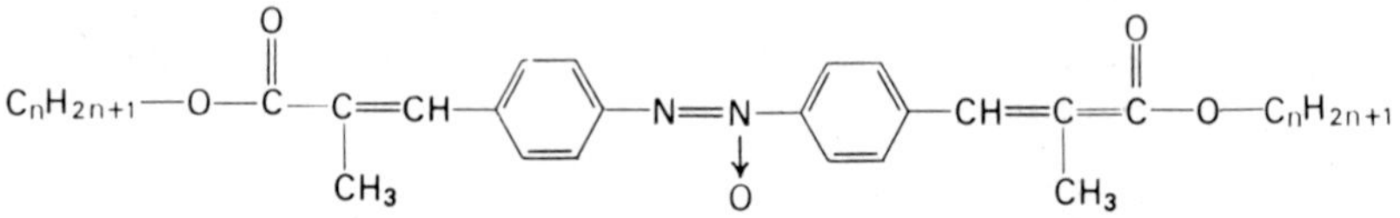

$$C_nH_{2n+1}{-}O{-}\overset{O}{\overset{\|}{C}}{-}\underset{CH_3}{\underset{|}{C}}{=}CH{-}C_6H_4{-}N{=}\underset{O}{\underset{\downarrow}{N}}{-}C_6H_4{-}CH{=}\underset{CH_3}{\underset{|}{C}}{-}\overset{O}{\overset{\|}{C}}{-}O{-}C_nH_{2n+1}$$

with n = 2–9 and 11. These compounds represent a more symmetrical form of the Schiff's bases derived from esters of *p*-aminocinnamic acid described by Leclercq *et al.* [36] with a much increased dipole due to the azoxy bridge. This should be comparable with that in the previously discussed 4,4′-di-n-alkoxyazoxybenzenes [5]. The calorimetric data (Table 10.8) are from references [27, 41], where unfortunately, most of the mesophase temperatures and heats were presented graphically.

The di-ethyl and dipropyl esters have two mesophases: smectic *A* and nematic. The nematic mesophase is not exhibited by the next four members of the series; only smectic *A* appears. From the di-octyl to didodecyl esters two mesophases occur: smectic *A* and smectic *C*. Complete heat capacity curves for the mesophase ranges of the didecyl and didodecyl esters have been presented [39].

The entropy increase for this series is less rapid with increasing chain length than with the dialkoxyazoxybenzenes. However, it is more regular. The additional chain length in the cinnamic acid link probably does not contribute to the smectic order. If a contribution were made, the transition entropies of the didodecyl compounds of azoxy-α-methylcinnamates and

TABLE 10.8. *Thermodynamic properties of a series of di-n-alkyl p-azoxy-α-methylcinnamates* [27, 41]

(Average purity ≈98 to 99 mole %)

Compound	Transition	T (°C)	ΔH (kcal/mole)	ΔS (cal/mole/°K)
Ethyl	solid → smectic A	110	7·89	20·6
	smectic A → nematic	125	0·26	0·65
	nematic → isotropic	141	0·29	0·70
Propyl	solid → smectic A	72	6·69	19·4
	smectic A → nematic	119	0·26	0·66
	nematic → isotropic	131	0·29	0·72
Butyl	solid → smectic A	62	9·08	27·1
	smectic A → isotropic liquid	103	0·69	1·8
Pentyl	solid → smectic A	61	9·56	28·6
	smectic A → isotropic liquid	100	0·81	2·2
Hexyl	solid → smectic A	55	5·98	18·2
	smectic A → isotropic liquid	91	0·96	2·6
Heptyl	solid → smectic A	52	6·21	19·1
	smectic A → isotropic liquid	90	1·14	3·14
Octyl	solid → smectic A	63	13·9	41·4
	smectic A → isotropic liquid	89	1·39	3·84
	smectic C → smectic A	55	?	?
Nonyl	solid → smectic C	67	6·45	19·0
	smectic C → smectic A	78	?	?
	smectic A → isotropic liquid	88	1·57	4·35
Decyl	solid → smectic C	66	8·93	26·3
	smectic C → smectic A	73	≈0·24	≈0·69
	smectic A → isotropic liquid	88	1·72	4·76
Undecyl	solid → smectic C	68	9·80	28·7
	smectic C → smectic A	77	?	?
	smectic A → isotropic liquid	87	2·15	5·97
Dodecyl	solid → smectic C	79	15·7	44·6
	smectic C → smectic A	83	≈0·24	≈0·71
	smectic A → isotropic liquid	87	2·01	5·58

azoxybenzenes should be comparable, with the cinnamate the larger. This is not the case.

Esters of Cholesterol

Since 1963 the n-alkyl esters of cholesterol have been studied in detail from the formate to the nonadecanoate. Heats and entropies of transition accurate to ±5% or better are available from at least three laboratories. Reasonably accurate heat capacity data are also available [14]. Indeed, these esters have been explored as completely as any series of liquid crystal-forming materials. In addition, data have been published on eleven other esters, including aryl and unsaturated esters.

Cholesterol

This material forms no mesophases. A sample, 98·0% pure by DSC, melted with an endotherm at 147·0°C and required 6·55 kcal/mole to go from the solid to the amorphous liquid. This is an entropy change of 14·7 cal/mole/°K [42]. The chloride, bromide and iodide of cholesterol are reported to form mesophases. Cholesteryl chloride melts to the amorphous isotropic liquid at 96·5°C with a heat of fusion of 4·9 kcal/mole (entropy-change, 13·3 cal/mole/°K). On cooling the melt the cholesteric mesophase forms at 66·5°C and liberates 0·09 kcal/mole (entropy change, 0·3 cal/mole/°K). These data were obtained by DSC with a stated precision of ±6% on a sample of ≈99% purity [24].

Cholesteryl Formate

This material forms a monotropic cholesteric mesophase. Due to complex solid polymorphism it is possible to miss the mesophase under calorimetric conditions. Under the polarizing microscope or in a narrow melting point tube the cholesteric "colours" are easily demonstrated on *rapid* cooling. On slow cooling the high melting solid nucleates at about 82°C in very pure samples and no mesophase is observed. In less pure samples the solid phase supercools easily and the mesophase is more easily obtained.

Using a sample with a DSC purity of 99·86% (98·67% using the method of Davis *et al.* [42]) the following transitions were noted using a polarizing microscope and hot stage [43].

1. On heating, conversion of the solid to the isotropic liquid was noted at 96·7°C.
2. The stage temperature was lowered to 95·7°C and a prismatic solid formed.
3. Rapid cooling of the melt produced a cholesteric.
4. By trial and error, a stage temperature, 57·2°C, was found at which the mesophase would neither pass to the amorphous isotropic liquid nor the above solid form rapidly.
5. After several changes of the "moss-like" cholesteric texture, the mesophase became isotropic at 59·6°C.
6. If the mesophase was cooled to below 50°C a new, spherulitic crystal formed.

The above observations suggest the following path of phase transition:

Solid I ⇄ Isotropic Liquid (→ 96·7°C; ← >60°C, <96·7°C)
Isotropic Liquid ⇅ Cholesteric (59·6°C)
Solid II* ← Cholesteric (≃50°C)

* Solid II gives the isotropic liquid at 97·8°C

The enthalpy difference between solid I and solid II is at present not known. The present authors suggest that it is less than 1 kcal/mole.

TABLE 10.9. *Thermodynamic properties of cholesteryl formate**

Solid → melt			Melt → mesophase			
T (°C)	Δ*H* (kcal/mole)	Δ*S* (cal/mole/°K)	*T* (°C)	Δ*H* (kcal/mole)	Δ*S* (cal/mole/°K)	Ref.
98	5·40	14·5	[59·9	−0·089	−0·27]	44
97·0	5·24	14·2	[60·1	−0·083	−0·25]	45
96·6	5·22	14·5	*	—	—	46
96	—	—	[57	—	—]	47
97·5	—	—	[60·5	—	—]	48
97·0	5·29	14·4	59·4	−0·086	−0·26	

* Davis *et al.* did not observe a mesophase in this compound.
[ ] Monotropic transition.

Three sets of calorimetric data are available and are given in Table 10.9. Agreement amongst three independent studies of the solid → amorphous isotropic liquid transition is excellent—2% in terms of Δ*S*. The difference between the two measurements of the very small mesophase transition entropy is ≈9% in terms of Δ*S*.

Cholesteryl Acetate

The issue raised about the formate ester has been raised about the acetate ester. The majority of workers agree that the acetate, on cooling, will give a monotropic cholesteric. In very pure ester, supercooling is difficult and a high-temperature solid phase forms *before* the formation of the cholesteric mesophase is accomplished [43]. Thermodynamic data from several sources are shown in Table 10.10.

Microscopy of the ester indicates that there are at least two and probably three solid phases—needle-like habit, spherulitic with well-defined cross and spherulitic with poorly defined cross, respectively. No significant information is available concerning the crystal class or unit cell dimensions.

TABLE 10.10. *Thermodynamic properties of cholesteryl acetate*

Solid → Melt			Melt → mesophase			Ref.
T (°C)	ΔH (kcal/mole)	ΔS (cal/mole/°K)	T (°C)	ΔH (kcal/mole)	ΔS (cal/mole/°K)	
114	6·77	12·5	[95·4	−0·14	−0·37]	44
110·9	4·95	12·0	—	—	—	45
114·6	4·89	12·6	—	—	—	46
115	—	—	[92	—	—]	47
116·5	—	—	[94·5	—	—]	48
116·0*	4·78	12·3*	[94·5*	—	—]	41
114·5	5·35	12·4	94·1	−0·14	−0·37	

* Estimated from graphical literature presentation.

[ ] Monotropic transition.

Cholesteryl Propionate

This ester is the first of the series that *all* workers agree forms a mesophase—and that it does so on heating as well as cooling. Precise scanning calorimetry demonstrates the existence of two solid phases with different melting points [49], and microscopy has verified two solid forms— solid I, spherulitic and solid II, needle-like. The following recrystallization path was suggested [49]:

$$\text{Solid I} \underset{84\cdot2°\text{C}}{\overset{95\cdot2°\text{C}}{\rightleftarrows}} \text{Cholesteric} \underset{111\cdot6°\text{C}}{\overset{111\cdot8°\text{C}}{\rightleftarrows}} \text{Isotropic Liquid}$$

$$\text{Cholesteric} \underset{98\cdot0°\text{C}\ (\uparrow)}{\overset{\simeq 88°\text{C}\ (\downarrow)}{\rightleftarrows}} \text{Solid II}$$

No direct conversion from solid I to solid II without the mesophase was observed.

Table 10.11 shows five sets of thermodynamic data and two sets of transition temperatures. The variation in transition temperature is probably due to differing amounts of solids I and II in individual samples.

Cholesteryl Butyrate

The transition heat and temperature data [41, 44, 45, 47, 48] indicate that there are probably several solid phases for this ester depending on recrystallization (solvent) and impurity content. The cholesteric mesophase is easily formed and persists for at least 10°C.

Cholesteryl Pentanoate (Valerate)

Some large variations in thermodynamic data for this ester have been observed [41, 44, 45, 47, 48]. In consideration of the reasonable agreement

TABLE 10.11. *Thermodynamic properties of cholesteryl propionate*

Solid → cholesteric			Cholesteric → isotropic liquid			Ref.
T (°C)	ΔH (kcal/mole)	ΔS (cal/mole/°K)	T (°C)	ΔH (kcal/mole)	ΔS (cal/mole/°K)	
98	5·61	15·1	114·1	0·12	0·30	44
101·6	5·76	15·4	115·2	0·10	0·26	45
97·2	5·89	15·9	113·2	0·075	0·20	46
96	—	—	112	—	—	47
102	—	—	116	—	—	48
100	5·98	16·0	115	0·16	0·42	19
96	5·5	14·9	114	0·13	0·33	24
98·7	5·75	15·5	114	0·12	0·30	

between workers on many of the other esters, it is impossible to account for the differences on the basis of instrumentation alone. Probably, traces of solvent and unreacted acid are responsible. Residual solvent depresses the mesophase → amorphous liquid transition more than the solid → mesophase transition in most cases. This could account for the apparent monotropic behaviour observed by Barrall *et al.* [45].

Cholesteryl Hexanoate (*Caproate*)

The entropy data [41, 44, 45, 47, 48] agree better than the transition temperature data. This is probably due to solid phase polymorphism and to impurities. The cholesteric mesophase is probably *not* monotropic in very pure material.

Cholesteryl Heptanoate

This ester [41, 44, 45, 47, 48] exhibits a monotropic cholesteric mesophase. This forms in some cases from 10°C to 20°C below the melting point of the solid. Problems are usually encountered when supercooling a melt by such a large amount, but some transition heat data have been obtained.

Cholesteryl Octanoate (*Caprylate*)

This ester, as with the heptanoate, shows [41, 44, 45, 47, 48] a monotropic cholesteric mesophase. This ester is the highest n-alkyl ester to form only a cholesteric mesophase. Repeated efforts at several laboratories have been unsuccessful in obtaining the smectic mesophase from this ester.

Cholesteryl Nonanoate

A large amount of calorimetric data have been obtained with this ester. All workers agree that the ester forms a well-defined cholesteric mesophase on heating. In addition, the solid → mesophase transition involves somewhat *less* energy than would be predicted from the lower members of the series. The nonanoate, in the opinion of the present authors, marks the beginning of a new type of order in both the solid and the mesophase. This is demonstrated graphically in Figs. 10.14 and 10.15. At first glance it appears that a smooth curve, without a discontinuity, could be drawn. However, such a smooth curve, neglecting the C_9 point, would contravene

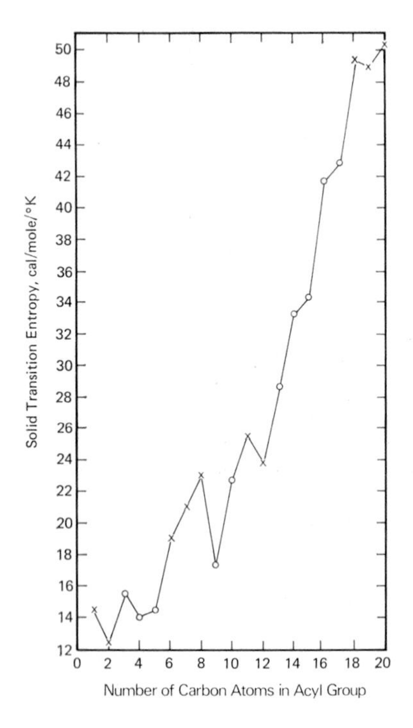

FIG. 10.14. Effect of acid carbon number on solid → mesophase (○) or solid → isotropic liquid (×) transition entropies for cholesteryl esters.

the measurements of five separate workers. The C_9 ester is the first of the series that exhibits a smectic mesophase (monotropic). There is, obviously, a new chain order possible when the ester tail reaches C_9. This is further reflected by the change in the percentage of the total transition entropy involved in the mesophase transitions for the cholesteryl esters as shown in Table 10.12.

Cholesteryl Decanoate

Several sets of calorimetric data [24, 41, 44, 45, 47, 48] are available for this ester. Like the C_9 ester, this ester exhibits a cholesteric on heating and a smectic (monotropic) on cooling. The entropy of the cholesteric → amorphous isotropic liquid transition is larger than would be anticipated, but is well substantiated by four independent studies.

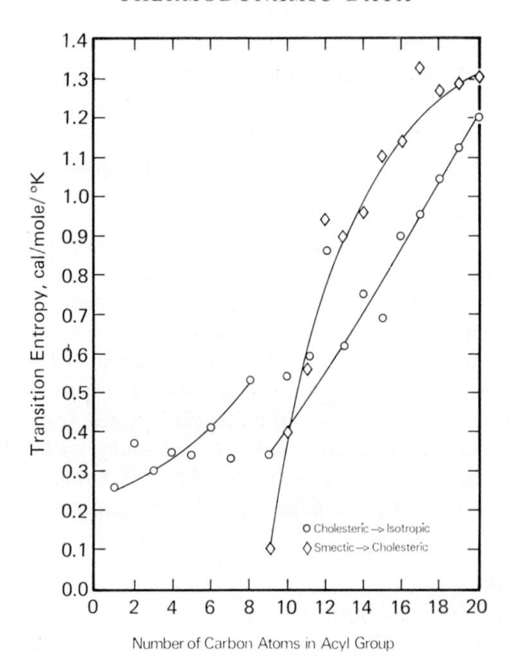

FIG. 10.15. Effect of acid carbon number on mesophase transition entropies for cholesteryl esters.

TABLE 10.12. *Variation of the percentage of the total entropy change due to the mesophase transitions in a series of cholesteryl esters*

Compound	ΔS percentage due to mesophase
Cholesteryl formate	1·8
Cholesteryl acetate	2·9
Cholesteryl propionate	1·9
Cholesteryl butyrate	2·4
Cholesteryl pentanoate	2·3
Cholesteryl hexanoate	2·1
Cholesteryl heptanoate	1·5
Cholesteryl octanoate	2·3
Cholesteryl nonanoate	3·5
Cholesteryl decanoate	4·0
Cholesteryl undecanoate	4·3
Cholesteryl dodecanoate	7·1
Cholesteryl tridecanoate	5·0
Cholesteryl tetradecanoate	4·9
Cholesteryl pentadecanoate	5·0
Cholesteryl hexadecanoate	4·7
Cholesteryl heptadecanoate	5·1
Cholesteryl octadecanoate	4·5
Cholesteryl nonadecanoate	4·7
Cholesteryl eicosanoate	5·0

Cholesteryl Undecanoate

This ester [41, 44–46] exhibits mesophases only on cooling. However, these (cholesteric and smectic) are well defined. Microscopy has indicated the presence of two solid forms which could account for some of the variations noted in transition entropy of the solid phase.

Cholesteryl Dodecanoate (Laurate)

This ester exhibits two solid phases on heating. These are separated by an exothermal recrystallization. This behaviour is reproducible and observable with the polarizing microscope. Davis and Porter [50] have prepared a sample which exhibits only one solid phase using n-pentanol as the recrystallization solvent. In addition, in Table 10.13, the four sets of data for the monotropic mesophases are not as reproducible as for the other esters. A trace impurity or metastable crystal form could account for this variation. Additional work is necessary with this ester. In Table 10.13 the entries on the left are the ΔH values for the change stable solid to liquid with the heat of the exothermal solid–solid change deducted [45].

Cholesteryl Tridecanoate

This ester [44–46] forms both the smectic and cholesteric mesophases on heating. Agreement amongst three sets of data is excellent. The simple mesophase order and well-spaced transitions make this ester and the myristate ideal inter-laboratory calibration standards for mesophase-forming compounds.

Cholesteryl Tetradecanoate (Myristate)

Five sets of data for this ester are shown in Table 10.14; agreement is reasonably good. The myristate ester is almost ideal for inter-laboratory and extended studies of mesophase transitions and properties. It is, probably, one of the easiest esters to purify due to the favourable solubility relationships of cholesterol and myristic acid. The authors have noted that a single ethanol recrystallization of freshly produced cholesteryl myristate is as effective in removing impurities as three of such recrystallizations of the decanoate or stearate [49].

Cholesteryl Pentadecanoate

Data from three laboratories [44–46] indicate that this ester exhibits both a smectic and cholesteric mesophase on heating; agreement is good.

Cholesteryl Hexadecanoate (Palmitate)

The thermodynamic properties of this ester have been measured by at least five workers (Table 10.15). Most studies agree that the cholesteric is formed and that a monotropic smectic forms on cooling. However, there are notable exceptions. Vogel *et al.* uncovered one reason for the unusual behaviour of the smectic mesophase [52]. A trace of antioxidant added by the manufacturer to the final product can depress the solid $\rightarrow$ mesophase transition without altering the mesophase transition temperatures or heats. This is reasonable if it is assumed that the antioxidant is

TABLE 10.13. *Thermodynamic properties of cholesteryl dodecanoate (laurate)*

Solid → liquid			Smectic → cholesteric			Cholesteric → isotropic liquid			
T (°C)	ΔH (kcal/mole)	ΔS (cal/mole/°K)	T (°C)	ΔH (kcal/mole)	ΔS (cal/mole/°K)	T (°C)	ΔH (kcal/mole)	ΔS (cal/mole/°K)	Ref.
92	7·82	21·4	[82·1	0·30	0·83	89·2	0·26	0·71]	44
99·0	9·96*	26·8	[80·7	0·40	1·13	87·4	0·49	1·36]	45
91·3	7·62	20·9	[80·2	0·23	0·64	87·2	0·18	0·49]	46
76	—	—	78	—	—	85	—	—	47
93	—	—	[83·5	—	—	90	—	—]	48
92*	9·32	25·5	[82	0·41	1·15	89	0·32	0·88]	19
91	8·68	23·7	81·1	0·34	0·94	88·0	0·31	0·86	

* Last of two peaks, see text.

[ ] Monotropic transition.

Table 10.14. *Thermodynamic properties of cholesteryl tetradecanoate (myristate)*

Solid → smectic			Smectic → cholesteric			cholesteric → isotropic liquid			
T (°C)	ΔH (kcal/mole)	ΔS (cal/mole/°K)	T (°C)	ΔH (kcal/mole)	ΔS (cal/mole/°K)	T (°C)	ΔH (kcal/mole)	ΔS (cal/mole/°K)	Ref.
71·5	10·4	30·1	79·9	0·31	0·87	85·2	0·25	0·70	44
73·6	—	—	80·0	—	—	85·6	—	—	45
70·5	11·1	32·6	77·8	0·33	0·95	83·2	0.24	0·69	46
71	—	—	81	—	—	86·5	—	—	48
72	13·4	38·8	78	0·39	1·1	84	0·33	0·93	19
70	11	32·1	78·5	0·32	0·91	83·5	0·25	0·70	24
73·6	11·2	32·3	79·5	0·34	0·97	85·5	0·26	0·73	51
71·7	11·4	33·2	79·2	0·34	0·96	84·8	0·27	0·75	

TABLE 10.15. *Thermodynamic properties of cholesteryl hexadecanoate (palmitate)*

Solid → cholesteric			Smectic → cholesteric			Cholesteric → isotropic liquid			Ref.
T (°C)	ΔH (kcal/mole)	ΔS (cal/mole/°K)	T (°C)	ΔH (kcal/mole)	ΔS (cal/mole/°K)	T (°C)	ΔH (kcal/mole)	ΔS (cal/mole/°K)	
77·5***	14·2	40·5	78·1	0·39	1·1	82·6	0·29	0·83	44
79·6****	14·2	40·3	[64·0	0·36	1·1]	[70·0	0·28	0·82]	45
77·3	14·0	40·0	[76·5	0·36	1·0]	81·6	0·28	0·78	46
77	—	—	[75	—	—]	80	—	—	47
79	—	—	[78·5	—	—]	83	—	—	48
76*	17·0	*48·7	[72*	0·44*	1·3*]	79*	0·41*	1·2*	41
77·51	13·69	39·1	[76·4	0·410	1·21]	82·0	0·32	0·89	52**
77·7	14·6	41·7	74·3	0·39	1·14	79·7	0·32	0·90	

* From graphically presented data.
** See text.
*** Solid → smectic.
**** Solid → isotropic
[ ] Monotropic transition.

insoluble in the solid and equally soluble in the mesophases and amorphous isotropic liquid. Such selective impurities have been noted in nematic systems [53, 54]. The unusual data shown in ref. [45] are due to a trace of ethanol remaining in the ester after final drying. The data from Vogel's work are based on a 99·97 mole % sample synthesized under very carefully controlled conditions with toluene-*p*-sulphonic acid as catalyst; see ** in Table 10.15. No antioxidant was present in any of the reactants or the product.

Cholesteryl Heptadecanoate

This material exhibits a cholesteric and a monotropic smectic mesophase. Reasonably good agreement exists for three sets of data [17, 44, 45].

Cholesteryl Octadecanoate (*Stearate*) [41, 44–48]

Both the smectic and cholesteric mesophases are monotropic. The smectic mesophase is particularly difficult to observe on DTA or DSC due to the formation of the solid phase.

Cholesteryl Nonadecanoate

This ester, as with the stearate, exhibits only monotropic mesophases [44–46].

Cholesteryl Eicosanoate

Only one complete set of calorimetric data [44] is available for this ester. Both mesophases are monotropic. Impurities are particularly difficult to remove from this ester. However, it is interesting to note that even as high as C_{20}, the cholesteric and smectic mesophases are exhibited although the hydrocarbon tail is dominating the fusion process (very high ΔS). As yet, no upper limit to the chain length–mesophase formation relation has been clearly defined. Davis *et al.* have suggested on the basis of smectic-cholesteric range (the difference between T smectic and T cholesteric), that C_{20} could mark the end of cholesteric mesophase formation [46]. At C_{21} or C_{22}, it is proposed that the two transitions merge to give only a smectic mesophase. However, this is conjecture at present.

Other Cholesteryl Esters

Thermodynamic data have been obtained on a number of esters of cholesterol, derived from aromatic, substituted and unsaturated acids (Table 10.16). The entropy of fusion (solid → mesophase or amorphous isotropic liquid) appears to be most closely related to the length in Ångstroms of the fully extended acyl group derived from the acid. This is shown in Fig. 10.16 for the solid → mesophase transition. The relationship is also true of the mesophase → amorphous isotropic liquid transition [55]. The correlation is extremely good; many materials which contain the cholesteryl ring structure, but lack the oxygen bridge, also lie on the curves with a high degree of agreement. This is described later.

Table 10.16. *Thermodynamic properties of some aromatic, substituted and unsaturated esters of cholesterol****

Ester	Solid → mesophase****			Mesophase → isotropic liquid			Ref.	Code
	T (°C)	ΔH (kcal/mole)	ΔS (cal/mole/°K)	T (°C)	ΔH (kcal/mole)	ΔS (cal/mole/°K)		
Cholesteryl anisoate	179·2	8·44	18·7	264·9	0·29	0·54	55	1
Dicholesteryl adipate	195·5	7·00	14·9	225·5	0·87	1·75	55	2
Cholesteryl cinnamate	162·6	6·87	15·8	215·2	0·17	0·35	55	3
Cholesteryl benzoate	145·8	5·30	12·7	180·7	0·17	0·38	55	4
Cholesteryl crotonate	112·8	5·91	15·3	162·0	0·30	0·69	55	5
Dicholesteryl sebacate	179·1	10·6	23·4	[175·8	0·88	1·96]	55	6
Dicholesteryl phthalate	186·4	9·48	20·6	**	—	—	55	7
Cholesteryl heptafluorobutyrate	114·2	6·35	16·4	**	—	—	55	8
Cholesteryl *p*-phenylazobenzoate	192·4	7·08	15·2	185·3	0·36	0·79	55	9
Cholesteryl hydrogen phthalate	169·8	9·78	22·0	[153·4 [89·1	0·080 0·49	0·19Ch] 1·35S]	55	10
Cholesteryl chloroformate	120·6	6·34	16·1	**	**	—	55	11
Cholesteryl oleate	50·0	7·31	22·6	[46·4 [41·8	0·18 0·24	0·56Ch] 0·76S]	56	12
Cholesteryl linoleate	42·6	7·17	22·7	[35·1 [34·0	0·13 0·35	0·42Ch] 1·14S]	56	13
Cholesteryl linolenate	34·8	*	*	[34·8 [33·0	0·10 0·44	0·33Ch] 1·43S]	56	14

* Heat could not be calculated due to overlapping exotherm.
** Mesophase not detected calorimetrically, but visible by microscopy.
*** The first nine esters exhibit only a cholesteric mesophase.
**** Except where mesophase is monotropic—then solid → amorphous isotropic liquid.
[ ] Monotropic transitions.

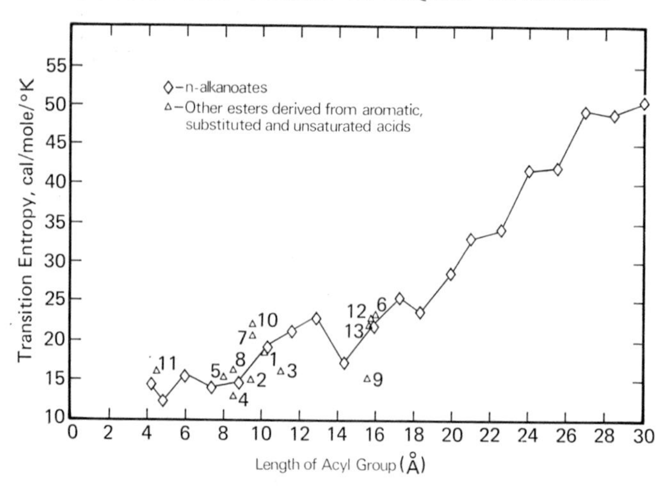

FIG. 10.16. Effect of acid chain length on solid → mesophase or isotropic liquid transition entropies for a group of cholesteryl esters. (Readings in the graph refer to Table 10.16.)

General Comments on Calorimetry of Cholesteryl Esters

The total order in the solids and mesophases of cholesteryl esters appears to be based only on geometrical considerations, i.e. the length of the substituent ester tails. There is some evidence that for n-alkyl esters above C_9, even–odd effects appear in the solid → mesophase transition entropy. The entropy break between C_8 and C_9 esters strongly indicates that a new type of geometrical symmetry becomes possible when the ester tail becomes longer than ≈13 Å. This same symmetry causes the appearance of the smectic mesophase in all esters longer than C_9. The *whole* cholesteryl ester molecule contributes to the order in the solid, smectic and cholesteric mesophases. This is in contrast to nematic mesophase-forming materials discussed previously. In the nematic, only the central core of the molecule appears to affect the total order of the system in the nematic range. The entropy of transition (cholesteric → amorphous isotropic liquid) for the dicholesteryl esters is unusually large, about twice the expected value (Table 10.16). This indicates that *both* cholesteryl ring systems are participating in the helical order of the cholesteric mesophase. The introduction of a double bond in the ester chain apparently forces the part of the chain beyond the first double bond from the carbonyl group from participating in the order of all phases (solid, smectic and cholesteric). This is amply demonstrated for the C_{18} esters in Table 10.16.

From thermodynamic data alone it is apparent that the esters of cholesterol form mesophases in a very unique fashion. The conditions controlling the order in the smectic and cholesteric cannot be deduced from a consideration of known nematogenic materials that do not contain the cholesteryl ring system. The esters of cholesterol reflect more than any other materials studied the importance of pure geometrical factors in complex phase order.

RELATED SYSTEMS DERIVED FROM STEROLS

Elser and Ennulat have prepared a number of interesting carbonates, thio-esters and thiocarbonates with the following formulae [44, 57, 58, 59]:

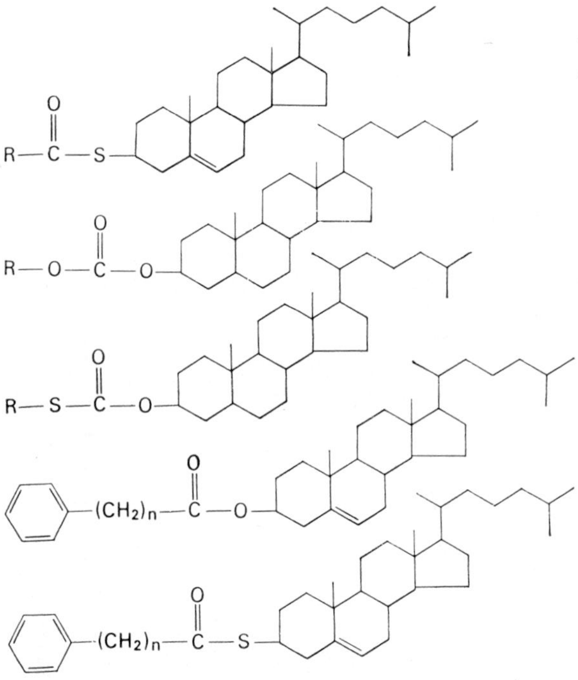

Thiocholesteryl alkanoates

Cholestanyl n-alkyl-carbonates

Cholestanyl *S*-alkyl thiocarbonates

Cholesteryl ω-phenylalkanoates

Thiocholesteryl ω-phenylalkanoates

Several other series were made [57], but transition heat data have not been presented. The thermodynamic data for two of these series are presented in Tables 10.17 and 10.18. Since data originating from this laboratory (U.S. Army Electronics Command, Night Vision Laboratory, Fort Belvoir, Va.) are in close agreement with those of other workers with respect to the cholesteryl n-alkanoates, it may be supposed that the data in these tables are of equal reliability. However, in the experience of Elser and Ennulat, as well as of the present authors, the compounds formulated above are much more difficult to purify than the cholesteryl n-alkanoates. As the original authors have stated, purity differences have undoubtedly altered some of the results, but with reasonably complete homologous series, these differences do not disturb the trends significantly.

Thiocholesteryl Esters (Table 10.17)

The replacement of the alcohol oxygen by sulphur produces some unique changes in the enthalpy relationships. The entropy of the transition from the solid to the mesophase or amorphous isotropic liquid is shown in Fig. 10.17. The broken line in this figure represents the relationship previously obtained for the esters of cholesterol. The solid line shows the trend for

TABLE 10.17. *Thermodynamic data for a series of thiocholesteryl alkanoates* [44]

Alkyl chain	Solid → smectic*			Smectic → cholesteric			Cholesteric → isotropic liquid		
	T (°C)	ΔH (kcal/mole)	ΔS (cal/mole/°K)	T (°C)	ΔH (kcal/mole)	ΔS (cal/mole/°K)	T (°C)	ΔH (kcal/mole)	ΔS (cal/mole/°K)
Formate	118	7·06	18·0				—	—	—
Acetate	126	6·53	16·4				[119·5	0·071	0·18]
Propionate	112	5·88	15·3				[111·1	0·059	0·15]
Butyrate	100	5·64	15·1				117·6	0·085	0·22
Pentanoate	91	5·70	15·7				104·7	0·058	0·15
Hexanoate	95	6·27	17·0				107·7	0·10	0·27
Heptanoate	107	7·78	20·5	[73·1	0·025	0·073]	[102·1	0·077	0·20]
Octanoate	97	6·93	18·7	[76·6	0·076	0·22]	100·8	0·16	0·41
Nonanoate	69	5·34	15·6	84·0	0·14	0·39	97·5	0·12	0·34
Decanoate	80	5·59	15·8	87·4	0·15	0·42	98·3	0·18	0·47
Undecanoate	81	6·25	17·7	88·1	0·25	0·70	95·0	0·15	0·41
Dodecanoate	84	8·34	23·5	86·7	0·36	1·0	92·3	0·19	0·53
Tridecanoate	80	8·33	23·7	86·5	0·31	0·87	91·4	0·15	0·42
Tetradecanoate	73	7·08	20·5	85·5	0·34	0·95	90·4	0·19	0·52
Pentadecanoate	62	8·73	26·1	84·3	0·37	1·0	88·3	0·20	0·56
Hexadecanoate	54	7·97	24·1	83·0	0·38	1·1	87·3	0·12	0·33
Heptadecanoate	66	10·2	29·9	81·2	0·40	1·1	85·0	0·20	0·57
Octadecanoate	64	9·77	28·9	80·2	0·41	1·2	84·2	0·13	0·36
Nonadecanoate	74	9·94	28·7	78·3	0·40	1·1	82·4	0·13	0·37
Eicosanoate	71	11·3	32·8	77·1	0·42	1·2	81·3	0·24	0·67

* For the first eight compounds, this transition is either solid → amorphous isotropic liquid or solid → cholesteric depending upon the monotropy indicated.

[ ] Monotropic transition.

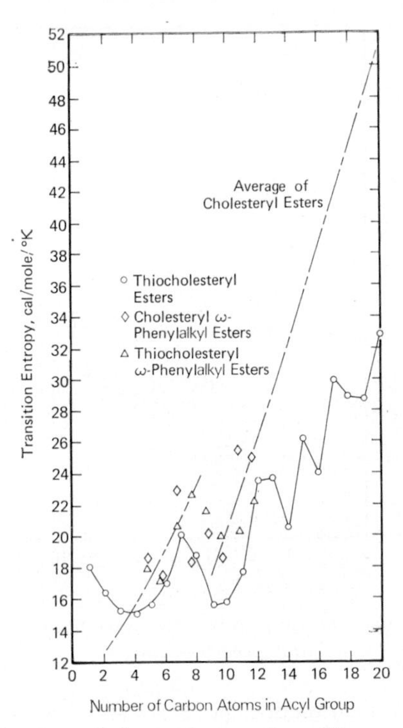

FIG. 10.17. Effect of carbon chain number on the solid → mesophase or isotropic liquid transition entropies for cholesteryl esters, thiocholesteryl esters, and ω-phenylalkyl analogues.

the thio-esters. A break in the smoothly increasing entropy occurs between carbon numbers 7 and 9. The entropies of transition for the thio-esters below C_7 first decrease and then increase. Above C_9 the entropy of transition increases almost rectilinearly with carbon number. Unlike the simple esters, the thio-esters above C_9 show clear evidence of even–odd alternation with increasing ester chain length. As with the simple esters, the break in the solid transition entropy curve is at the same carbon number as the first appearance of the smectic mesophase. The total order, as defined by the entropy of the solid phase of the thio-ester, is lower than the total order of the solid phase of the simple ester.

The mesophase entropy curves, Fig. 10.18, are similar to the previously given curves for the simple esters (Fig. 10.15). Below C_7, only the cholesteric mesophase appears. There is clear evidence of even–odd alternation up to C_{13}. The entropy range from thio-acetate to thioheptanoate is lower than for the simple esters. Mesomorphism apparently begins in the series with the thio-acetate. However, the cholesteric mesophase of cholesteryl thioformate may have been overlooked. The thio-ester smectic → cholesteric transition entropy curve is almost identical (super-imposable) with the simple ester curve. The result for the last member, C_{20}, lies only 0·1 cal/mole/°K below that for C_{20} in Fig. 10.15. The entropy of the

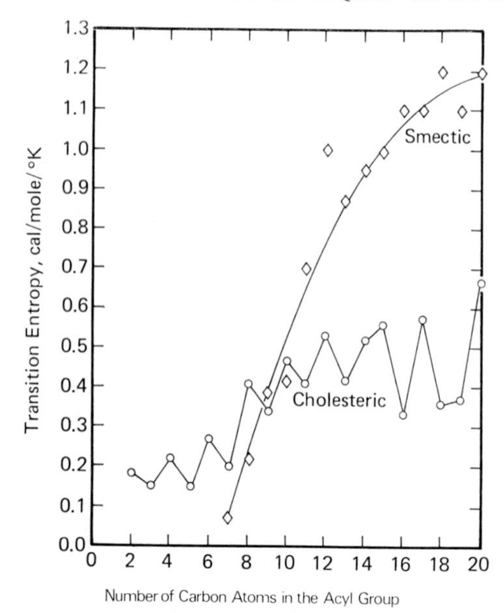

FIG. 10.18. Effect of acid carbon number on mesophase transition entropies for thiocholesteryl esters.

smectic mesophase transition does not show signs of even–odd alternation. The cholesteric → amorphous isotropic liquid transitions of the thio-esters above C_8 show much less increase in entropy with increasing chain length, i.e. are less ordered, compared with the simple esters. However, there is *clear* evidence of even–odd alternation up to C_{13}.

Sulphur appears to cause the following changes:

1. Decreases the order in the solid phase.
2. Produces almost the same order in the smectic mesophase.
3. Causes smectic order to appear at C_7 as opposed to C_9 in n-alkyl esters of cholesterol.
4. Introduces a clear odd–even effect in transition temperatures and transition entropies with the exception of the smectic mesophase.
5. Decreases the transition entropy of the cholesteric mesophase.

All of the above suggest that sulphur increases lateral molecular interactions and decreases terminal attractions [44].

Cholestanyl n-*alkyl Carbonates and* S-*alkyl Thiocarbonates*

Ennulat *et al.* noted that the carbonates of cholestanol and n-alkanols did not produce a smectic mesophase at any chain length. The entropy of the solid → mesophase or solid → amorphous isotropic liquid transition increases almost linearly with increasing alkyl carbon number (Fig. 10.19). There is no apparent break in the curve (allowing for data scatter). The total entropy increase in going from the C_1 to C_{20} alkyl chain is almost exactly equal for cholestanyl carbonates and cholesteryl esters.

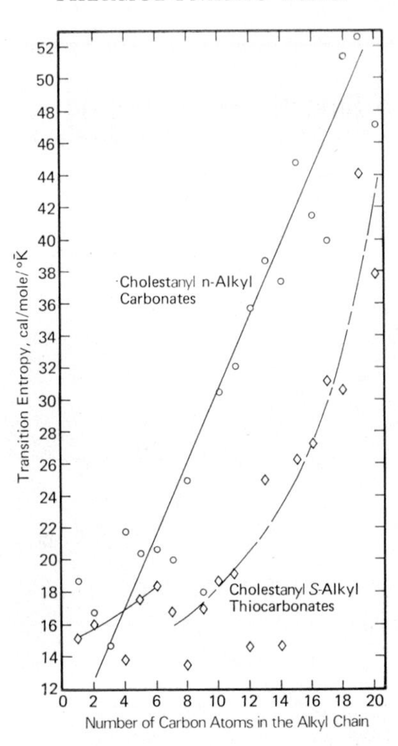

FIG. 10.19. Effect of alkyl carbon number on the solid → mesophase or solid → isotropic liquid transition entropies for cholestanyl n-alkyl carbonates and thiocarbonates.

The replacement of oxygen by sulphur at the alkyl group produces a smectic mesophase at C_7 (Fig. 10.20). There is an apparent break in the curve of the solid phase transition entropy at this point. Some values for the higher thiocarbonates, C_8, C_{12} and C_{14}, are anomalously low. The cholesteric mesophase appears with the first member and continues to C_9 giving a smoothly increasing curve in Fig. 10.20. Above C_9 a new relationship is established for the cholesteric mesophase. The break is much smaller than that observed for the esters and thio-esters. The smectic mesophase appears at C_7 and gives an unbroken curve up to C_{12}. Above this carbon number another relationship is established, and the smectic → cholesteric and cholesteric → amorphous isotropic liquid transition entropies become equivalent. However, the data exhibit a broad scatter.

Elser and Ennulat state, from extrapolation of data, that a smectic mesophase should form at C_5 [59]. Their optical tests did confirm the existence of a metastable smectic mesophase in rapidly cooled C_5 and C_6 thiocarbonates. In addition, however, the same authors concluded that many of their thiocarbonates contained significant impurities [59]. The final clarification of the transition heats of the alkyl thiocarbonates of cholestanol awaits the development of an efficient method of purification.

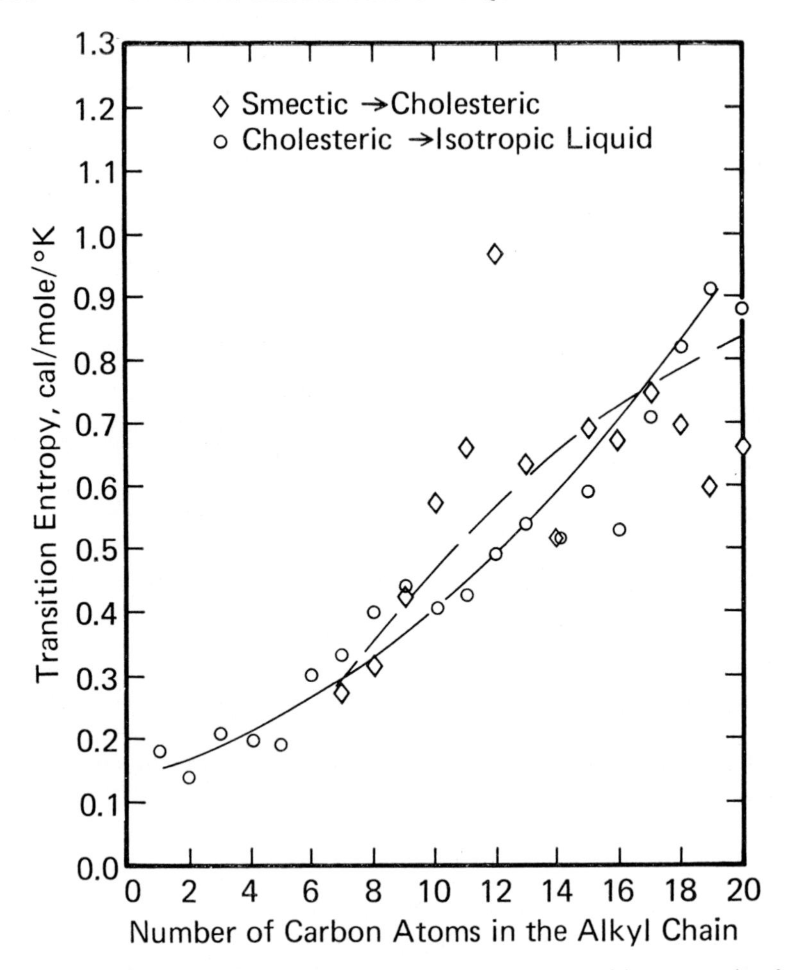

FIG. 10.20. Effect of carbon number on mesophase transition entropies for cholestanyl *S*-alkyl thiocarbonates.

Cholesteryl and Thiocholesteryl ω-phenylalkanoates

Calorimetry of the ω-phenyl esters and thio-esters [58, 59] (Table 10.18) indicates that the phenyl ring does participate in the solid as well as the mesophase order. This conclusion is based on two points:

1. If the phenyl ring did not participate, ΔS values for the solid →mesophase and mesophase → amorphous isotropic liquid transitions should be comparable with the corresponding values for the cholesteryl and thiocholesteryl n-alkanoates of the same alkyl chain length. This is not so. If we assume that the phenyl group is equivalent in length to about C_5 (actually between 8·5 and 8·85 Å) a reasonably good fit is obtained, not only with the corresponding n-alkanoate, but also with the corresponding thio-ester (Fig. 10.17). Thus, unlike the unsaturated C_{18} esters,

TABLE 10.18. *Thermodynamic data for a series of cholesteryl ω-phenyl alkanoates* [44]

Alkyl chain	Solid → cholesteric**			Smectic → cholesteric			Cholesteric → isotropic liquid		
	T (°C)	ΔH (kcal/mole)	ΔS (cal/mole/°K)	T (°C)	ΔH (kcal/mole)	ΔS (cal/mole/°K)	T (°C)	ΔH (kcal/mole)	ΔS (cal/mole/°K)
ω-Phenylformate***	145·4	7·82	18·7	—	—	—	187·6	0·16	0·35
ω-Phenylacetate	120·5	6·85	17·4	—	—	—	—	—	—
ω-Phenylpropionate	111·5	8·77	23·2	—	—	—	114·1	0·12	0·32
ω-Phenylbutyrate	90·9	6·74	18·5	—	—	—	[26·1	0·02	0·07*]
ω-Phenylpentanoate	98·3	7·50	20·2	—	—	—	[91·3	0·15	0·42]
ω-Phenylhexanoate	81·3	6·63	18·7	—	—	—	[45·2	0·07	0·21]
ω-Phenylheptanoate	97·4	9·46	25·5	—	—	—	[82·7	0·19	0·53]
ω-Phenyloctanoate	91·2	9·12	25·0	[34·7	0·14	0·47]	[56·6	0·84	0·26]

* Reported to be very inaccurate.
** Due to monotropy, as indicated, this transition may also be solid → isotropic liquid.
*** or benzoate—difference from data in Table 10.16 may be due to solid phase polymorphism.
[ ] Monotropic transition.

entire chains of the ω-phenylalkyl esters participate in the solid and cholesteric order.

2. There is no smectic mesophase reported; data are not available on esters with alkyl chains longer than C_8.

Structural Analogues

Data are available on Marker's Acid [60]. This is a chain analogue of cholesteryl formate.

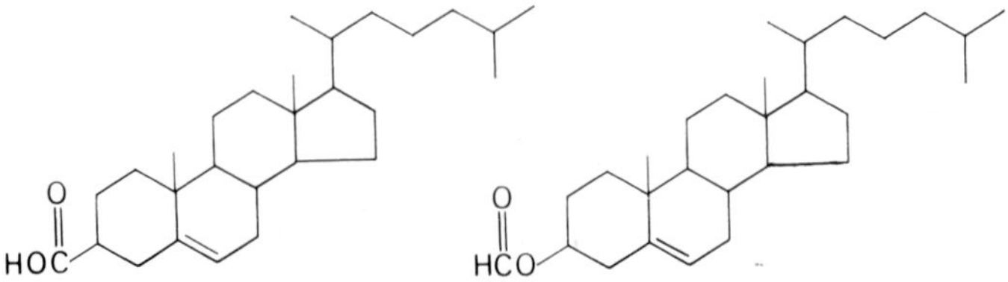

Solid Marker's Acid converts to the cholesteric at 224·5°C and the cholesteric changes to the amorphous isotropic liquid at 258·6°C. The first transition requires 6·73 kcal/mole or 13·5 cal/mole/°K. This agrees well with the value of ≈14 cal/mole/°K predicted by analogy with cholesteryl formate. The formate group is ≈4·2 Å long and the acid group is ≈4 Å. The mesophase → amorphous isotropic liquid transition requires 0·744 kcal/mole or 1·40 cal/mole/°K. This is much too high by comparison with the formate analogue (expected ≈0·25 cal/mol/°K). However, it is probable that the acid is dimerized in the mesophase. Thus, two cholesteryl ring systems are available separated by about 5·5 Å. If this is assumed, agreement is excellent with other dicholesteryl materials studied, i.e. the adipate and sebacate [55]. This evidence from Marker's Acid further strengthens the argument that the *length* and *geometry* of the group substituted on the cholesterol system play a major rôle in determining the order found in the solid and the mesophase. In addition, the question is raised as to how *two* cholesterol rings can participate in the helical stacking order of the cholesteric mesophase.

Effect of Purity on Mesophase Transitions

Mesophase-forming materials are extremely sensitive to very small amounts of impurity. In addition, this sensitivity can be very specific for certain types of impurity. The effects of residual solvents [49, 50], residual reactants [16, 17] and other impurities [52] have been explored.

Prior to any discussion of the effect of purity on mesophase and solid properties, methods for determining purity must be considered. Two general methods are available: thermodynamic and specific.

The thermodynamic approach to purity involves the application of the van't Hoff equation to a DSC curve of the appropriate phase transition. A number of assumptions are made in this application. The method is far from simple, since a number of instrument limitations must be considered. These have been described in detail elsewhere [61]. Two mathematical approaches have been described [62, 63]. The method of Davis *et al.* generally gives lower purity results than that of Gray and others [63]. A method somewhat similar in concept to the above has been described by Casey *et al.* [64]. In any case, the DSC purity determinations generally represent the lower limit of purity, i.e. the "worst case" condition.

The specific purity methods are many and varied. These methods involve an analysis for the impurity directly. Spectrophotometric methods are generally not sufficiently sensitive for the formidable job at hand; 0·01 mole % of impurity is important in the thermodynamic order of mesophase-forming material. Chromatographic methods are much more sensitive. Gas chromatography [44] and thin layer chromatography [65] have been applied with good results. However, it must be remembered that these methods measure specific impurities. It is very easy to overlook impurities of major importance due to a lack of search or close correspondence of the properties of the impurity to those of the compound of interest [52].

Which method of purity determination is employed depends on the system under study. However, *some* impurity measurement should be presented with every thermodynamic study. The DSC method is probably the most general and convenient.

Residual Solvent Effects

Recrystallization solvent can be one of the most persistent and serious impurities. Many mesophase-forming materials are highly retentive of specific solvents. A study of cholesteryl esters has indicated that benzene and carbon tetrachloride may be retained up to 8% in solid esters exposed to a high vacuum for several days [49]. The retained solvent lowered the transition heats and temperatures seriously. In several cases the mesomorphic behaviour vanished. For the esters of cholesterol, recrystallization solvents must be selected with great care. A trace of acid or base usually results in cleavage of the ester on standing for a prolonged period in the solid phase at room temperature. Preferred solvents for the cholesteryl esters are ethanol, pentan-1-ol, acetone and light petroleum. Ethanol is somewhat suspect due to reported cases of transesterification and cleavage in the presence of traces of residual catalyst [66]. In addition, the solvent can determine the morphology of the solid phase when a compound can exhibit two or more solid polymorphs. The technique of high vacuum sublimation has recently received some attention and appears to be an excellent method for purification [49].

Unreacted Starting Materials and Oxidation

The general effect of unreacted acid and of oxidation of cholesterol on cholesteryl esters has been explored [16, 17]. Figs. 10.21 and 10.22 show typical thermograms of the peak broadening and translation which these impurities induce. Oxidation is a serious source of impurities in many mesophase-forming materials. In general, all peaks are broadened. The mesophase transitions exhibit "shoulders" as if the transition occurs in two steps. The broadening of *all* transitions in this case indicates that the impurities have different solubilities in the solid and each mesophase.

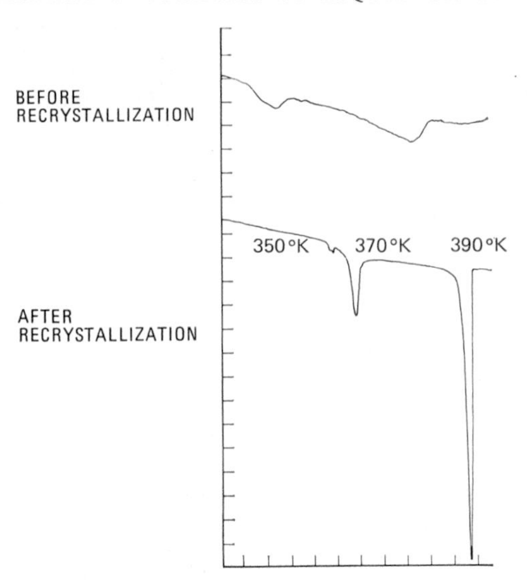

FIG. 10.21. Effect of impurities on the mesophases of ethyl 4-(*p*-methoxybenzylidene)aminocinnamate [16].

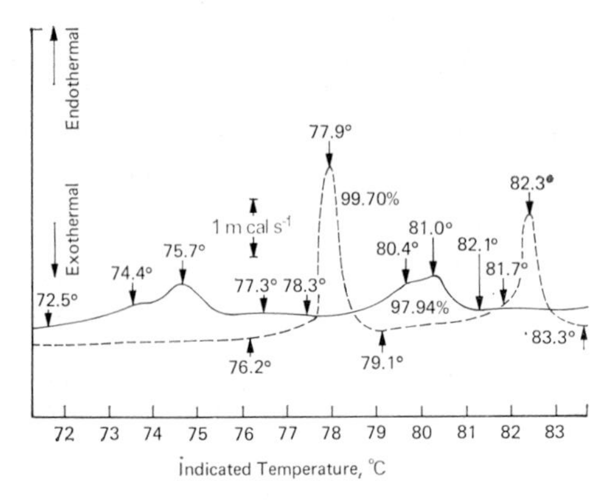

FIG. 10.22. Effect of impurities on the mesophases of cholesteryl heptadecanoate [17].

This is in agreement with studies using mesomorphic materials as a stationary phase in gas liquid chromatography.

SELECTIVE IMPURITIES

These impurities selectively depress one or more phase transitions while leaving the remaining transitions undisturbed. This effect has been explored in some detail when liquid crystal-forming materials are intentionally made into a binary mixture [53, 54]. The solid → mesophase transition is usually selectively depressed. This results in mesophases, which are monotropic in the pure material, forming on both heating and cooling. Vogel *et al.* (see p. 286) have found an impurity which converts the normally monotropic smectic mesophase of cholesteryl palmitate into a mesophase reversible with respect to the solid [52]. Such effects in both nematic and cholesteric systems indicate that the van't Hoff analysis should be applied to the solid → mesophase transition for absolute purity measurement.

Conclusions

Obviously, many profound and useful conclusions may be derived from the thermodynamic data presented in this chapter. However, the authors have refrained from making all but the more obvious statements of fact for reasons of space and the purpose of this chapter. Table 10.19 is presented as a concluding summary of general mesophase entropy considerations and presents many obvious relationships and exceptions.

TABLE 10.19. *Entropies of mesomorphic transitions for homologous series*

Mesomorphic series	Increment in: Nematic → isotropic: Average ΔS for n ⩽ 5 (cal/mole/°K)	Increment in: Nematic → isotropic: ΔS/chain carbon for n ⩾ 8 (cal/mole/°K)	Increment in: Smectic → isotropic: ΔS/chain carbon (cal/mole/°K)
$CH_3O\phi{-}CH{=}\overset{O\atop\uparrow}{N}{-}\phi OC_nH_{2n+1}$	0·41	0·13	0·15
$C_nH_{2n+1}O\phi{-}\overset{O\atop\uparrow}{N}{=}N{-}\phi OC_nH_{2n+1}$	0·62	0·32	0·54
Cholesteryl—$O_2C{-}C_nH_{2n+1}$	0·30	0·073[a]	0·20
$C_nH_{2n+1}O\phi{-}CH{=}N{-}\phi\overset{O\atop\Vert}{C}CH_3$	0·34	—	0·23
$(C_nH_{2n+1}O\phi{-}CO_2H)_{dimer}$	2·8	—	0·48

[a] These data refer to cholesteric → isotropic liquid transitions.

It is hoped that the data presentation in this chapter will be of general use in the construction of a more coherent theory of liquid crystal formation. For the convenience of future tabulators and researchers, we would make the following requests to all who report thermodynamic data on liquid crystals:

1. Present data somewhere in the text in *tabular form*. Graphical presentations are useful in showing relationships, but are usually not a satisfactory way of reporting numerical data. (Footnote ● on p. 304).

2. Present thermodynamic data consistently in °C, kcal/mole, cal/mole/°K, and cal/g/°C for T, ΔH, ΔS and Cp, respectively, or in corresponding SI units.* Heats reported in cal/g usually require multiplication by the molecular weight prior to any calculation. Transition entropy is a useful term when comparing various transition types.
3. Give some estimate of the purity of the compounds studied and the precision of the measurements.

REFERENCES

[1] Brown, G. H. and Shaw, W. G. *Chem. Rev.* **57,** 1049 (1957).
[2] Gray, G. W. *Molecular Structure and the Properties of Liquid Crystals*, Academic Press, Inc., London and New York (1962), pp. 131–3.
[3] Fergason, J. L., Taylor, T. R. and Harsh, T. B. *Electro-Technology* **85,** 41 (1970).
[4] Arnold, H. *Z. phys. Chem.* **225,** 45 (1964).
[5] Arnold, H. *Z. phys. Chem.* **226,** 146 (1964).
[6] Furukawa, G. T. and Douglas, T. B. *Bur. Stand. J. Res.* **57,** 67 (1956).
[7] Kreutzer, K. and Kast, W. *Naturwissenschaften* **25,** 233 (1937).
[8] Barrall, E. M. and Johnson, J. F. *Techniques and Methods of Polymer Evaluation* (edited by P. E. Slade and L. T. Jenkins), M. Dekker, New York (1966), Vol. I, pp. 1–40.
[9] Barrall, E. M., Porter, R. S. and Johnson, J. F. *J. phys. Chem.* **68,** 2801 (1964).
[10] Barrall, E. M. and Johnson, J. F. *Fractional Solidification* (edited by M. Zief), M. Dekker (1969), Vol. 2, p. 77.
[11] E.I. DuPont Instrument Products sales literature.
[12] Thermal Analysis Newsletter, No. 9, Perkin-Elmer Corp., Norwalk, Conn. (1970), p. 3.
[13] Torgalkar, A., Porter, R. S., Barrall, E. M. and Johnson, J. F. *J. chem. Phys.* **48,** 3897 (1968).
[14] Barrall, E. M., Porter, R. S. and Johnson, J. F. *J. phys. Chem.* **71,** 895 (1967).
[15] Arnold, H. *Z. phys. Chem.* **226,** 146 (1964).
[16] Ennulat, R. D. *Analytical Calorimetry* (edited by R. S. Porter and J. F. Johnson), Plenum Press, New York (1968), Vol. I, p. 219.
[17] Barrall, E. M. and Vogel, M. J. *Thermochem. Acta* **1,** 127 (1970).
[18] McCullough, J. P. *Pure appl. Chem.* **2,** 221 (1961).

● Regrettably considerations of space have necessitated the omission of some tables from this chapter, but the numerical data are available in the papers to which references are made.

* We are at a turning point in the presentation of thermodynamic data. The adoption of the SI system is at hand. In the future, enthalpy data should probably be expressed in joules rather than calories. The present authors have used calories out of familiarity.

[19] Arnold, H. and Roediger, P. *Z. phys. Chem.* **239,** 283 (1968).
[20] Schenk, R. *Kristallinische Flüssigkeiten und flüssige Kristalle,* W. Engelmann, Leipzig (1905), pp. 84–9.
[21] Martin, H. and Müller, F. H. *Kolloidzeitschrift* **187,** 107 (1963).
[22] Kreutzer, K. *Annln Phys.* **33,** 192 (1938).
[23] Chow, L. C. and Martire, D. E. *J. phys. Chem.* **73,** 1127 (1969).
[24] Leclercq, M., Billard, J. and Jacques, J. *C. r. hebd. Séanc. Acad. Sci., Paris* **264**, 1789 (1967).
[25] Bondi, A. *Chem. Rev.* **67,** 565 (1967).
[26] Saupe, A. *Molec. Crystals Liqu. Crystals* **7,** 59 (1969).
[27] Arnold, H., El-Jazairi, E. G. and König, H. *Z. phys. Chem.* **234,** 401 (1967).
[28] Herbert, A. J. *Trans. Faraday Soc.* **63,** 555 (1967).
[29] Gray, G. W. *J. chem. Soc.* 4179 (1953); 2556 (1954).
[30] Young, W. R., Haller, I. and Aviram, A. *Molec. Crystals Liqu. Crystals* **13,** 357 (1971).
[31] Bogojawlensky, A. and Winogradow, N. *Z. phys. Chem.* **64,** 229 (1908).
[32] Dave, J. S. and Dewar, M. J. S. *J. chem. Soc.* 4305 (1955).
[33] Glasstone, S. and Lewis, D. *Elements of Physical Chemistry*, D. Van Nostrand Co., New Jersey (1966), 2nd edn, p. 393.
[34] Young, W. R., Haller, I. and Williams, L. *Liquid Crystals and Ordered Fluids* (edited by J. F. Johnson and R. S. Porter), Plenum Press, New York (1970), p. 383.
[35] Gray, G. W. and Jones, B. *J. chem. Soc.* 683 (1954).
[36] Leclercq, M., Billard, J. and Jacques, J. *Molec. Crystals Liqu. Crystals* **10,** 429 (1970).
[37] Barrall, E. M. *Analytical Calorimetry* (edited by R. S. Porter and J. F. Johnson), Plenum Press, New York (1968), Vol. II, p. 121.
[38] Barrall, E. M., Porter, R. S. and Johnson, J. F. *Molec. Crystals* **3,** 299 (1968).
[39] Arnold, H., Jacobs, J. and Sonntag, O. *Z. phys. Chem.* **240,** 177 (1969).
[40] Eastman Kodak Distillation Products Catalogue of Liquid Crystal-forming Materials.
[41] Arnold, H., Demus, D., Koch, H. J., Nelles, A. and Sackmann, H. *Z. phys. Chem.* **240,** 185 (1969).
[42] Davis, G. J., Porter, R. S. and Barrall, E. M. *Molec. Crystals Liqu. Crystals* **10,** 1 (1970).
[43] Barrall, E. M. and Porter, R. S., work in progress.
[44] Ennulat, R. D. *Molec. Crystals Liqu. Crystals* **8,** 247 (1969).
[45] Barrall, E. M., Porter, R. S. and Johnson, J. F. *J. phys. Chem.* **71,** 1224 (1967).
[46] Davis, G. J., Porter, R. S. and Barrall, E. M., *Molec. Crystals Liqu. Crystals* **10,** 1 (1970).
[47] Sell, P. J. and Neumann, A. W. *Z. phys. Chem.* (*Frankfurt*) **65,** 19 (1969).
[48] Gray, G. W. *J. chem. Soc.* 3733 (1956).

[49] VOGEL, M. J., BARRALL, E. M. and MIGNOSA, C. P. Second Symposium on Ordered Fluids and Liquid Crystals (edited by R. S. Porter and J. F. Johnson), Plenum Press, New York, p. 333 (1970).
[50] DAVIS, G. J. and PORTER, R. S. *Molec. Crystals Liqu. Crystals* **6**, 377 (1970).
[51] GRAY, A. P., private communication used later in [45].
[52] VOGEL, M. J., BARRALL, E. M. and MIGNOSA, C. P. *I.B.M. J. Res. Dev.* **15**, 52 (1971).
[53] DAVE, J. S. and LOHAR, J. M. *Chemy. Ind.* **37**, 597 (1959).
[54] DAVE, J. S. and DEWAR, M. J. S. *J. chem. Soc.* 4305 (1955).
[55] BARRALL, E. M., JOHNSON, J. F. and PORTER, R. S. *Molec. Crystals Liqu. Crystals* **8**, 27 (1969).
[56] DAVIS, G. J., PORTER, R. S., STEINER, J. W. and SMALL, D. M. *Molec. Crystals Liqu. Crystals* **10**, 331 (1970).
[57] ELSER, W. *Molec. Crystals Liqu. Crystals* **8**, 219 (1969).
[58] ELSER, W. *Molec. Crystals* **2**, 1 (1967).
[59] ELSER, W. and ENNULAT, R. D. *J. phys. Chem.* **74**, 1545 (1970).
[60] YOUNG, W. R., BARRALL, E. M. and AVIRAM, A. *Analytical Calorimetry* (edited by R. S. Porter and J. F. Johnson), Plenum Press, New York (1968), Vol. II, p. 113.
[61] BARRALL, E. M. and DILLER, R. D. Symposium on Recent Advances in Thermal Analysis, 159th ACS National Meeting at Houston, Texas, 1970. See also *Chem. Engng News* **48**, No. 10, 43 (1970); also *Thermochem. Acta* **1**, 509 (1970).
[62] Thermal Analysis Newsletter, No. 4, Perkin-Elmer Corp., Norwalk, Conn. (1966), p. 5.
[63] DAVIS, G. J. and PORTER, R. S. *Molec. Crystals Liqu. Crystals* **6**, 377 (1970).
[64] CASEY, D. L. and LEVY, P. F. 117th Meeting of the American Pharmaceutical Association, Washington, D.C., 1970.
[65] KAUFMANN, H. P., MAKUS, Z. and DEIKE, F. *Fette Seifen Anstr-Mittel* **63**, 235 (1961).
[66] HALLER, I., private communication, April 1970.

Index

M

N

O

P